# Research in Collegiate Mathematics Education. VI

# CBMS

Conference Board of the Mathematical Sciences

## Issues in Mathematics Education

Volume 13

# Research in Collegiate Mathematics Education. VI

Fernando Hitt
Guershon Harel
Annie Selden
Editors

Shandy Hauk, *Production Editor*

**American Mathematical Society**
Providence, Rhode Island
in cooperation with
**Mathematical Association of America**
Washington, D. C.

2000 *Mathematics Subject Classification*. Primary 00-XX, 97-XX.

ISBN-13: 978-0-8218-4243-0
ISBN-10: 0-8218-4243-9
ISSN 1047-398X

# Contents

**Preface**                                                                vii
*Fernando Hitt, Guershon Harel, and Annie Selden*


**An Image of Calculus Reform: Students' Experiences
of Harvard Calculus**                                                        1
*Jon R. Star and John P. Smith III*


**Effects of Concept-Based Instruction on Calculus Students'
Acquisition of Conceptual Understanding
and Procedural Skill**                                                      27
*Kelly K. Chappell*


**Constructing a Concept Image of Convergence of Sequences
in the van Hiele Framework**                                               61
*Maria Ángeles Navarro and Pedro Pérez Carreras*


**Developing and Assessing Specific Competencies
in a First Course on Real Analysis**                                       99
*Niels Grønbæk and Carl Winsløw*


**Introductory Complex Analysis at Two British Columbia
Universities: The First Week - Complex Numbers**                          139
*Peter Danenhower*


**Using Geometry to Teach and Learn Linear Algebra**                      171
*Ghislaine Gueudet-Chartier*

**Investigating and Teaching the Processes Used
    to Construct Proofs**                                        197
*Keith Weber*


**The Transition to Independent Graduate Studies**              233
    **in Mathematics**
*Janet Duffin and Adrian Simpson*

# Preface

Welcome to the sixth volume of *Research in Collegiate Mathematics Education (RCME VI)*. The present volume, like the previous volumes in this series, reflects the importance of research in mathematics education at the collegiate level. In 1994 the first volume of *RCME* appeared, and volume V was published in 2003; thus, we are commemorating more than a decade of *RCME* with volume VI. In *RCME. I,* Ed Dubinsky, Alan Schoenfeld, and Jim Kaput stated:

> We hope to serve two audiences. *RCME* is for researchers in collegiate mathematics education, and for the wider community of mathematicians who may be interested in these issues both for fundamental intellectual reasons and because of applications to their instruction. (p. vii)

Since that first volume, *RCME* has continued the aim of serving these two audiences, providing a bridge of communication between two academic communities, both of which are interested in improving the teaching and learning of mathematics at the college level. Research in collegiate mathematics education is a complex task that requires the study of phenomena from a variety of theoretical perspectives and methodologies, with special attention given to the implementation of didactical situations in the classroom. That is why, since its beginning, *RCME* has published studies based on different theoretical approaches to the problems of learning and teaching mathematics at the college level.

Mathematics education is a relatively new discipline and the research tradition at the college level is even more recent. In their article about research on undergraduate teaching and learning in *RCME. III*, Annie and John Selden cited Deborah Ball's call to action: "We need to know more about the kinds of mathematical understanding that matter in teaching, how to help teachers develop those understandings, and how to help teachers learn mathematics in, and from, their daily practice." From its beginnings, *RCME* has addressed this need and has become an important forum for the publication of studies related to the issues of learning and teaching at the college level. Just as importantly, it is a venue for new ways of dealing with teaching and learning topics in collegiate mathematics practice.

## Overview of this Volume

The eight papers in this volume come from researchers in six different countries: the United States, Spain, Denmark, Canada, France, and the United Kingdom. They deal with five general themes: calculus and real analysis, complex analysis, linear algebra, proofs, and the transition to graduate work in mathematics. The first four studies, coming from three different countries and four different approaches, address curriculum and the learning and teaching of calculus and real analysis.

The first paper is an analysis of calculus reform as implemented at a large U.S. university. The second, also conducted at a U.S. university, addresses students' acquisition of conceptual understanding and procedural skills in calculus. The third, conducted at a Spanish university, deals with students' construction of a viable concept image linked to the formal definition of convergence for sequences. The fourth article involves the development and assessment of student competencies in a first course in real analysis at a Danish university. The fifth paper focuses on the teaching and learning of complex analysis at two Canadian universities and uses APOS theory to consider difficulties students experience when shifting from one representation of a problem to another. The sixth study, completed at a French university, concentrates on the teaching and learning of linear algebra and considers whether an emphasis on the construction of geometrical intuition aids students' understandings in generalizing linear algebra concepts from $\mathbb{R}^2$ and $\mathbb{R}^3$ to $\mathbb{R}^n$. The seventh paper presents an instructional approach, carried out with U.S. undergraduates, for improving students' performances in constructing group theory proofs. Finally, there is a study of mathematics Ph.D. students in the United Kingdom that describes several different styles of learning in mathematics and identifies obstacles that students might experience when adopting a particular one.

## The Curriculum and the Learning and Teaching of Calculus and Real Analysis

This volume begins with a study by Jon Star and John Smith III about calculus reform at a large U.S. university: "An Image of Calculus Reform." In this paper, Star and Smith investigate the implementation of a calculus course based on the Harvard Consortium Calculus. The authors explore what happened when the faculty decided to transition from a large lecture format to cooperative learning in small groups. They document how the instructors taught and how the students learned. The authors consider the mathematical performance of nineteen students from several points of view, including interviews and student journals. A main claim that emerges from their study is that "the 'taught' curriculum did not differ much from traditional practice in calculus." One problem that becomes apparent when teaching mathematics in a cooperative learning environment is that the instructor needs to be aware of the important role that each member of a small group can play in the learning process. Also, the instructor often must pay attention to the positions taken by each group and lead the class to a general consensus – something that is very difficult to arrive at in every session for every concept. In addition, the instructor needs to develop a certain disposition to teach with cooperative small groups and should be prepared to resolve difficulties that might arise during these sessions. An instructor who is unfamiliar with this method would probably experience difficulties during its implementation and, as a consequence, affect students' performances. Even when an instructor has developed some experience with the use of cooperative small groups, this does not guarantee the complete success of students. This may be so particularly if they are participating in this environment for the first time, as was the case in this study.

The second paper, written by Kelly Chappell and titled "Effects of Concept-Based Instruction on Calculus Students' Acquisition of Conceptual Understanding and Procedural Skill," compares student achievement in two teaching formats. Chappell compares concept-based and procedure-based instructional environments, and documents students' calculus performance in each. Research using such comparisons is well established. Richard Skemp (1978) in his article "Relational Understanding and Instrumental Understanding," pointed out the importance of analyzing the development of two kinds of understanding during the construction of mathematical concepts, referring to them as relational and instrumental understanding. Similarly, Hiebert and Lefevre (1986) in their chapter "Conceptual and Procedural Knowledge," argued that

> Mathematical knowledge, in its fullest sense, includes significant, fundamental relationships between conceptual and procedural knowledge. Students are not fully competent in mathematics if either kind of knowledge is deficient or if they both have been acquired but remain separate entities. When concepts and procedures are not connected, students may have a good intuitive feel for mathematics but not solve the problems, or they may generate answers but not understand what they are doing. (p. 9)

In this volume, Chappell focuses on these connections between procedure and concept in calculus learning situated in technological environments (graphing calculators and computer-based algebra systems). She provides additional evidence at the post-secondary level that concept-based instruction can effectively foster the development of student understanding without sacrificing skill proficiency.

In the third paper "Constructing a Concept Image of Convergence of Sequences in the van Hiele Framework," Maria Ángeles Navarro and Pedro Pérez Carreras discuss students' difficulties in understanding the limit concept. They propose first teaching this concept from a visual approach using particular software, then in a formal way. Navarro and Carreras frame their study in terms of concept image and concept definition. As Tall and Vinner (1981) originally stated,

> We shall use the term *concept image* to describe the total cognitive structure that is associated with the concept, which includes all the mental pictures and associated properties and processes. It is built up over the years through experiences of all kinds, changing as the individual meets new stimuli and matures. (p. 152)

In contrast, a concept definition is the conventional linguistic formulation of a mathematical object – its mathematical if-and-only if definition. Navarro and Carreras tried to promote the development of a rich concept image that would provide students with a solid background for constructing the concept definition. In order to understand students' performances, they found it useful to interpret students' productions using a van Hiele model of learning. As the authors point out, the van Hiele model was initially applied to geometric concepts at more elementary levels of education; but in this study they apply it in higher education. The van Hiele model asserts that the learner moves sequentially through five levels of understanding (see Teppo, 1991) and provides a description of the five phases through which one can help students move, from one level to the next. The levels are: Basic level (or Level 0), Visualization; Level 1, Analysis; Level 2, Informal deduction; Level 3, Deduction; Level 4, Rigor. Mainly used to promote understanding in geometry, this model was modified by Navarro and Carreras to examine learning of the concept of convergence. In their visual approach with computers, the authors used a somewhat peculiar notation; instead of the usual notation for a sequence $(n, s_n)$, they used $(1/n, s_n)$. The authors' experimental results show that this notation did not impede the construction of a formal definition. Their study led them to a characterization of a methodology for introducing the notion of convergence:

- It should allow the introduction of the concept without requiring the student to have any kind of logical or operational skill.
- It should convey the essence of the processes of reasoning of an infinite kind.
- An identifiable hierarchy of levels of reasoning must be present.
- It must prepare the student for the "jump" to the formal definition.

In the fourth paper, Niels Grønbæk and Carl Winsløw deal with the important issue of developing and assessing specific student competencies in a first course in real analysis. The article demonstrates, in a practical way, how to implement a series of activities to develop students' understanding and assess student competencies. The authors followed the French theoretical structure of didactic engineering (see, for example, Artigue, 1988, 1990). Didactic engineering appeared in France in the 1980s as a response to understanding the complexity of classroom phenomena. It is about the relationships between research and action in the educational system. This research methodology takes into account the theory of didactical situations in mathematics of Guy Brousseau (1997). It consists of: (a) an epistemological analysis of the mathematical content in the study (some authors also consider it important to analyze the historical development of the particular mathematical concept), (b) an analysis of students' conceptions about that concept, (c) an *a priori* analysis of the didactical variables that may play a significant role in the study, (d) the actual teaching experiment along with *a posteriori* analysis thereof, and (e) a process of validation. Using this methodology, Grønbæk and Winsløw report on *a priori* analyses of the curriculum, the students in the sense of learning in the curriculum, and teaching and assessing, in that order. In their *a posteriori* analyses, they follow the reverse order. Finally, the authors reflect on their first cycle of using the didactical engineering approach to revise both the curriculum and teaching of real analysis.

## Teaching and Learning Complex Analysis

Peter Danenhower's article, "Introductory Complex Analysis at Two British Columbia Universities," documents some of the problems students face when learning complex analysis. Very little previous research has been conducted in this area. Danenhower completed a large research study and, in this article, he focuses on the introductory topic of complex numbers and their representations. We know that any representation of a mathematical object only partially represents that object. Yet, we expect students to apply a concept flexibly in a variety of mathematical situations, shifting from one representation to another as appropriate. Thus, it is important to understand how students articulate representations to arrive at the construction of a mathematical object and to understand how they deal with several representations of the same mathematical object. Danenhower's theoretical approach is based on the Action-Process-Object-Schema (APOS) framework (see, for example, Asiala et al., 1996). Danenhower focuses on the depiction of problems when shifting from one representation to another and describes different levels of understanding in accordance with the APOS framework.

## Teaching and Learning Linear Algebra

The paper "Using Geometry to Teach and Learn Linear Algebra," written by Ghislaine Gueudet-Chartier, considers geometrical intuition as an aid to learning linear algebra. This mixed-methods quantitative and qualitative study has a theoretical framework based on Fischbein's work, *Intuition in Science and Mathematics* (1987). To build a definition of geometrical intuition, Gueudet-Chartier uses Fischbein's definition of model: "A system $B$ represents a model of system $A$ if, on the basis of a certain isomorphism, a description or a solution produced in terms of $A$ may be reflected consistently in terms of $B$ and vice versa." In the first part of her article, the author defines the meaning of geometrical intuition as "the use of models stemming from a geometry." After a review of the historical development of linear algebra, she discusses responses to a questionnaire about geometry and linear algebra given to 28 French university linear algebra teachers and presents an examination of geometry and drawing use in linear algebra texts. Gueudet-Chartier continues by reporting on the results of following one university linear algebra course and interviewing the teacher and eight students. The teacher believed that by developing a geometrical model when working in $\mathbb{R}^2$ and $\mathbb{R}^3$, students would generalize more or less easily to $\mathbb{R}^n$. Indeed, in the interview, the teacher said: "For quadratic forms, all the phenomena already happen in three-dimensional spaces. It is necessary to understand how to move from 2 to 3. After that, there is nothing new." Contrary to this teacher's belief, the author's analysis reveals problems that students face when passing from $\mathbb{R}^2$ and $\mathbb{R}^3$ to $\mathbb{R}^n$.

## Teaching and Learning Proofs

Many researchers have addressed the theme of mathematical proof, but it is always interesting because of its complexity. The paper by Keith Weber, "Investigating and Teaching the Processes Used to Construct Proofs," considers this theme in relation to understanding group homomorphisms. His review of the literature asserts three causes of student difficulties with proof. The first cause is that students often possess an inaccurate conception of what constitutes a mathematical proof. The second cause is that students may lack an understanding of a theorem or a concept and systematically misapply it. However, even if students have accurate conceptions of proof and the ability to derive logical inferences, as necessary conditions for proof-writing competence, these skills alone are not sufficient. The third cause is that students who cannot construct proofs often do not have effective strategies for doing so. In conducting two studies, Weber addressed this third cause of students' difficulties with proofs in the context of group homomorphisms. He describes a prescriptive procedure for constructing proofs that he tested with students and analyzes the results. The high percentage of students constructing valid proofs supports his claim that learning a powerful procedure to prove certain kinds of statements about group homomorphisms improved students' performance.

## Learning Graduate Mathematics

The paper by Janet Duffin and Adrian Simpson, "The Transition to Independent Graduate Studies in Mathematics," documents changes in students' learning styles when going from undergraduate to graduate study. The authors conducted a qualitative study in the United Kingdom, interviewing thirteen Ph.D. students in mathematics. They describe their methodology as a "conversation with a purpose" to explore these students' ways of learning new areas of mathematics. Building on their literature review, the authors develop an expanded version of their previous theory of learning styles and explore the changes in those styles that their graduate student participants experienced. They identify four categories of learning styles: (a) *Natural*, where a student sees new knowledge connected to old; (b) *Alien*, where a student constructs new knowledge that is separated from the old, but that continues to exist at graduate level in a radically different form akin to mathematical formalism; (c) *Coherence*, where a student constructs new knowledge from old, with a separation of new knowledge from old, but where the new knowledge needs to have a clear internal structure; and, (d) *Flexible*, where a student constructs an ability to adapt her or his learning style to the perceived value to be gained from the learning. Duffin and Simpson state that,

> The transition to independent graduate study in mathematics thus seems to be stable only for natural learners. Alien learners predominately respond in two ways: by altering their learning style towards a natural stance by more clearly seeking analogical links or by developing a learning style (which we had not previously seen at other levels) in which they retain the separateness of a new area of mathematics, but seek internal coherence and structure for that new area.

Finally, the editors express deep gratitude and appreciation to Cathy Kessel for her professional work and the special attention she gave to the past issues of *RCME* for which she served as Managing Editor. Cathy did her best to strengthen authors' contributions to the two most recently published volumes, and we thank her for that wonderful work. We are also grateful to Shandy Hauk, who has generously given her time to take on a similar role as Production Editor for this volume. We wish to acknowledge all those who have given of their time and expertise to review manuscripts for this and previous *RCME* volumes. The production of this volume has been greatly enhanced due to their significant contributions.

Fernando Hitt
Guershon Harel
Annie Selden

## References

Artigue, M. (1988). Ingénierie didactique. *Recherches en Didactique des Mathématiques, 9*(3), 281–308.

Artigue, M. (1990). Epistémologie et didactique. *Recherches en Didactique des Mathématiques, 10*(2, 3), 241–286.

Asiala, M., Brown A., DeVries, D. J., Dubinsky, E., Mathews, D., & Thomas, K. (1996). A framework for research and curriculum development in undergraduate mathematics education. In J. Kaput, A. H. Schoenfeld, & E. Dubinsky (Eds.), *Research in collegiate mathematics education. II.* (pp. 1–32). Providence, RI: American Mathematical Society.

Brousseau, G. (1997). *Theory of didactical situations in mathematics.* (N. Balacheff, M. Cooper, R. Sutherland, & V. Warfield, Eds. and Trans.) Dordrecht, The Netherlands: Kluwer.

Dubinsky, E., Schoenfeld, A. H., & Kaput J. (1994). Preface. *Research in collegiate mathematics education. I.* (pp. vii–xi). Providence, RI: American Mathematical Society.

Fischbein, E. (1987). *Intuition in science and mathematics: An educational approach.* Boston: D. Reidel.

Hiebert, J., & Lefevre, P. (1986). Conceptual and procedural knowledge in mathematics: An introductory analysis. In J. Hiebert (Ed.), *Conceptual and procedural knowledge: The case of mathematics* (pp. 1–28). Hillsdale, NJ: Erlbaum.

Selden, A., & Selden, J. (1998). Questions regarding the teaching and learning of undergraduate mathematics (and research thereon). In A. H. Schoenfeld, J. Kaput, & E. Dubinsky (Eds.), *Research in collegiate mathematics education. III.* (pp. 308–313). Providence, RI: American Mathematical Society.

Skemp, R. (1978). Relational understanding and instrumental understanding. *Arithmetic Teacher, 26*(3), 9–15.

Tall, D., & Vinner, S. (1981). Concept image and concept definition in mathematics with particular reference to limits and continuity. *Educational Studies in Mathematics, 12,* 151–169.

Teppo, A. (1991). Van Hiele levels of geometric thought revisited. *Mathematics Teacher, 84,* 210–221.

CBMS Issues in Mathematics Education
Volume **13**, 2006

# An Image of Calculus Reform: Students' Experiences of Harvard Calculus

Jon R. Star and John P. Smith III

ABSTRACT. This paper examines students' experiences of calculus reform at one major research university that has implemented and maintained the Harvard Consortium calculus program for many years. Our focus is on students' experiences with the new course, particularly on which features students noticed as different from traditional expectations and assumptions about the teaching and learning of mathematics. Nineteen first-year students who were enrolled in reform calculus courses at the University of Michigan participated in this study. These students were interviewed multiple times, kept mathematics journals, and reported their grades in all of their courses. In addition, research staff observed their mathematics classes. Our results indicate that the "taught" curriculum did not differ much from traditional practice in calculus. However, students did notice significant differences between the traditional mathematics courses they had taken in high school and Harvard calculus, most notably the presence of group homework assignments, the accompanying requirement to provide written explanations for one's problem-solving steps and an increase in the percentage of application or story problems assigned. We did not find any relationship between repeated mentioning of these differences and changes in students' achievement in mathematics. The relevance of our results to those implementing or considering implementing reform calculus programs is discussed.

## 1. Introduction

This paper examines students' experiences of calculus reform at one major research university. The University of Michigan has implemented and maintained the Harvard Consortium calculus program for many years. After nearly two decades of design, development, and implementation, fundamental changes have taken place in both the content and the pedagogy of calculus at many colleges and universities. Mathematicians and educators continue to reflect upon the success of such reform efforts, with an eye toward further improving learning and engagement in this pivotal course.

The Tulane Conference in 1986 is commonly recognized as the beginning of the reform calculus movement in the U.S., although its roots can be traced to the 1950's (Tucker & Leitzel, 1994). In general, reform calculus is characterized by a scaling back in the number of topics covered and an enrichment of the pedagogy used in instruction, including attention to graphical, symbolic, and numeric representations;

a move to smaller classes rather than large lectures; increased use of cooperative learning groups, technology, and extensive writing; and more generally an increased focus on conceptual understanding and decreased attention to symbol manipulation (Ganter, 2001; Tucker & Leitzel, 1994).

There continues to be a need for evaluation of calculus reform efforts (Darken, Wynegar, & Kuhn, 2000; Ganter & Jiroutek, 2000). Despite a recent increase in the number of evaluation studies and a growing body of evidence of the effect of reform calculus programs on student learning, additional research is needed (Darken et al., 2000; Ganter & Jiroutek, 2000; Schwingendorf, McCabe, & Kuhn, 2000).

In particular, many evaluations of calculus reform focus exclusively on academic performance, particularly students' grades on course examinations. While grades are certainly one way to assess a students' success in a course, other measures provide important complementary information (Armstrong, Garner, & Wynn, 1994; Bonsangue, 1994). Evaluation efforts have moved beyond relying only on students' grades to include measures such as continued course-taking (Ganter & Jiroutek, 2000), surveys of students' opinions (Edwards, 1993), students' performance on other examinations or tasks (Beidleman, Jones, & Wells, 1995), and students' overall GPAs[1] and high school grades (Schwingendorf et al., 2000).

One perspective on calculus reform that has not been thoroughly examined is students' experiences with the new course (Keith, 1995). Students' voices typically enter the dialogue indirectly, as anecdotes shared by their instructors. For example, the Assessing Calculus Reform Efforts survey, that was administered in 1993-94 by the MAA, asked faculty to comment on students' attitudes and changes in students' understandings as a result of the reform calculus experience (Tucker & Leitzel, 1994). Keith (1995) notes that anecdotal reports of students' experiences are potentially problematic for at least two reasons. First, student anecdotes are often selected to make particular, often favorable, points. And second, anecdotal reports of students' experiences are often accompanied by explanations from an instructor's point of view that serve to rationalize and dismiss students' concerns. Our reading of the literature on calculus reform indicates that only rarely have students' views of reform been directly and systematically surveyed (e.g., Walker, 1999).

**1.1. Student Voices.** The research we report here is distinctive in its detailed examination of students' achievement and their assessments and judgments at one implementation site. Students' judgments about the content they study are important because engagement and connection with mathematics mediates and directly influences learning and achievement, in calculus as much as any subject matter. In fact, a desire to address concerns about students' not being engaged was one of the issues motivating initial reforms in calculus (Tucker & Leitzel, 1994). Since it is still the case that most students come to college from twelve years of more traditionally designed and taught mathematics courses, it is reasonable to expect they would find that reform calculus courses violate some of their assumptions about the ways that mathematics should be taught and learned. Indeed, developers of reform materials argue that significant differences do exist between traditional and reform curricula (Star, Herbel-Eisenmann, & Smith, 2000). Our particular interest

---

[1]GPA, or grade point average, indicates a student's average grade for some period of course-taking (e.g., all of high school, or all of college). GPAs are typically reported using a 4-point scale with 4 being best.

was to study what students themselves noticed and reported as different as they moved into and came to terms with one reform calculus curriculum – in our case, the materials developed by the Harvard Consortium. (Hughes-Hallett et al., 1994)

Questions concerning what students notice as different between traditional and reform mathematics curricula rarely have been asked of reform calculus students. As noted above, the predominant focus in prior studies has been on students' academic performance. We feel that expanding the focus of current evaluative research on calculus reform to include students' own voices is important for several reasons. First, although authors of reform curricula clearly articulate the ways that their textbooks and recommended pedagogies differ from more traditional approaches (e.g., Harvard Consortium's "rule of four"), there is no evidence that students see the new curricula in the same light as they were intended. In fact, our experience in grades K through 12 indicates that students' perceptions of what is new in reform curricula are quite different from those of curriculum designers and researchers (Smith, Star, & Hoffmann, 2002). How students perceive what it is different in reform curricula is an open, empirical question – one that seems clearly related to efforts to evaluate the success of such curricula.

Second, viewing a curriculum through students' eyes allows potential adopters of reform materials to know what students may react to or against. Universities considering the adoption of a reform curriculum not only need to consider whether or not it is "successful;" they also need to consider the resources that must be made available to support its implementation. Knowing what students find especially noteworthy and challenging can assist in decisions about such resources. For example, how should instructor training be changed to prepare to teach a particular reform curriculum? How should instructor trainers be trained differently? These decisions should depend, at least in part, on what students find easy or especially difficult and whether instructors (and instructor trainers) have the resources or background to scaffold students through such challenges. Furthermore, decisions about other kinds of external support for students (extra help sessions, technology support persons or facilities, etc.) are better made when potential adopters have access to what students' find different, noteworthy, easy, or difficult.

Third, students' perceptions of reform curricula can provide explanations for why an implementation was, or was not, successful. Although grades are often used to determine success, they provide little or no information to account for or explain outcomes. When faced with findings from studies of students' academic performance, such as the growing body of evidence that reform calculus students perform no worse, and often somewhat better, than students in more traditional classes (e.g., Ganter & Jiroutek, 2000), researchers are left to speculate on the "why" underlying these results. Students' voices, particularly their impressions of what they found different, easy, or challenging, are one way to get beneath top-level findings from performance data and to understand more deeply why some implementations of reform curricula are successful.

**1.2. Calculus Reform at the University of Michigan.** Locating our inquiry in the mathematics department at the University of Michigan [U-M] was appropriate for a number of reasons (other than proximity). First and foremost, the department has maintained a stable implementation of an explicit and well-regarded reform curriculum for a long period. Small-scale preparations began in 1991, with full adoption of the Harvard Consortium materials in 1994. Second, the department is a highly regarded community of research mathematicians who also actively support the "introductory" calculus program, despite the costs of small classes and graduate student training. Third, the department was interested in our inquiry and facilitated our work with students and instructors.

The introductory program at U-M, primarily designed for incoming first-year students, consists of three semester courses, Pre-Calculus [Math 105] (Connally et al., 1998) and Calculus I [Math 115] and II [Math 116] (Hughes-Hallett et al., 1994). These courses constitute the calculus sequence for all U-M majors, including mathematics, engineering, physical sciences, pre-med, business, statistics, and nursing.[2] In these three courses, large lectures with small recitation sections were eliminated in favor of many small sections (25-30 students) that meet three hours each week (in two 90-minute or three 60-minute sessions). Each year approximately 35 sections of Pre-Calculus and Calculus II and approximately 60 sections of Calculus I are offered. Each section of these courses is taught independently by a single instructor, though all sections share common course homework assignments, unit tests, and final exams. Group work is required in and out of class.

The overall goal of the Harvard calculus program, as stated by the curriculum's developers, is "to provide students with a clear understanding of the ideas of calculus," focusing on deep understanding of key concepts, rather than coverage (Hughes-Hallett et al., 1994, p. v). Two main principles oriented the design of these texts. First, problems in each section are varied so that students will be less likely to solve novel problems by rote pattern-matching to the steps of example problems or highlighted procedures. Many problems, especially in Pre-Calculus and Calculus I, require making sense of a quantitative situation and explaining the meaning of the numerical or symbolic result in terms of that situation. Second, topics are presented and developed in multiple representations (in their terms, using the "the Rule of Four") – geometrically, numerically (that is, tables), algebraically, and verbally – to build meaning for symbolic expressions and avoid mindless manipulation.

In class, U-M students typically sit at tables with three other students and are regularly encouraged to work on problems with their tablemates. Out of class, they must complete group homework assignments each week, also in groups of four, and are encouraged to complete weekly individual homework assignments in groups as well. Group homework is graded (all members of the group receive the same score); individual homework typically is not. Group homework problems are usually more difficult than test or individual homework problems; the goal is to encourage greater effort in the group.

In general, group homework is designed to force students to interact, explain their thinking, and write coherent explanations, rather than simply calculate and circle answers. Students are expected to look over the problems on their own before their groups meet to discuss the problems and write up their solutions. They are also asked to assume particular roles within their group ("scribe," "reporter," "clarifier," and "manager"), and these roles are intended to rotate among group members week to week. Though the roles and procedures are clearly proscribed, students are free to enact them (or not) and often organize their groups to function in very different ways. It is important to note here that while the Harvard program actively encourages group work, group homework was a site-specific feature of U-M's

---

[2] U-M has an honors calculus sequence that does not use reform curricula. Almost all mathematics majors take the honors courses.

implementation. The particular instantiation described above was not mandated by the written curriculum. It should be re-emphasized here that the goal of our research is to understand students' experiences in one example of reform calculus, not to evaluate the effectiveness of the Harvard Consortium materials in general or as implemented by U-M.

Given the large number of sections in each course, the introductory program demands a large number of instructors, most of whom are graduate students in the department. Many who become graduate student instructors ("GSIs") have no prior teaching experience. To train and support them, the department carries out a mandatory one-week instructor professional development in the summer, publishes and teaches from a detailed instructor's guide, requires attendance at regular course-specific GSI meetings during the semester, and conducts formal evaluations by trained staff from the campus teaching and learning center. This training and support emphasizes increased use of cooperative groups and extensive student writing. The Instructor's Guide (Shure, Brown, & Black, 1999) indicates that students should be "encouraged to experiment and conjecture, to describe and discuss" (p. 5) by working together, writing, and solving real-world problems. Instructors are asked to reduce the amount of time spent lecturing, give up some control of the classroom flow, and work hard to listen to students' responses and questions. More detailed sections of the guide are devoted to using cooperative groups in the classroom, different kinds of classroom activities, and questioning techniques.

## 2. Method

Given this description of the implementation of reform calculus at U-M, as well as prior research on students' experiences in high school and college mathematics (e.g., Schoenfeld, 1989; Walker, 1999), we found it reasonable to expect that U-M first-year students would experience something of a mathematical discontinuity in their calculus or pre-calculus classes. Students would have developed expectations about what it meant to think, know, and do mathematics from their K-12 experiences. But these expectations would likely be jarred (and perhaps forced to change) as they worked through the Harvard materials and associated U-M teaching practices.

Our study focused on students' perceptions and adaptations in this "new" mathematical context. Specifically, we wanted to know what students noticed as different between the U-M program (curriculum and teaching) and their pre-college mathematics experiences as well as how students performed in these classes and what instruction actually looked like. In fact, this work was part of a larger project exploring students' reactions and adjustment to fundamental changes in mathematics curriculum and pedagogy (see Smith, Star, & Hoffmann, 2002). In addition to the work at U-M, our project examined the shift from a reform middle school to a traditional high school program and vice versa, and the move from a reform high school curriculum to a traditional college calculus program.

**2.1. Participants.** Nineteen U-M first-year students volunteered for the study (10 females; nine males) in response to flyers posted where their mathematics classes met. To be eligible, students had to be enrolled full-time, over the age of 18, and graduates of Michigan high schools. In what follows we will refer to each of the students in our study with a two-letter code.

We gathered grade and standardized test score information from participant self-reports. All 19 students had been quite successful in high school. Like most first-year students at U-M who elect to take a mathematics course in their first semester, all participants had taken four years of mathematics in high school, had a high GPA in high school, and had high standardized test scores. Their mean high school GPA was 3.84 (on a 4-point scale), the mean of their ACT mathematics scores was 29 out of 36 (at or above the 95th percentile), and the mean grade in their 12th grade mathematics course was 3.43.

All but one participant used "traditional" mathematics curricula during at least three years of high school. By "traditional," we refer to a diverse variety of mathematics textbooks written prior to the development, in the 1990s, of reform curricula sponsored by the National Science Foundation. For the purposes of this study, we also chose to classify the high school materials from the University of Chicago School Mathematics Project (UCSMP) as traditional. The one exception (SB) took 1.5 years of a reform mathematics program during her first two years of high school and then switched to the traditional sequence of courses thereafter (see the Appendix for more detail on the sequences of mathematics courses that participants took during high school). All volunteers were compensated for their participation.

At many U.S. high schools, students may have the option of taking Advanced Placement, or AP, courses in a variety of subjects. After taking an AP course, students may then take a national AP exam. By scoring well on the AP exam (AP exams are graded on a five-point scale, with 5 being the best possible score), students can earn college credit for the AP course. In mathematics, AP courses are offered in Calculus and Statistics. There are two versions of the AP Calculus course, which are referred to as AB Calculus and BC Calculus. Each is a year-long course; AB Calculus corresponds roughly to one semester of college Calculus, while the much faster-paced BC Calculus course corresponds roughly to a full year of college Calculus.

Sixteen of the 19 students took AB Calculusin their senior year of high school, two students took Pre-Calculus, and the remaining student took AP Statistics (but not the Advanced Placement examination). Of the 16 who completed AB Calculus, 10 took the Advanced Placement examination. One student received a score of 5, one received a 4, three received 3's, three received 2's, and two students received a score of 1.

In their first semester at U-M (Fall 1999), all 19 students took a mathematics course because one or two semesters of mathematics were required for their intended major. Most U-M degree programs do not require a mathematics course. Almost all students who enroll in the introductory calculus sequence (and all 19 in our sample) intend to major in engineering, business, statistics, or pre-med.    Five of the 19 participants enrolled in Pre-Calculus, 10 in Calculus I, and four in Calculus II. One student (CA, in Calculus I) dropped mathematics about halfway through the semester. None planned to major in mathematics. In recent department history, very few mathematics majors have taken these courses; instead they generally place into Calculus III or the Honors calculus sequence, both of which are taught with more traditional texts.

In the next semester (Winter 2000), 13 of the 19 students enrolled in mathematics again. All five Pre-Calculus students took Calculus I; seven of the 10 Calculus

I students took Calculus II; and one of the four students in Calculus II enrolled in Calculus III. The six who did not take mathematics cited various reasons for their choice, including a lack of interest in the subject and that only one semester of mathematics was required for their intended major.

**2.2. Data Collection.** We sought information about students' mathematical experiences in multiple ways. Project staff observed participants' classes and homework groups. We observed all participants' classes at least once per semester, except in two cases where particular instructors preferred not to be observed. All observations were documented in detailed written field notes, taken in a structured format.

In addition, we gathered a broad range of data on participants' individual experiences of their mathematics classes and work. First, students kept "math journals," in which they wrote about their experiences in their mathematics classes. They were asked to write in their journals twice each week, and they submitted their entries to us by e-mail. Second, participants reported all their mathematics grades, including scores for homework, quizzes, tests, midterms, and final exams, as well as their high school grades and college entrance exam scores. Third, students were interviewed two or three times each semester they were enrolled in mathematics. The students who did not enroll in mathematics in the second semester were interviewed once. All interviews were semi-structured and were tape-recorded.

In the interviews, students were asked to talk about their experiences in high school and college mathematics. In particular, we asked students directly what they perceived as different between their high school and collegiate mathematics courses. We used both broad, open-ended questions like, "What did you find different between high school math and Calc I?" and more pointed questions like, "Was there anything about the instruction that you found different between high school and college math?" Students were also asked to confirm or revise descriptions from previous interviews. For example, "In the first interview, you said that you felt that there was no difference between the instruction you received in high school and college math. Do you still feel that this is true?"

It should be noted here that although our sample size ($n = 19$) is quite small relative to other studies evaluating calculus reform, the number of interviews (a total of 80) we conducted is large. For example, Ganter and Jiroutek's (2000) study looked at the performance of approximately 300 students and interviewed only 5 (see also Bonsangue (1994), who interviewed 22 of his approximately 850 participants, each once).

## 3. Results

Before describing how students saw the differences between their traditional and reform-oriented mathematics experiences, we look carefully at two factors that cannot be separated from and indeed likely influenced our students' experiences of the Harvard calculus. First, we summarize what we observed in our visits to mathematics classrooms for a portrait of students' experiences. Since curricula influence but do not define the teaching that students receive or the experiences they walk away with, we present the "taught" mathematics curriculum and contrast it with the "written" curriculum. Second, we analyze how students' grades in their mathematics courses changed from high school, both alone (mathematics only) and in relation to their overall GPAs. Given U-M students' focus on achievement, it is reasonable to expect that their relative success in a "new" mathematics program would affect their overall experience of that program, and in particular, their perceptions of how it differed from their previous school mathematics.

**3.1. Typical U-M Class.** We observed three Pre-Calculus classes (all in the Fall semester), 14 Calculus I classes (nine in the Fall and five in the Winter semester), and six Calculus II classes (three in the Fall and three in the Winter semester). To analyze our field notes from these observations, we first characterized the structural features of each class, including its composition (number and gender), the physical arrangement of the classroom (e.g., the position of the tables and chairs), the start and end time, and attendance. Next, we created and coded for categories of classroom actions, including whether or not the class started or ended late, when and for how long students worked in groups, what they worked on in those groups, when and for how long the instructor lectured (on old or new material), and when and how students were assessed. Finally, the coded action sequences were tabulated.

Although there was some variation, our analysis shows a great deal of uniformity in instructional activity.[3] The classes typically began and ended roughly on time; all of the classes we observed were approximately 80 minutes long (90 minutes minus a 10-minute passing period). Male instructors outnumbered female instructors by a factor of two, while the gender balance was about equal for students. Classes typically consisted of the following activities undertaken in the order listed (see Table 1): (a) announcements and collecting/passing out of student work, (b) instructor's review of the homework/quiz/exam problems, (c) group work to practice material covered in a previous class, for an upcoming exam, or new material, and (d) lecture on new material. This analysis excludes atypical classes, such as quiz days (four classes; three Calculus I classes and one Calculus II class) and full review days for upcoming assessments (three classes; one for each course).

Overall, Table 1 shows a progressive transformation in a typical day from an emphasis on procedural matters and homework review in Pre-Calculus to a focus on group work and lecture on new material in Calculus II. In Calculus I a rough balance was achieved between paper passing/review and group work and lecture. Similarly, Pre-Calculus classes were more likely to end early, while Calculus II instructors used the entire allotted class time.

The group work we observed in most classes was not very collaborative. In many cases, it was group work only in the sense that students were sitting in a group (at tables that seated four). At some point in the class, the instructor put a problem or two on the board and asked students to work in groups to solve it (them). Then the instructor circulated around the room assisting students who asked for help. In this activity, students rarely worked collaboratively. Instead, they worked on the problems individually and only occasionally asked a group member for help. Instructors typically did not indicate dissatisfaction with this mode of group work. Occasionally we observed instructors reminding students that the point of group work was for them to work together and talk to each other, but this was rare. So instead of group work, it would be more accurate to call this activity "problem solving practice while sitting with three other students."

---

[3]Though we observed more than a few classes, we also admit that our sample was relatively small in relation to the total number of sections of each course.

TABLE 1. Mean Time in Minutes by Class Activity in a Typical U-M Mathematics Class

|                                    | Pre-Calc (n=2) | Calc I (n=10) | Calc II (n=4) |
|------------------------------------|----------------|---------------|---------------|
| Start time                         | 0 (on time)    | 0 (on time)   | -1 (late)     |
| Announcements/Hand out papers      | 8              | 7             | 1             |
| Review of homework                 | 32             | 23            | 6             |
| Group work to practice material    | 13             | 27            | 33            |
| Lecture on new material            | 15             | 13            | 34            |
| End time                           | -9 (early)     | -6 (early)    | 0 (on time)   |

Beyond group work, the sequence of activities in Table 1 indicates that typical class sessions in the introductory program did not look very different from sessions in a more traditional Calculus or Pre-Calculus course taught in small sections. Though, we did find significantly less class time was devoted to instructor lecturing than has been found in prior research on traditional calculus teaching (e.g., Friedman, 1993).

Notably absent in the observed class meetings were reform practices such as students' presentations of their solutions, instructors' calls for multiple solutions to already-solved problems, and alternative forms of assessment (e.g., partner quizzes). Department documents given to GSIs and discussed in the summer professional development described a quite different classroom environment from what we observed. The 40-page *Instructor's Guide* (Shure, Brown, & Black, 1999) oriented our expectations, so we were somewhat surprised (initially at least) to find that the "taught" curriculum did not differ much from traditional practice in calculus. Our observations indicated that written directions (including scripted lessons) and intensive short-term training had been largely ineffective in orienting graduate students' teaching practices (see also Speer, 2001).

What factors might underlie this marked difference between the "intended" and "taught" curriculum? Certainly, the explanation might begin with the basic fact that most sections were taught by GSIs who had very limited (if any) teaching experience. For most, this was their first experience with classroom teaching and their comfort level was very low, especially at the start of the semester. This inexperience included little or no awareness of reform teaching principles and practices. Many had attended high schools in foreign countries and did not have a sense of current issues in pre-college mathematics education in the U.S. Those who had attended U.S. high schools did so prior to the implementation of many reform curricula and teaching practices, so the chance of having experienced reform teaching was quite low. Moreover, many GSIs, especially the least experienced, translated the department's commitment to reform teaching practices to being exclusively about the addition of group work and group homework. They felt that their support of group homework and their regular use of group work in class satisfied this mandate. The overall result was that if the group homework assignments were removed and in-class group work time reduced, what would be left would have been a very traditional calculus class, despite the use of a reform text.

**3.2. Changes in Student Achievement.** Another facet of students' experience that is helpful in understanding their perceptions of differences between traditional and reform mathematics programs is their academic performance in their college courses relative to high school. As we and others (e.g., Tucker & Leitzel,

TABLE 2. First-Year Achievement Data, U-M site

| | | High School 12th grade math | | First semester U-M | | | | Second semester U-M | | |
| | GPA | Course | Grade | GPA | Math Course | Grade | Change Math[a] | GPA | Math Course | Grade | Change Math |
|---|---|---|---|---|---|---|---|---|---|---|---|
| BD | 3.8 | APC | 3.0 | 3.1 | CI | 4.0 | 1.0 | 2.7 | CII | 2.7 | -1.3 |
| BE | 4.0 | APC | 4.0 | 3.7 | CII | 3.7 | -0.3 | 3.8 | CIII | 3.3 | -0.4 |
| BL | 3.6 | APC | 2.5 | 2.1 | Pre | 1.7 | -0.8 | nd | CI | 1.7 | 0 |
| BM | 3.9 | APS | 4.0 | 3.8 | Pre | 4.0 | 0 | 3.1 | CI | 2.7 | -1.3 |
| CA | 3.5 | APC | 0.7 | 3.3 | CI | nd | nd | nd | | nd | nd |
| CM | 4.o | APC | 4.0 | 2.9 | CI | 2.7 | -1.3 | 2.9 | CII | 2.3 | -0.4 |
| DD | 3.9 | APC | 4.0 | 2.7 | CI | 2.0 | -2.0 | 2.8 | CII | 2.7 | 0.7 |
| DJ | 4.0 | Pre | 4.0 | 3.2 | CI | 2.7 | -1.3 | nd | | nd | nd |
| FD | 4.0 | APC | 4.0 | 2.3 | CI | 2.0 | -2.0 | nd | nd | nd | nd |
| GD | 3.9 | APC | 4.0 | 3.5 | CII | 3.3 | -0.7 | nd | | nd | nd |
| JC | 4.0 | APC | 4.0 | 2.6 | CI | 2.3 | -1.7 | nd | nd | nd | nd |
| KK | 3.9 | APC | 4.0 | 4.0 | CII | 4.0 | 0 | nd | | nd | nd |
| LS | 3.7 | APC | 3.5 | 3.9 | Pre | 3.7 | 0.2 | 2.8 | CI | 3.7 | 0 |
| MM | 4.0 | APC | 4.0 | 3.6 | CII | 3.3 | -0.7 | nd | | nd | nd |
| MT | 3.9 | Pre | 4.0 | 3.5 | Pre | 2.3 | -1.7 | nd | CI | 3.0 | 0.7 |
| PJ | 4.0 | APC | 4.0 | 3.9 | CI | 4.0 | 0 | 3.2 | CII | 2.7 | -1.3 |
| SB | 3.8 | APC | 3.7 | 3.7 | Pre | 3.7 | 0 | 3.0 | CI | 3.3 | -0.4 |
| TM | 3.2 | APC | 1.5 | 1.9 | CI | 1.0 | -0.5 | nd | | nd | nd |
| VJ | 3.8 | APC | 2.5 | 2.7 | CI | 3.3 | 0.8 | 2.8 | CII | 2.0 | -1.3 |

*Note*: Course abbreviations APC = Advanced Placement Calculus; APS = Advanced Placement Statistics; Pre = Pre-Calculus; CI = Calculus I; CII = Calculus II; CIII = Calculus III; nd = no data; blank if no mathematics course taken.

[a]Change Math columns are calculated by subtracting the old mathematics grade from the new mathematics grade; in this case, (first semester grade) minus (high school grade). Thus, change values less than 0 indicate a drop in mathematics grade in the more recent course.

1994) have noted, students who struggle in calculus can be expected to have different attitudes and perceptions than those who do not. However, few prior studies have looked explicitly at the change in students' grades from high school to college and its value in understanding students' experiences in reform calculus.

The most basic pattern that emerged from our analysis of students' grades was that almost all students' overall GPAs dropped in their first semester of collegiate mathematics relative to high school (see Table 2). Participants' mean first semester GPA was 3.17, a drop of 0.67 points from the high school mean GPA of 3.84. Individually, 17 of the 19 students' GPAs dropped. The largest drops were 1.8 (FD) and 1.5 (BL) points. The GPAs of only two students (LS and KK) rose, and those went up only slightly (LS, 0.2; KK, 0.1). Of course, this drop in grades was not unexpected. Many first-year students enter U-M with high GPAs, and the general pattern across years is for students' GPAs to fall when they enter U-M and other prestigious universities (Kirst, 2004; Venezia, Kirst, & Antonio, 2003).

As shown in Table 2, a similar drop took place in mathematics. The mean grade for students' senior year of high school mathematics was 3.43 and the mean for the first semester at U-M was 2.98 – a drop of 0.45 points. Eleven of the 19 students had lower grades in the first semester compared to 12th grade; the largest drop was 2.0 points (DD and FD). Four students' first semester grades were the same, and three students' first semester grades were higher (the highest rise was 1.0 points, for BD). One student, CA, dropped her first semester mathematics course.

This pattern continued for the 13 students who also took mathematics the following semester. The 11 of those 13 students who reported grades to us(whose mean for their first semester mathematics grade was 3.19 and for 12th grade mathematics was 3.56) had a second semester mathematics mean grade of 2.74. This represented a 0.45-point drop from the first semester and a 0.82-point drop from high school. Seven of those 11 students' grades dropped (1.3 points was the largest:

PJ and VJ). Two students' grades did not change; two others' grades rose (0.7 points; MT and DD).

This result – students' grades generally dropping in college in mathematics and in all subjects – is not surprising. Incoming first-year students' grades are likely to drop due to (1) a ceiling effect (there is no room for grades to move in any direction except down) and (2) college courses are generally more difficult than high school courses. However, our interest was in the relationship between a marked change in subject-matter approach (true for mathematics at U-M but not other subjects) and the grade drop in that subject. Determining whether such an effect took place required considering how individual students performed in mathematics *relative* to their overall GPA. In particular, a student whose overall GPA dropped but whose mathematics grade rose presents a very different picture than a student with a similar drop in overall GPA and an equal drop in mathematics.

For example, consider the cases of JC and MT (see Table 2). Both JC and MT did very well in high school, in mathematics (4.0 for both) and in their overall GPA (4.0 and 3.9, respectively). And both struggled in the first semester of mathematics at U-M, with each earning a 2.3, or a drop of 1.7 points. However, JC's overall GPA fell to 2.6, while MT's GPA only dropped to 3.5. From our perspective, JC suggests a case of a student adjusting to college generally; most of his grades, including mathematics, suffered a major drop. In contrast, in MT's case something unusual appeared to happen in mathematics because she did relatively well in all of her classes except for mathematics.

Students like MT, whose grades indicated some particular challenge or difficulty in mathematics, merited further investigation, because of a mismatch between over-all achievement and achievement in mathematics. We measured such *mismatches* by computing the difference between the change in a student's mathematics grade and the change in her term GPA, from high school to the first semester of college (and also from first to the second semester). This "relative change" score (or $R\Delta$) was calculated as follows:

$$R\Delta = \Delta_{\text{Mathematics Grade}} - \Delta_{\text{Overall GPA}}$$

As the examples of MT and JC illustrate, the $R\Delta$ score is a more sensitive measure than using mathematics grades alone. $R\Delta$ gives us control over a poten-tially confounding variable in our analysis – namely, that college is generally harder than high school in all subjects (this observation was explicitly mentioned to us by many of our participants).

The $R\Delta$ scores for all U-M students are given in Table 3 below. Interpreting the $R\Delta$ score requires the identification of a threshold value (or $\delta$) above which this score is considered to be a "meaningful" difference. For example, a $R\Delta$ of 0.1 probably does not indicate a mismatch, while a $R\Delta$ of 0.9 probably does. We chose our working value of $R\Delta$ to be 0.5. Students whose $R\Delta$ score was greater than 0.5 or less than -0.5 were considered to have achievement mismatches. This decision was somewhat arbitrary, based more on face validity than empirical evidence. In particular, values smaller than 0.5 did not generally seem large enough to be signif-icant to students. As it happened, and as described more below, a threshold value of 0.5 divided the sample into halves: nine students experienced a mismatch during at least one term and the other nine did not (there was no data for CA).

TABLE 3. First-Year Relative Change ($R\Delta$) Achievement Scores

| | High School to Fall semester | | | Fall semester to Winter semester | | |
|---|---|---|---|---|---|---|
| | $\Delta$Math | $\Delta$GPA | $R\Delta$ | $\Delta$Math | $\Delta$GPA | $R\Delta$ |
| | | | Mismatch | | | |
| BD | 1 | -0.7 | **1.7** | -1.3 | -0.4 | **-0.9** |
| BL | -0.8 | -1.5 | **0.7** | 0 | nd[a] | nd |
| BM | 0 | -0.1 | 0.1 | -1.3 | -0.7 | **-0.6** |
| DD | -2 | -1.2 | **-0.8** | 0.7 | 0.1 | **0.6** |
| LS | 0.2 | 0.2 | 0 | 0 | -1.1 | **1.1** |
| PJ | 0 | -0.1 | 0.1 | -1.3 | -0.7 | **-0.6** |
| TM | -0.5 | -1.3 | **0.8** | nd | nd | nd |
| MT | -1.7 | -0.4 | **-1.3** | 0.7 | nd | nd |
| VJ | 0.8 | -1.1 | **1.9** | -1.3 | 0.1 | **-1.4** |
| | | | No mismatch | | | |
| BE | -0.3 | -0.3 | 0 | -0.4 | 0.1 | -0.5 |
| CA | nd | -0.2 | nd | nd | nd | nd |
| CM | -1.3 | -1.2 | -0.1 | -0.4 | 0 | -0.4 |
| DJ | -1.3 | -0.8 | -0.5 | nd | nd | nd |
| FD | -2 | -1.8 | -0.2 | nd | nd | nd |
| GD | -0.7 | -0.4 | -0.3 | nd | nd | nd |
| JC | -1.7 | -1.4 | -0.3 | nd | nd | nd |
| KK | 0 | 0.1 | -0.1 | nd | nd | nd |
| MM | -0.7 | -0.4 | -0.3 | nd | nd | nd |
| SB | 0 | -0.1 | 0.1 | -0.4 | -0.7 | 0.3 |

[a]nd = no data; student did not take math, dropped the course, or was dropped from the project.

As Table 3 shows, there were 12 cases of mismatches (indicated in bold) for nine students; four of these nine (VJ, BD, DD, and LS) had mismatches twice. The 12 cases are of two types: Those where the mathematics grade change was greater than the overall GPA change ($R\Delta$ values that were negative), and those where the mathematics grade change was less ($R\Delta$ values that were positive). Mismatches of the first type ($R\Delta < 0$) occured when $\Delta_{\text{Math}} - \Delta_{\text{GPA}} < -\delta$. For example, in the first term, DD was one of two students (with MT) who fell into this category. Both his mathematics grade and his GPA fell in the first semester, but his mathematics grade drop was more severe than his GPA drop. BD, BM, PJ, and VJ fell into this category in the second semester. Mismatches of the second type ($R\Delta > 0$) occured when $\Delta_{\text{Math}} - \Delta_{\text{GPA}} > \delta$. In the first semester, BD, VJ, BL, and TM all experienced this type of mismatch; their mathematics grade change was not as severe as the change in their GPA. In the second semester, LS and DD fit this pattern.

Given that our choice of a 0.5 cutoff was somewhat arbitrary, it is reasonable to ask how our analysis would have changed if we had chosen a different threshold value. In exploring this issue, we found that if the cutoff score changes by 0.1, the number with mismatches only changes by one or two students. For example, increasing the threshold to 0.6 results in eliminating two students from the set of mismatches, while decreasing the threshold to 0.4 results in two more students with mismatches. Given that such small changes in the threshold do not have a large impact on the number of mismatches (there is gradual change only), we chose to stay with the value that we find to have the greatest face validity, which is 0.5. In addition, changing the threshold value up or down slightly did not significantly affect the comparison we make toward the end of this paper between differences noticed and mismatches.

Although the $R\Delta$ measure (a difference of differences) is somewhat complex and difficult to interpret, it does permit sorting students' achievement patterns into useful categories. Indeed, other studies have used composite quantitative measures such as $R\Delta$ as a way to more carefully interpret students' experiences before, during, and after taking a calculus course. For example, see Bonsangue's (1994) CAR or Course Attempt Ratio variable.

Half of our sample did not experience any mismatch. Of the nine who had achievement mismatches, Table 3 shows which students had mismatches in both terms and in which direction. We will return to these nine students and this measure, as a way to further interpret students' experiences.[4]

**3.3. What Students Reported as Different.** With the preceding as background, we now consider the issue that is the main focus of this paper – what students noticed as different between their traditional high school mathematics courses and their reform calculus class(es) at U-M. The results described below were arrived at via close reading and analysis of the transcripts of students' interviews, using the following method. First, students' interviews were transcribed. Next, a list of possible differences was generated in brainstorming sessions with project staff. This list was developed and refined multiple times over a period of several months. Once this list and a framework for differences were formalized, they were used to code interviews and journals. Each kind of difference was given a alphanumeric code, and two independent coders went through all interviews and journals, coding for differences noted. The two coders subsequently met and resolved all disagreements (see Smith & Berk, 2001, for a more detailed discussion of our methodology).

For each student, a distinction was made between differences that were mentioned and those that were repeatedly mentioned. A mentioned difference was one that was expressed by a student at least once in an interview or in a journal entry. For a difference to qualify as repeatedly mentioned, it had to be mentioned a total of at least six times, in at least three different interviews. Because two students only had two interviews (CA and GD), they were excluded from this analysis. The remaining 17 students were interviewed from three to five times (4.5 times, on average).

Table 4 shows the percentage of students who mentioned or repeatedly mentioned each dimension of difference. We use the remainder of this section to discuss and interpret these results, starting with the differences that were mentioned most frequently.

**Group Homework.** We found the difference that was repeatedly mentioned by the most students was group homework. Recall that students were placed by their instructors into groups of four and were assigned a weekly problem set to be done in groups. It was expected that individuals would look over the problem set

---

[4]Given our focus on making students' own voices a part of efforts to evaluate reform calculus, one interesting opportunity that we did not take advantage of concerned students' perceptions of their own grades. Exploring in more depth how students themselves perceived the changes in their own grades would have been a useful complement to our analysis.

TABLE 4. Percentage of Students who Mentioned or Repeatedly
Mentioned Differences

| Category of "difference" | Repeatedly mentioned | Mentioned at least once |
|---|---|---|
| Group homework | 59% | 100% |
| Typical problems | 59% | 100% |
| Teacher/student relationship | 29% | 94% |
| Assessment | 24% | 88% |
| Basic communication | 12% | 76% |
| Content | 12% | 71% |
| Lesson format | 12% | 71% |
| Textbook | 12% | 65% |
| Pace | 0% | 94% |
| Deeper mathematical knowledge | 0% | 65% |
| Classroom management | 0% | 41% |
| Independence | 0% | 41% |
| Class size | 0% | 29% |

on their own and then the group would convene (outside of class time) to discuss
the problems and to write up a formal solution. None of the students in our sample
had participated in formal group work in high school, and all made mention of this
difference at least once. While it was common in high school for informal groups
of friends to complete problem sets and to study together for exams, formal group
problem solving with classmates who were not friends was a new experience.

The students in our sample had a lot to say about group homework. Some
students felt that participating in group homework sessions was a positive aspect
of the course and aided their learning. For example, MM found that her peers'
contributions counteracted what she felt were deficiencies in instruction and the
text.

> I had trouble figuring out what was going on, so I had to rely on
> either my classmates or the book, which I did. I also felt the book
> didn't really clearly explain some of the material. So I think the
> group study homework sessions helped a lot, because then you had
> four guys who were in the same, or three other guys in the same
> situation as me. Not really knowing and just going on minimal,
> you know, what you can pick up in the book, and each person can
> pick up a different type of concept, and together we could make
> one person who actually knew what they were doing. It helped
> out. (MM, 3/26/00 Interview)

However, other students disagreed with MM's assessment of group work, particularly lamenting the fact that the homework groups sometimes did not work together. For these students, the subverting of proscribed group homework practices
seemed to follow from irresponsible behavior from at least some group members.
As SB wrote in her journal,

> My team homework group is the exact antithesis of a team. 'There
> is no I in team'; well, in this team there are about four. We met

yesterday and worked on the problems for about an hour, then
finally decided that it would be a lot easier to just split them up.
We decided that if you need help on your problem, go to the math
lab. (SB, 2/10/00 Journal)

By far the most prevalent aspect of group homework that students commented
on concerned the requirement to explain verbally or in writing one's solution steps
while problem-solving (referred to by students as having to provide "explanations").
Having to provide explanations was mandated by the U-M mathematics department, the Harvard Consortium curriculum, and each instructor. Students were
expected to write several sentences explaining exactly what steps were taken in
solving particular problems on group homework and exams; instructors were responsible individually for communicating these expectations to students and also
for grading students' explanations. No student had had experience writing explanations in their high school mathematics courses and, for many, this feature of
the U-M program was the most salient (and first mentioned) difference that they
raised during interviews. They often felt that this writing requirement was tedious,
unnecessary, and inconsistently implemented. Many questioned the benefits, and
claimed that instructors rarely motivated or justified the requirement explicitly in
class.

Some students did comment on the benefits of writing. For example, PJ commented that writing forced him to focus on "deep understanding of the concepts"
(PJ, 10/26/99 Interview) rather than on merely calculating answers. In fact, even
students who expressed strong dislike for having to write explanations, such as BD,
saw the value of this requirement:

> If you have to explain something to another person or you're in a
> work environment where you have to show a colleague how to do
> something, you have to be able to get across without just putting
> numbers on paper. (BD, 10/8/99 Interview)

**Typical Problems.** Along with group homework, the most frequently mentioned difference was our category "typical problems." It captures students' observations about how the Harvard calculus problems differed from those typically
done in high school. Students indicated that their work in high school mathematics
courses mostly involved "pure" symbol manipulation, such as calculating derivatives and integrals. Recall that almost all of these students took AP Calculus in
high school (see Table 2). However, at U-M the focus was much more on application or "story" problems where expressions or equations had to be derived from
the problem setting before any manipulation was possible. Particularly for group
homework, but also for exams, students commented that most problems were very
wordy, situated the relevant mathematics in the context of a real-world example,
and required some initial sense-making before they could use the mathematical
procedures they knew.

Although the students had solved similar "word" problems in high school, story
problems appeared much more frequently in their U-M courses. As CM commented,

> In high school we really didn't, I mean, we did a few story problems
> but mostly it was just like, you know, problems you know. These,
> the ones that we do now, like even for individual [homework] are

> more like story problems and it's just they're, like, different from
> what I used to do I guess. (CM, 11/4/99 Interview)

PJ drew this contrast even more clearly and explicitly.

> But then depending on the section we did, we would have some
> word problems, like I remember specifically where you got, you
> know, metal plate placed under water or the pressure on certain
> points, and those would be more word problems and there were
> certain sections where it would be all word problems. But I'd say
> typically about somewhere between 50 and 75% [would] be all just
> basic problems and no wording. (PJ, 9/30/99 Interview)

Some students saw the usefulness of doing lots of application problems; students
gave varied evaluations of this substantial shift in the kinds of problems they were
asked to solve. For all students, however, the prevalence of word problems made
the course more difficult. For example, BL felt that the word problems obscured
the content that she was trying to learn:

> Well I don't know if it was the differences in the kind of math, but
> I always had a book where there was a lot of number problems.
> Something simple, but it would be, they'll be like factor this or
> something. Just, I don't know, find the derivative, and now, it's
> like a lot of word problems. My book, I don't know we would have
> like pages and pages of number problems, and my book now isn't
> like that, and I think that those kind of things help. That helped
> me in high school cause then you know how to do it, and with
> word problems, if you don't know how to do what you need to do
> first, then adding all the other stuff in is just going to confuse you
> more. (BL, 12/05/99 Interview).

**Discussion.** As Table 4 indicates, group homework as well as typical problems
(story problems that required understanding first and then explaining of solution
methods) were differences repeatedly mentioned by a majority of our students and
were mentioned at least once by every student. Although other differences emerged
(see below), our data suggest that the introduction of group homework, particularly with the focus on written explanations, and the change in typical textbook
and exam problems were the most significant changes effected by the shift from
traditional high school mathematics programs to the Harvard calculus program at
U-M. We find this to be an interesting result – for what it includes as well as excludes. As we have argued elsewhere (Star, Herbel-Eisenmann, & Smith, 2000),
K-16 reform mathematics curricula are different from more traditional ones along
many dimensions. Those dimensions include differences in typical problems, the
form of expected solutions to those problems, and the shape of typical lessons (e.g.,
group work). But our work at U-M suggests that other dimensions are much less
salient to students. The increased emphasis on conceptual understanding and more
balanced reliance on multiple representations, to name two, are both central features of the Harvard curriculum, yet the U-M students either had difficulty naming
those differences (the focus on conceptual understanding, see below) or never did
(multiple representations).

**3.4. Less Frequently Mentioned Differences.** In addition to the two differences discussed above, Table 4 lists many other differences – the majority of

which have little to do with written curricula. Indeed many participants named features of classroom life in their collegiate mathematics courses that are either independent of the calculus reform movement or are not influenced by the choice of a particular reform calculus curriculum.

**General Differences.** Many of the differences listed in Table 4 are typical, if not inherent aspects of college classes that can be especially problematic for first-year students. For example, 94% of our sample noted the increasing impersonality of the teacher-student relationship, and 88% complained about the implications of this altered relationship for assessment. Also, 76% expressed, on at least one occasion, dissatisfaction with the basic communication patterns of the instructor (e.g., the ability to be understood by students). In addition, 71% noted that the format of college lessons was different; in particular, instructors allowed less time for individual questions. What is more, 94% of students noted that the pace of instruction was much faster than in their high school mathematics courses. Though the frequency of mention of these differences was high, repeated mention was not. We think that the most likely explanation of this pattern is that students came to U-M expecting these differences, based on information they received from parents, alumni, and older peers. Though they were not welcome features of college life, they were anticipated, were not specific to mathematics, and therefore were mentioned less frequently.

**Deeper Mathematical Knowledge.** One final difference that was mentioned by the majority of participants, though not repeatedly, was likely related to design features of the Harvard curriculum. We refer to this difference as a stronger emphasis on gaining deeper mathematical knowledge. On at least one occasion, 65% of participants expressed the belief that their U-M mathematics class expected them to learn the material in a different, deeper way than was necessary in high school. But they struggled to find language to express this judgment. Some, like DD, expressed it as an increase in the expected level of abstraction.

> Right now it is a lot more abstract than what we did in high school, but it is getting to the point the long way. In high school we cut right to the chase. I think that in the end this way of learning will be better. (DD, 12/6/99 Journal)

For PJ, the difference was between knowing how to execute a procedure (only) versus also understanding what you are doing and why – essentially the distinction that Skemp (1976) has drawn between "instrumental" and "relational" understanding. When asked to advise prospective high school seniors on how the U-M program was different from high school, PJ said,

> I'd probably just tell them to make sure that you understand the concepts fully of calculus, not just the sort of problem you're going to be given, but understand why, why and how you do each problem. Not just being able to go out and do say a logarithm problem, it's, it's harder than just being able to do it. You got to know what you're doing and why you're doing it. (PJ, 9/30/99 Interview)

MT linked her instructor's focus on understanding the central concepts to the greater depth he brought to his treatment of calculus.

> Yeah, well, he's focusing on the details, but if we get them wrong,
> like little minor stuff, he doesn't really worry about as long as we
> understand, he wants to make us understand the concept. A lot of
> times he even goes farther then what's in the book, oh yeah, and
> they got this from some guy 500 years ago, and etc. We learn about
> a whole bunch of different stuff, and then it makes us understand
> the concept we're working on even better. So that's what he's
> worried about more than just, oh yeah, you didn't get the right
> answer, that's wrong. (MT, 3/26/00 Interview)

SB expressed this shift as a move from the high school focus on the product to an
understanding of the process at U-M.

> They take what you're learning like in high school and go and just
> take it a step further, make it a little more in depth, you know,
> like you understand what's going on, not just what the results are
> but the process of getting the result. (SB, 12/16/99 Interview)

These students struggled to find words to convey a similar feature of their collegiate mathematics experience: that their U-M mathematics courses aimed to develop deeper, richer, more conceptual knowledge than did their high school courses.
Despite the fact that it was not repeatedly mentioned, we found this difference significant for two reasons: it corresponded to a central design feature of the Harvard
materials, and it named classroom and cognitive phenomena that students found
difficult to describe but worthy of report. As is evident in the series of quotes, students had difficulty pinpointing the features responsible for this difference. Some
attributed it to the orientation of their instructor; some saw connections between
other differences, such as group homework or the focus on explanations, and the
emphasis on deeper understanding. Most merely noted it as a feature of their mathematics classes and were unable to express the ways that the course, their instructor,
or the U-M mathematics department instantiated this different emphasis.

**3.5. Connecting Differences and Achievement.** If citing differences could
be seen as evidence of "complaint," then it would be reasonable to expect that
students who struggled to achieve good grades in their mathematics classes would
mention more differences. On this hypothesis, struggling students would have more
reason to complain. We explored this conjecture in a number of ways.

First, we sorted students into two groups – those who experienced a "mismatch," (as defined above) and those who did not. Then we examined the differences mentioned or repeatedly mentioned by each group. We were especially
interested in looking at students who had a mismatch of the first type ($R\Delta < -\delta$),
as these students struggled particularly in their mathematics courses. Recall that
for the six students who experienced this type of mismatch (DD, MT, BD, BM,
PJ, and VJ), their mathematics grade change was more severe or negative by at
least 0.5 points as compared to their overall GPA (e.g., the mathematics grade
dropped by 1.0 points, while the GPA only dropped by 0.5 points). Row 1 of Table
5 displays the results of this analysis.

Students who experienced this type of mismatch mentioned, on average, 9.3
differences and repeatedly mentioned 2.8 differences. These average values did not
significantly differ from those students who did not experience such a mismatch
(9.3 differences mentioned and 1.8 differences mentioned repeatedly). This result

TABLE 5. Comparisons of Achievement and Noticed Differences

| Row | Criteria for comparing differences mentioned | n | Repeatedly mentioned M(SD) | | Different? | Mentioned M(SD) | | Different? |
|-----|-----|-----|-----|-----|-----|-----|-----|-----|
| | | | Yes | No | p-value | Yes | No | p-value |
| 1 | $R\Delta < -0.5$ (math grade change 0.5 points more severe than overall GPA in either term) | 6 Yes; 11 No | 2.8 (2.4) | 1.8 (1.1) | 0.24 | 9.3 (1.6) | 9.3 (2.1) | 0.95 |
| 2 | $R\Delta < 0$ (math grade change more severe than overall GPA in either term) | 13 Yes; 4 No | 2.3 (1.8) | 1.8 (1.0) | 0.58 | 9.3 (2.1) | 9.7 (1.2) | 0.96 |
| 3 | Math grade dropped more than 0.5 points in any term as compared to previous term | 12 Yes; 5 No | 2.3 (1.9) | 1.8 (1.1) | 0.56 | 8.9 (1.8) | 10.2 (1.9) | 0.21 |
| 4 | Math grade dropped more than 0.5 points in the fall term as compared to high school | 8 Yes; 9 No | 2.6 (2.1) | 1.8 (1.2) | 0.31 | 9.0 (2.0) | 9.6 (1.9) | 0.56 |
| 5 | Math grade dropped more than 0.5 points in the winter term as compared to the fall term | 4 Yes; 7 No | 1.8 (1.5) | 3.1 (1.9) | 0.24 | 8.8 (1.7) | 10.4 (1.4) | 0.11 |
| 6 | Math grade dropped by any amount in any term as compared to the previous term | 15 Yes; 2 No | 2.2 (1.7) | 2.0 (1.4) | 0.88 | 9.1 (2.0) | 10.5 (0.7) | 0.35 |
| 7 | Math grade dropped by any amount in the fall term as compared to high school | 10 Yes; 7 No | 2.5 (1.9) | 1.7 (1.3) | 0.36 | 9.3 (2.2) | 9.3 (1.5) | 0.99 |
| 8 | Math grade dropped by any amount in the winter term as compared to the fall term | 6 Yes; 5 No | 1.8 (1.2) | 3.6 (1.9) | 0.11 | 9.7 (2.2) | 10.0 (1.0) | 0.76 |

indicates that students who struggled more in their mathematics classes as compared to their overall GPA did not notice significantly more differences than their non-struggling peers.

To further explore this result, we reduced our threshold value for this type of mismatch from $\delta = 0.5$ to $\delta = 0$. In other words, we looked at those students whose grade change in mathematics was more severe by as little as 0.1 points as compared to their GPA. As Row 2 of Table 5 indicates, we again failed to find significant differences between the groups.

To confirm this non-relationship, we performed several additional analyses involving students' mathematics grades alone, without considering overall GPA. We looked at students whose mathematics grade dropped by more than 0.5 points in any term (Table 5 Row 3), in the Fall term only (Row 4), and in the Winter term only (Row 5); we also looked at students whose mathematics grade dropped by any amount, in any term (Row 6), in the Fall term only (Row 7), and in the Winter term (Row 8). For none of these comparisons did we find significant differences, either in differences mentioned or in those mentioned repeatedly.

We believe that these analyses provide strong evidence that there was not a strong relationship between noticing differences between high school and reform calculus courses and academic difficulties in calculus. This finding has important implications. First, contrary to the belief that students are "ruled" by their grades, achievement patterns did not determine students' experiences more broadly. Other factors were at play. Second, one such candidate factor was how students saw their "new" program as different from their "old" program. And here the conceptual categories they cited (e.g., Group Homework and Typical Problems) and the language they used to describe those categories were important.

## 4. Discussion

These results raise many issues of current concern to collegiate mathematics educators, especially those who have implemented or are considering implementing reform of their calculus program. In this section we identify some of these issues and consider how our data and results bear on them. Where appropriate, we also point to questions for further study, both for our own research team and for other researchers who seek to understand the mathematical experiences of collegiate calculus students.

**4.1. What Changes When Reforms are Implemented?** Our study suggests that the easiest thing to change is the written curriculum (and the associated assessments), and fundamentally different (written) calculus curricula can make a deep impact on students' behavior and thinking. As we have seen, the focus in the Harvard Consortium materials on problems embedded in realistic contexts, clear written explanations, and deeper conceptual understandings forced the students in this study to adjust how they did mathematics, if not what they thought was mathematically appropriate.[5] Work on word problems in traditional high school programs does not seem to prepare students for these foci. Our students generally did not have ready competency for this work. What appears to be much harder to change is instruction, or how instructors orchestrate students' activity and learning during class time (we return to this issue below). Somewhere in the middle lies U-M's mandated feature of group homework, which is not a part of the written Harvard curriculum. Like the differences listed above, required group homework also represented a real change for students, even as they acted against its principles and found more traditional ways of completing these assignments. In addition, our study finds that not all top-level features of a written curriculum will necessarily make a deep impact on students. In other words, the differences that might be identified as most salient by curriculum developers and researchers (e.g., the presence of multiple representations and the "rule of four" in the Harvard Consortium materials) differed from what students noticed as most prominent (e.g., a difference in the typical problems). Students did not notice (or did not find important enough to merit reporting) a marked increase in tabular and graphical representations in the Harvard materials.

**4.2. Group Work Collides With Students' Beliefs.** One of the most common features of the diverse collection of reform calculus programs is intentional support for collaborative learning, in and outside of the classroom. The U-M program attempted to support group work in both settings, by seating students in groups of four in class, sometimes actively encouraging their interaction, and by carefully structuring required group homework away from class. In both settings, we found evidence that socially responsible and fully collaborative problem-solving violates most students' views of mathematics learning (and perhaps school learning more generally) as an individual enterprise. Collaboration in high school was

---

[5]Here we see our data as offering a mixed result: Some students were convinced of the value of the Harvard approach, while some clearly were not.

usually a matter of getting help when needed, and group work was working with friends. For most, understanding was achieved on one's own time and terms. This contradiction between instructional practice and students' beliefs about learning accounts for the frequent mention of group work, especially group homework, as an important difference from high school. But it constitutes a significant challenge to any reform calculus program that seeks to emphasize and support collaboration.

**4.3. Where Will Good Instructors Come From?** The main result from our classroom observations was that instruction differed little from traditional instruction in small sections. This must be considered in light of the U-M mathematics department's challenge to staff so many sections with qualified instructors, whether new faculty, post-docs, or graduate students. The department is keenly aware of the gap between candidate instructors' knowledge and experience and the demands of teaching this program and has devoted substantial resources to training new instructors. But as the research literature on K-12 mathematics teachers and teaching shows (e.g., Borko, et al., 1992), reorienting teachers' fundamental beliefs about mathematics, teaching, and learning is difficult, if not impossible, with short-term interventions. We do not foresee the department resolving this problem completely as long as the teaching that most instructors have experienced (and therefore tend to re-enact) is so different from the intended U-M model. More generally, the U-M case suggests that changes in teaching practice may be the most difficult dimension of calculus reform programs to put in place (see also Speer, 2001). This "non-change," however, makes the impact of the curriculum (and group homework practices) more significant. Students felt their impact even without a supporting change in instruction.

**4.4. Calculus Reform and Achievement.** Did calculus reform affect student achievement? Though the U-M program clearly affected how students experienced freshman collegiate mathematics relative to high school, there is no clear evidence in our data that it changed or altered their mathematics achievement. We will never know, first of all, how our 19 students would have fared in more traditional calculus courses. More importantly, our students struggled more with their U-M coursework generally (again, relative to high school) than they struggled with the Harvard calculus. Judged against the generalization that "college is more difficult than high school," there was no clear pattern of change in mathematics achievement. In the Fall semester, mathematics grades generally followed GPAs downward, but with more cases of rosier change in mathematics achievement than in overall GPA. The Winter semester reversed this order (with two more cases of significant decline in mathematics achievement relative to GPA change than significant increase) but small numbers restrict conclusions that might be drawn. Considered from another perspective, the felt impact of reform, as indicated by the number of differences mentioned, did not correlate with achievement. Students' reactions to and judgments about the U-M program were relatively independent of their grades.

This project explicitly set out to move beyond grades as an indicator of students' experiences, and our data confirm that patterns in achievement tell only a partial story of those experiences and fail to correlate with other dimensions of impact. Here we have analyzed and reported "impact" in terms of how students see their reform calculus courses as different from high school mathematics. But we think it would also be relevant to track and analyze other dimensions of students' experience, such as their attitudes towards mathematics, their beliefs about mathematics and themselves as learners, and potential changes in their actions and strategies for learning mathematics in their new courses. We have such data in hand and are currently undertaking its analysis (see Smith, Star, & Hoffmann, 2002).

**4.5. On Generalizing From Our Results.** The small number of participants in this study raises legitimate questions about whether our results can be generalized to all of U-M's introductory calculus program. We offer the following thoughts on this issue. First, a study with only 19 participants, which is only a small fraction of U-M students who take calculus each year, cannot claim to generalize in the traditional statistical sense.

However, we believe that our sample of 19 students is reasonably representative of the population of students who take introductory calculus at U-M. There were 9 males and 10 females in our sample; there were 8 engineering majors and 10 students who had not yet chosen a major; one of the 19 dropped her mathematics course in the first semester; one dropped out of college after a year; and one transferred to another institution after a year. The representation in our sample for each of these student characteristics is roughly equal to that of the entire student population taking introductory calculus. Although our sample was self-selected (students volunteered to participate in our study), we felt that we ended up with a sufficiently representative group. Had our sample not been representative of the population, we would likely have continued our recruiting efforts until we obtained a more representative sample. We believe that the issues and trends that we noticed, particularly the ones for which there was overwhelming unanimity, are likely to be represented in the general population of students enrolled in introductory calculus at U-M. For example, given that 100% of our sample mentioned group homework and typical problems as being different, we would expect to find similar perceptions within the population as a whole.

Furthermore, although we had only 19 participants, we know a lot about these 19 students. We understand these 19 students' experiences in great depth, and this allows us to feel cautiously confident in generalizing particular claims to the entire U-M introductory calculus population. These 19 students are not isolated cases but are typical ones; through these few students, we have been able to identify common concerns and issues that we would expect to see more generally.

**4.6. Conclusion.** Our project was grounded on the assumption that students' experience should be one factor considered in current debates about collegiate calculus reform and K-16 mathematics education reform more generally. Rather than quoting ideologically suitable but potentially isolated individual cases, views of students' experience from more sizable samples are needed. That said, we also consider it a serious mistake to equate or even closely associate "how students do" in reform classes with "the worth of reforms." The broader question of the desirability of pursuing current reform principles cannot be answered from portraits of students' experience alone. Where expectations from "traditional" practices do not match the demands of "new" curricula, tensions necessarily arise, especially in the short-term. But what should influence our national direction most is a shared view of what skills, competencies, and values we want for our students at the end of their mathematical educations.

## Acknowledgments

In addition to the authors, the rest of the Mathematical Transitions Project staff played vital roles in the work reported here. These other team members, in alphabetical order, were Dawn Berk (University of Delaware), Carol Burdell (Michigan State University), Beth Herbel-Eisenmann (Iowa State University), Amanda Jansen Hoffmann (University of Delaware), Violeta Lazarovici (Michigan State University), and Gary Lewis (Michigan State University). We also acknowledge the fine work of our supporting undergraduate student analysts at the University of Michigan: Alissa Belkin, Rachel Catt, Matthew Dilliard, Adrienne Frogner-Howell, Andrea Kaye, Abby Magid, Kelly Maltese, Brian My, William Nham, Sharon Risch, Samantha Spatt, Brian Walby, and Heidi Walson. Thanks as well to Natasha Speer and several anonymous reviewers for comments on drafts of this paper. We are grateful for the assistance and support we received from the faculty, graduate students, and staff of mathematics department at the University of Michigan, particularly Pat Shure. The project is supported by a grant from the National Science Foundation to the second author (REC-9903264). Any opinions, findings and conclusions or recommendations expressed in this material are those of the author(s) and do not necessarily reflect the views of the National Science Foundation.

## References

Armstrong, G., Garner, L., & Wynn, J. (1994). Our experience with two reformed calculus programs. PRIMUS, *14*, 301–311.

Beidleman, J., Jones, D., & Wells, P. (1995). Increasing students' conceptual understanding of first semester calculus through writing. PRIMUS, *5*, 297–316.

Bonsangue, M. V. (1994). An efficacy study of the calculus workshop model. In E. Dubinksy, A. H. Schoenfeld, & J. Kaput (Eds.), *Research in collegiate mathematics education. I.* (pp. 117–137). Providence, RI: American Mathematical Society.

Borko, H., Eisenhart, M., Brown, C. A., Underhill, R. G., Jones, D., & Agard, P. C. (1992). Learning to teach hard mathematics: Do novice teachers and their instructors give up too easily? *Journal for Research in Mathematics Education, 23*, 194–222.

Connally, E., Hughes-Hallett, D., Gleason, A., Avenoso, F., Cheifetz, P., Hillyer, J., et al. (1998). *Functions modeling change: A perspective for calculus* (preliminary ed.). New York: Wiley and Sons.

Darken, B., Wynegar, R., & Kuhn, S. (2000). Evaluating calculus reform: A review and a longitudinal study. In E. Dubinksy, A. H. Schoenfeld, & J. Kaput (Eds.), *Research in collegiate mathematics education. IV.* (pp. 16–41). Providence, RI: American Mathematical Society.

Edwards, H. C. (1993). Student surveys and calculus reform. *PRIMUS, 3*(2), 133–140.

Friedman, M. L. (1993). Research - The sparse component of calculus reform - Part II: An investigation of calculus instruction. *PRIMUS, 3*(2), 113–123.

Ganter, S. L., & Jiroutek, M. R. (2000). The need for evaluation in the calculus reform movement: A comparison of two calculus teaching methods. In E. Dubinksy, A. H. Schoenfeld, & J. Kaput (Eds.), *Research in collegiate mathematics education. IV.* (pp. 42–62). Providence, RI: American Mathematical Society.

Ganter, S. (2001). *Changing calculus: A report on evaluation efforts and national impact from 1988-1998.* Washington, DC: Mathematical Association of America.

Hughes-Hallett, D., McCallum, W., Gleason, A., Quinney, D., Flath, D., Osgood, B., et al. (1994). *Calculus.* New York: Wiley and Sons.

Keith, S. Z. (1995). How do students feel about calculus reform, and how can we tell? *UME Trends 6*(6), 3–31.

Kirst, M. W. (2004). The high school/college disconnect. *Educational Leadership 62*(3), 51–55.

Schoenfeld, A. H. (1989). Explorations of students' mathematical beliefs and behavior. *Journal for Research in Mathematics Education, 20*, 338–355.

Schwingendorf, K. E., McCabe, G. P., & Kuhn, J. (2000). A longitudinal study of the C4L calculus reform program: Comparisons of C4L and traditional students. In E. Dubinksy, A. H. Schoenfeld, & J. Kaput (Eds.), *Research in collegiate mathematics education. IV.* (pp. 63–76). Providence, RI: American Mathematical Society.

Shure, P., Brown, M., & Black, M. (1999). *Instructors' guide.* Ann Arbor: University of Michigan, Department of Mathematics.

Smith, J., & Berk, D. (2001, April). *The Navigating Mathematical Transitions Project: Background, conceptual frame, and methodology.* Paper presented at the annual meeting of the American Educational Research Association, Seattle, Washington.

Smith, J., Star, J. R., & Hoffmann, A. (2002, April). *Students' experiences moving between "traditional" and "reform" curricula: What are the implications for K-16 education?* Paper presented at the Research PreSession of the annual meeting of the National Council of Teachers of Mathematics, Las Vegas, Nevada.

Skemp, R. (1976). Relational understanding and instrumental understanding. *Mathematics Teaching, 77*, 20–26.

Speer, N. M. (2001). *Connecting beliefs and teaching practices: A case study of teaching assistants in reform-oriented courses.* (Doctoral dissertation, University of California, Berkeley, 2001). *Dissertation Abstracts International, 63*(02), 533A.

Star, J. R., Herbel-Eisenmann, B., & Smith, J. P., III. (2000). Algebraic concepts: What's really new in new curriculum? *Mathematics Teaching in the Middle School, 5*, 446–451.

Tucker, J. & Leitzel, A. (1994). *Assessing calculus reform efforts: A report to the community.* Washington, DC: Mathematical Association of America.

Venezia, A., Kirst, M. W., & Antonio, A. L. (2003). *Betraying the college dream: How disconnected K-12 and post-secondary education systems undermine student aspirations.* Palo Alto, CA: Stanford Institute for Higher Education Research.

Walker, R. (1999). *Students' conceptions of mathematics and the transition from a standards-based reform curriculum to college mathematics.* (Doctoral dissertation, Western Michigan University, 1999). *Dissertation Abstracts International, 61*(02), 887B.

## Appendix

Participants' High School Mathematics Course Sequences

| Student | Course Sequence | Additional notes |
|---------|-----------------|------------------|
| BL | Standard[a] | |
| BE | Standard | |
| BD | Standard | |
| BM | Standard | Used UCSMP; did not take Calculus; took AP Stats |
| CA | Standard | |
| CM | Standard | |
| DJ | Standard | Used UCSMP; did not take Calculus |
| DD | Standard | Used a reform curriculum in 9th grade |
| FD | Standard | |
| GD | Standard | |
| JC | Standard | |
| KK | Standard | Used UCSMP; took Calculus at a community college |
| LS | Standard | |
| MM | Standard | |
| MT | Standard | Used UCSMP; did not take Calculus |
| PJ | Standard | |
| SB | Standard | Used reform curricula in 9th and part of 10th grade |
| TM | Standard | |
| VJ | Standard | |

[a]Standard sequence is Algebra I, Geometry, Algebra II (Advanced Algebra), Pre-Calculus (Analysis, Trigonometry, and Analytic Geometry), and Calculus.

COLLEGE OF EDUCATION, 513C ERICKSON HALL, MICHIGAN STATE UNIVERSITY, EAST LANSING, MI, 48824.

*E-mail address*: jonstar@msu.edu

COLLEGE OF EDUCATION, 513H ERICKSON HALL, MICHIGAN STATE UNIVERSITY, EAST LANSING, MI, 48824.

*E-mail address*: jsmith@msu.edu

CBMS Issues in Mathematics Education
Volume **13**, 2006

# Effects of Concept-Based Instruction on Calculus Students' Acquisition of Conceptual Understanding and Procedural Skill

## Kelly K. Chappell

ABSTRACT. The quantitative component of this study involved 144 calculus students and 4 instructors and investigated the effects of instructional environment (concept-based or traditional) on student performance on common assessments that measured conceptual understanding, procedural skill, and ability to extend knowledge to unfamiliar problems. All four instructors were selected for the study based on documented teaching accomplishments. The two traditional instructors emphasized the learning and practice of procedures and de-emphasized concept development. The two concept-based instructors emphasized concept development and de-emphasized the learning and practice of procedures. The students enrolled in the concept-based learning environment scored significantly higher than the students enrolled in the traditional learning environment on the *Conceptual Understanding Subscale* of both the midterm examination ($p < .001$) and the final examination ($p = .01$). These students also scored significantly higher on the *Procedural Skill Subscale* of the midterm examination ($p = .005$). There was no significant difference in students' scores on the *Procedural Skill Subscale* of the final examination. An analysis of student responses to select examination items provided insight into the nature of student performance on tasks requiring them to extend knowledge to new situations. Interview and questionnaire data gave a lens through which to view the quantitative data and provided a framework for analyzing responses and patterns of explanation on extension questions. The present study goes beyond the quantitative, offering qualitative evidence that makes a case for the "why" behind the reported differences between the two groups.

## 1. Introduction

The National Council of Teachers of Mathematics (NCTM, 1989) and the Mathematical Association of America (MAA, 1988) have advocated curriculum and teaching that emphasize conceptual understanding rather than the practice of routine skills. These organizations have criticized the "traditional" curriculum for an over reliance on symbolism, manipulative skills, and the rote memorization of facts at the expense of conceptual development. They have argued that many students learn and retain mathematics best in a "reform" environment where mathematics is introduced in a context related to student experiences, where oral and written communication of mathematical ideas is emphasized, and where students are actively involved in the learning process. Such reform agendas stress that the role of the teacher is no longer to present mathematics as a collection of facts and procedures to be memorized. Instead, it is to prepare students to think critically about mathematical situations, analyze problems through a variety of lenses, and communicate conceptual understanding in addition to mastering procedures.

More than five hundred mathematics departments at post-secondary institutions nationwide have engaged in discussions about calculus reform, strategized reform efforts, and implemented components of reform (Ganter, 1997). Mathematics departments at secondary and post-secondary institutions nationwide are divided about the merits of reform (Alarcon & Stoudt, 1997; Armstrong & Hendrix, 1999; Dreyfus & Eisenberg, 1990; Wilson, 1997). The discussion is often reduced to a debate between "traditional" approaches and "experimental or reform" approaches to mathematics instruction (Schoenfeld, 1994). Many teachers of secondary and collegiate mathematics believe that in order to develop student understanding some degree of procedural competency must be sacrificed. Due to a perception that conceptual understanding and procedural skill are not strongly related, they fear that if time were devoted to the development of conceptual understanding, students would fail to master the procedures of mathematics.

A number of publications have described the classroom implementation of reform instruction but few have included empirical research results (Bookman & Friedman, 1994; Ganter, 1997; Heid, Blume, Flanagan, Iseri, Deckert, & Piez, 1998). The lack of empirical evidence evaluating the effects of calculus instructional reform has created an environment of uncertainty (Ganter, 1997). This uncertainty has resulted in the withdrawal of faculty support for reform calculus instruction (Alarcon & Stoudt, 1997; Armstrong & Hendrix, 1999; Wilson, 1997). Thus, calculus courses are ideal testing grounds for the effects of instructional environment on students' conceptual understanding and procedural knowledge. Both conceptual understanding and procedural skills are extremely important for success in higher-level mathematics courses. In fact, it is hard to imagine an instructor who teaches procedures with no explanation of the related concepts. It is equally hard to imagine an instructor who develops concepts without ever connecting them to the related procedures. Therefore, it is particularly important to investigate how best to combine the two types of instruction.

The quantitative component of the study was designed to investigate the effects of instructional environment (concept-based or traditional) on the performance of calculus students on common assessments that measure conceptual understanding, procedural skill, and ability to extend knowledge to unfamiliar problems. The initial goal was to answer the following two questions: If an instructor chooses to emphasize the learning and practice of procedures and de-emphasize concept development, will students be able to perform well on examinations that require the extension of knowledge to new situations and that require students to explain verbally or graphically why a particular procedure makes sense? If an instructor chooses to emphasize concept development and de-emphasize the learning and practice of procedures, will students be able to perform well on examinations that require the recall and use of procedures and that require the extension of knowledge to new situations? Interview and questionnaire data provided a lens through which to view the quantitative data and provided a framework for analyzing student responses and patterns of explanation on extension questions.

**1.1. Review of Related Literature.** At the high school and post-secondary levels, many published studies that have empirically evaluated the effects of various instructional approaches on student achievement have compared the achievement of students enrolled in technology-enhanced courses to the achievement of students enrolled in traditional courses. In technology-enhanced courses, use of technology has decreased the amount of instructional time required for skill development and increased the amount of instructional time devoted to concept development.

With few exceptions (Meel, 1998), students from technology-enhanced courses have performed significantly better on measures of conceptual understanding than their traditionally taught counterparts (Bookman & Friedman, 1994; Heid, 1988; Hollar & Norwood, 1999; O'Callaghan, 1998; Palmiter, 1991; Park & Travers, 1996; Schwarz & Hershkowitz, 1999; Stiff, McCollum, & Johnson, 1992; Tall & Thomas, 1991). Some of these studies also compared the achievement of students in traditional courses to the achievement of students enrolled in technology integrated courses on measures of procedural skill, finding no significant difference in the computational capabilities of the two groups of students (Heid, 1988; Hollar & Norwood, 1999; Meel, 1998; O'Callaghan, 1998; Palmiter, 1991; Park & Travers, 1996; Stiff et al., 1992).

In a study conducted by Heid (1988), the first twelve weeks of two experimental classes were devoted to conceptual development through graphical and symbol-manipulation computer programs. Only the last three weeks of these experimental classes were devoted to skill development. In the control class, the instructor taught algorithmic skills the entire fifteen weeks. Heid found that the students enrolled in the experimental classes demonstrated better conceptual understanding than their traditionally taught peers. In addition, the conceptually taught students performed almost as well on a final examination of routine skills. Heid suggested that large differences in class size could have contributed importantly to differences in understanding. Fewer than 20 students were in each of the experimental classes, whereas 100 students were in the traditional large lecture class.

Palmiter (1991) investigated the effects of instructional environment on the performance of two groups of university students enrolled in an integral calculus course. The same conceptual material was presented to both the experimental and control groups. The experimental group ($n = 40$) was taught to compute antiderivatives and definite integrals using a computer algebra system (MACSYMA). The control group ($n = 41$) used paper-and-pencil methods to compute antiderivatives and definite integrals. The experimental group scored significantly higher than the traditionally taught group on a conceptual knowledge test and a computational exam. Palmiter suggested that the study could be biased by several underlying factors: first, the students enrolled in the experimental class may have performed better because they were fully aware of their participation in the study; second, the superior performance of the experimental group could be attributed to the fact that this group used the computer algebra system to perform the computations required on the computational test, whereas the control group did not; and third, the study involved relatively small sample sizes.

Bookman and Friedman (1994) compared the problem solving performance of second semester calculus students enrolled in a Project CALC section ($n = 23$) with the performance of second semester calculus students enrolled in two traditionally taught sections ($n = 42$). The traditional course, which met for three hours per week, emphasized the acquisition of paper-and-pencil computational skills. The Project CALC course, which met for three 50-minute sessions and one two-hour laboratory session per week, emphasized the development of conceptual understanding via interactive computer laboratories, investigation of real-world calculus problems, cooperative learning, open book tests, and extensive student writing.

The authors reported that, at the end of two semesters of calculus, students enrolled in a Project CALC section performed significantly better than students enrolled in the two traditionally taught sections on a set of five problems that required students to mathematically interpret non-routine verbal problems. The five problems were novel to both groups, were based on content that had been taught to both groups, and were to be done with open book, open notes, paper, pencil, and a scientific calculator.

The results of a replication study, in which a revised version of the test was administered to three sections of project CALC students ($n = 65$) and eleven sections of traditionally taught students ($n = 205$), were consistent with the results of the first study. The authors suggested that the test was skewed toward Project CALC's goals, since Project CALC placed greater emphasis on real-world problem solving. The test was not designed to evaluate the degree to which students from the different learning environments had mastered the procedures of calculus.

Park and Travers (1996) investigated the relative performance of students enrolled in a *Calculus and Mathematica* course and students enrolled in a standard course. *Calculus and Mathematica*, a computer laboratory calculus course with weekly discussion sections and no lectures, de-emphasized procedures and closed-form mechanical exercises and emphasized concept development, visualization of ideas, and solving open-ended realistic problems. The standard course was lecture oriented and did not involve the use of computers. Each course met for the same number of hours per week. Their "show all work" achievement test was comprised of eight items designed to measure conceptual understanding and eight items designed to measure computational proficiency. The results suggest that the *Calculus and Mathematica* group ($n = 26$) obtained a higher level of conceptual understanding than the standard group ($n = 42$) without loss of computational proficiency. In addition, concept maps and student interviews used to assess understanding of key concepts yielded findings favorable to the *Calculus and Mathematica* group. The authors suggested that the self-selection of students into the learning environments was a limitation of the study in that the *Calculus and Mathematica* group might have been more energetic and adventurous than the standard group.

Meel (1998) examined the differences in understandings of two groups of third semester honors calculus students in the areas of limit, differentiation, and integration on a technology-restricted written test, in a technology-permitted problem solving interview, and during an understanding interview. The experimental group ($n = 16$) was comprised of students enrolled in a *Calculus and Mathematica* course and the control group ($n = 10$) consisted of students enrolled in a traditional calculus course. The two groups were not significantly different in terms of overall performance on the technology-restricted written test, a paper-and-pencil instrument composed of ten open-ended items. However, the traditionally taught students performed significantly better than the *Calculus and Mathematica* students on items associated with the limit concept, items presented in a text-only format, and conceptually-oriented items. No significant differences were found between the two groups on the items associated with differentiation and integration, items presented in a text-and-pictorial format, and procedurally oriented items. In the problem solving interviews, the *Calculus and Mathematica* students were more successful and flexible in solving real-world problems than the traditionally taught students. The student responses to the understanding interview questions revealed that both groups had difficulties explaining limits and differentiation but displayed formalized understanding of integration.

Chappell and Killpatrick (2003) investigated the effects of instructional environment (concept-based or traditional) on students' conceptual understanding and procedural knowledge of calculus. The original study involved 305 college-level calculus students and eight instructors. A replication study involved 303 college-level calculus students and eight instructors. Though students were not allowed to use a computer algebra system on examinations, they were allowed to use a TI-82 or TI-83 graphing calculator on examinations.

In both the original study and the replication study, there were no significant differences in the measures of students' abilities to employ procedural skills. However, students enrolled in the concept-based learning environment scored significantly higher than students enrolled in the traditional learning environment on assessments that measured both conceptual understanding and procedural skills ($p < .001$).

These findings may be limited since only one concept-based instructor was involved in both the original and the replication study. The possibility exists that the concept-based instructor may have been a better teacher than the majority of the traditional instructors. The authors underscored the need for a replication study that involved a balance between the number of concept-based instructors and the number of traditional instructors, as well as a design where teaching ability was controlled for by selecting instructors of comparable effectiveness.

In all of the above comparative studies, the amount of instructional time devoted to skill development was decreased in the experimental course and the amount of instructional time devoted to conceptual development was increased. A preponderance of evidence has suggested that instructional programs emphasizing concept development fosters conceptual understanding without sacrificing skill proficiency among calculus students.

Darken, Wynegar, and Kuhn (2000) conducted a comprehensive literature review of quantitative investigations based on reform calculus textbooks. They found seven studies that compared reform and traditional student performance on common examinations or parts of examinations. Five studies (Hershberger & Plantholt, 1994; Holdener, 1997; Lefton & Steinbart, 1995; Penn, 1994; Tidmore, 1994) compared student performance on traditional problems and found no differences between the two groups, though only two studies conducting statistical tests to confirm this result (Holdener, 1997; Penn, 1994). Six of the studies compared graphical understanding, conceptual understanding, or problems solving ability; three found that the reform students performed significantly better than the traditional students (Holdener, 1997; Judson, 1994; Penn, 1994), two found results favorable to the reform students but did not report significance (Tidmore, 1994; Hershberger & Plantholt, 1994), and one found that traditional students performed significantly better (Brunett, 1995). Darken et al. (2000) concluded from their meta-analysis that students using reform calculus texts usually performed as well as traditional students on traditional test questions and perhaps better than traditional students on conceptual or graphical questions. Darken, et al. also indicated a need for larger in-depth investigations into the conceptual understanding of reform calculus students, noting that such in-depth evidence was hard to come by and that all of the studies that had investigated this key issue had involved small or even very small sample sizes.

**1.2. The Present Study.** The quantitative component of the study reported on here involved 144 calculus students and 4 instructors and investigated the effects of instructional environment (concept-based or traditional) on the performance of calculus students on common assessments that measured conceptual understanding, procedural skill, and ability to extend knowledge to unfamiliar problems. The two traditional instructors emphasized the learning and practice of procedures and de-emphasized concept development. The two concept-based instructors emphasized concept development and de-emphasized the learning and practice of procedures.

The study used a common midterm examination and a common final examination to measure the degree to which students from the two learning environments had mastered the concepts and procedures in first semester calculus. The following design features were incorporated into the study to maximize its contribution to the knowledge base regarding how instructional environment impacts students' understanding and skill in calculus courses:

- Instructors were selected based on documented teaching accomplishments to ensure that any effects would be due to the treatment, and not to teaching effectiveness.
- Although students self-selected into sections, they were unaware of their instructors' identities and of the learning environment into which they enrolled.
- Differences among class sizes were small and the difference between the number of students enrolled in each condition was small.
- The use of a computer-based algebra system, MATLAB, was consistent across learning environments.
- In order to address the concern that concept-based instruction "waters down" by-hand skills, students were not allowed to use a computer algebra system on examinations. However, students were allowed to use the capabilities of the TI-82 or TI-83 graphing calculator on examinations.

The study gathered evidence regarding the conceptual understanding of reform students through analysis of student responses to specially constructed *Conceptual Understanding Subscale* tasks. These tasks required the extension of knowledge to unfamiliar situations. Student responses were analyzed and coded to identify prevailing patterns. Additionally, written questionnaires and classroom interviews explored students' perspectives regarding the effects of concept-based instruction on their acquisition of procedural skills and on their ability to extend knowledge to unfamiliar problem situations. The major themes that emerged from these data are described and also illustrated by student quotations. The results offer qualitative evidence for "why" concept-based instruction is more effective than traditional instruction.

## 2. Quantitative Method

The setting for the study was a large state university in the western part of the U.S. serving more than 20,000 students.

**2.1. Course Description.** The calculus classes studied were primarily taken by students intending to major in engineering, a physical science, or mathematics. Historically, this had been a tightly controlled and traditionally taught course with common examinations that focused primarily on the recall and implementation of appropriate procedures. The textbook for the course was *Single Variable Calculus, Fourth Edition* (Stewart, 1999). According to Stewart, this textbook contains elements of reform but within the context of a traditional curriculum.

**2.2. Objectives.** In order to evaluate the effect of instructional environment on student achievement, hypotheses about student performance on the midterm examination's *Procedural Skill Subscale* and *Conceptual Understanding Subscale* and the final examination's *Procedural Skill Subscale* and *Conceptual Understanding Subscale* were formulated. How these subscales were derived is discussed below in the section on Subscale Development.

*Hypothesis One.* Students enrolled in the concept-based learning environment will score higher than students enrolled in the traditional learning environment on the *Conceptual Understanding Subscale* of both the midterm examination and the final examination.

*Hypothesis Two.* There will be no difference between the scores of the students enrolled in the concept-based learning environment and the scores of the students enrolled in the traditional learning environment on the *Procedural Skill Subscale* of both the midterm examination and the final examination.

*Hypothesis Three.* Students enrolled in the concept-based learning environment will be better able to extend knowledge to new situations than the students enrolled in the traditional learning environment.

**2.3. Participants.** The participants were 144 undergraduate students enrolled in one of four sections of first semester calculus during the Fall semester of 2000. Due to scheduling considerations, students self-selected into sections but were unaware of both the instructor's identity and the learning environment into which they enrolled. Students were unaware that they were in courses being researched. Permission to conduct the study without obtaining the consent of students was granted by the university's Human Research Committee.

Enrollment in each section was limited to 42 students. At the start of the study, the enrollments of the two traditional sections were 34 and 31 students, and the enrollments of the two concept-based sections were 42 and 37 students. All enrolled students had satisfied the prerequisites of the course.

**2.4. Instructors.** The instructors each taught according to learning environments that matched their instructional styles. Two instructors (one male tenured faculty member and one male graduate student instructor with four years of teaching experience) taught the two traditional sections of calculus. These two sections are referred to here as the traditional learning environment. Two instructors (one female fourth year, tenure-track faculty member and one female graduate student instructor with one year of teaching experience) taught the two concept-based sections of calculus. These two sections will hereafter be known as the concept-based learning environment. The first two instructors were characterized as traditional because they emphasized the learning and practice of procedures and de-emphasized concept development. The other two instructors were characterized as concept-based instructors because they emphasized concept development and de-emphasized the learning and practice of procedures. The author was one of the concept-based instructors.

All four instructors were selected for the study based on documented teaching accomplishments including teaching awards, student evaluations, and the achievement of their students on common examinations administered in previous years. Each of the two graduate student instructors recently had been awarded the Teaching Assistant Award for Excellence in Teaching. The faculty member assigned to the traditional learning environment had won several teaching awards throughout his career. The author, teaching in the concept-based learning environment, had been nominated for several teaching awards. Student evaluations of all four instructors had been consistently excellent.

**2.5. Instructional Practices in the Two Learning Environments.** All four instructors agreed that "conceptual" and "traditional" should not be thought of as disjoint terms, that there were elements of both types of instruction in any given class session, and that the boundary between conceptual instruction and traditional instruction was not well defined. However, for the purposes of this report, the terms "traditional" and "concept-based" refer to the instructional emphasis.

Each instructor shared methods for teaching specific mathematical content during weekly instructor meetings and the methods were recorded in the minutes of those meetings. The instructors of the traditional environment consistently reported that the emphasis in their sections was on the teaching of procedures, skills, and algorithms. The concept-based instructors consistently reported that the emphasis in their sections was on concept development, multiple methods of solving problems, and mathematical communication. A faculty member observed each of the four sections on randomly selected days (10 visits per section). She took copious notes documenting what was written on the board by each instructor, the wording each instructor used to explain mathematical content, and the homework assigned. To combat any bias that the author may unwhittingly bring to the interpretation of the classroom data, these notes were reviewed by the author, the observer, and one additional faculty member who was not otherwise involved in the study. This collaborative review was done to assure that the instructional methods were consistent with the designated learning environment and to empirically document that the self-reports of the instructional methods accurately reflected the observed classroom practice.

Data from the classroom observations and the weekly instructor meetings were used to identify the normative instructional practices in the two learning environments. While there was some variation among the instructional practices within each learning environment, there were general patterns of instruction. The following descriptions of the two learning environments are based on evidence gathered through instructor meetings and classroom observations.

2.5.1. *Concept-based learning environment.* The content, perspectives, and instructional strategies of the concept-based learning environment reflected the reform emphases articulated by the National Council of Teachers of Mathematics (NCTM,

1989) and the Mathematical Association of America (MAA, 1988). The three features that characterized the concept-based learning environment built upon the features presented in the earlier work of Chappell and Killpatrick (2003).

*2.5.1.a. Concept development preceded skill development in sequence as well as priority.* Instructional units were organized around the prior knowledge of students with the goal being to help students connect new ideas with prior knowledge. Students' entry knowledge and skills formed the basis for the development of more formal concepts and procedures. The majority of class time (approximately 75%) was devoted to developing conceptual understanding by linking the entry knowledge of students to more rigorous formulas. The instructors displayed a spiral approach to teaching by regularly reinforcing new concepts and procedures with the prior knowledge, skills, and intuition that had been used to develop the concepts and procedures. Since the amount of class time devoted to skill development was limited (approximately 25%), students enrolled in the concept-based learning environment were exposed to procedural examples that did not require advanced algebraic manipulation. Nightly homework was assigned out of the textbook and included basic procedural tasks.

*2.5.1.b. Concept-based instruction emphasized multiple methods of solving problems.* The philosophical underpinning of the concept-based learning environment was the cognitive view of understanding as the internal construction of connections or relations between representations of mathematical ideas. To understand, students must build connections among mathematical ideas and various representations of them (Brownell, 1935; Davis, 1984; Hiebert and Carpenter, 1992). In order to build such connections, topics were presented numerically, graphically, and algebraically and the importance of being able to work a problem in more than one way was emphasized in class. Nightly homework included tasks that required students to combine and compare graphical, numerical, and algebraic approaches and to explain the connections between different representations. A TI-83 graphing calculator and view-screen were used in class on a daily basis to help students make connections between algebraic, graphical, and numerical solution methods. Three examples of graphing calculator use in the concept-based learning environment are given below.

*Example 1.* When maximization or minimization problems were solved using the first derivative test, the instructors asked students to make connections between the algebraic test and the graph of the equation being maximized or minimized. From the graph, the first derivative test could be visualized graphically (why the derivative at the maximum is zero, why the derivative to the left of the maximum is positive, and why the derivative to the right of the maximum is negative).

*Example 2.* Before the derivative at a given point was calculated, the graph of the function was analyzed at the given point and it was first determined whether the derivative should be positive, negative, or zero. Next, the trace feature was used to determine another point close to the given point and the slope between the two points was calculated in order to approximate the derivative.

*Example 3.* Before definite integrals were evaluated, the graph of the integrand over the desired interval was used to determine whether the integral should be zero, negative, or positive.

*2.5.1.c. Students were expected to present and explain the variety of methods they employed as they solved a problem.* Nightly homework included tasks that required students to explain their answers and to explain the meaning of concepts and procedures. The correct answer was valued only if students could explain why they had applied certain mathematical concepts, methods, and procedures to solve problems.

2.5.2. *Traditional learning environment.* The features that characterized the traditional learning environment were in keeping with those that have characterized traditional instructional programs (Hiebert, 1999) and built upon the features presented in earlier work (Chappell, 2003; Chappell and Killpatrick, 2003):

*2.5.2.a. Skill development preceded concept development in priority.* The traditional class sessions most often began with a conceptual overview of the topic that involved the prior knowledge and skills of students (approximately 10% of class time). However, the teaching of procedures, skills, and algorithms was of primary emphasis. The instructors introduced definitions and formulas and demonstrated their use with textbook examples that required a wide range of basic and advanced procedural techniques (approximately 90% of class time). A linear approach to teaching was followed: once algebraic methods of problem solving had been presented, instructors did not revisit the ideas that had been presented in the conceptual overview.

Students in the traditional learning environment were exposed to considerably more procedural examples than the students in the concept-based learning environment. Basic procedural examples that did not require advanced manipulation were demonstrated first followed by procedural examples of a greater level of technical difficulty than those presented in the concept-based learning environment. For homework, students solved textbook problems that required an execution of the skills and techniques demonstrated in class.

*2.5.2.b. Traditional instruction emphasized algebraic solution methods over non-algebraic solution methods.* Instructors in the traditional learning environment presented an initial example numerically, graphically, and algebraically. However, once the algebraic method had been presented, students were not required or encouraged to work a problem in more than one way. The instructors and the students had access to graphing calculators but these were used primarily for numerical calculations. The graphing calculator's view-screen was not used in the traditional learning environment. Nightly homework did not include tasks that required students to combine, compare, or explain the connections between different representations.

*2.5.2.c. Students were not expected to present and explain the variety of methods they employed as they solved a problem.* Nightly homework did not include tasks that required students to explain their answers or to explain the meaning of concepts and procedures. Homework grades reflected the degree to which the students could accurately execute procedural skills.

**2.6. Two Treatments of Limit.** The traditional and concept-based treatments of limits (Section 2.3 of the textbook) are provided below to illustrate the normative instructional practices in the two learning environments and to distinguish between the two instructional environments. For an additional example that illustrates traditional and concept-based treatments of the definition of the derivative (Section 3.2 of the textbook), see Chappell and Killpatrick (2003).

2.6.1. *Limit in the concept-based learning environment.* The concept-based calculus instructors first introduced the definition of the limit of $f(x)$ as $x$ approached $a$ as the value that the function approached as $x$ got close to $a$. Then, the functions $f(x) = \frac{x^2+x-6}{x-2}$ and $g(x) = x+3$ were considered from graphical perspectives, concluding that the functions were not equal because $f(x)$ had a removable discontinuity at $x = 2$ and $g(x)$ was continuous everywhere. Students considered whether the limits of these two functions were equal as $x$ approached 2. The limits of both functions were evaluated in multiple ways: graphically, by generating the graphs of the functions on the TI-83 and view-screen and seeing what happened, through the trace feature, for each function as $x$ approached 2; numerically, by evaluating the functions at values slightly larger and smaller than 2 and inferring what value each function was approaching; and algebraically, by considering the various properties of the functions, such as continuity. Throughout this discussion, the instructor reinforced the idea that in finding the limit of a function as $x$ approached 2, one was only concerned with how the function was defined near 2, regardless of what happened at 2. The development of why it was valid to cancel common factors when evaluating limits comprised approximately 35 minutes of a 50-minute class period.

The remaining class time was devoted to finding limits, such as $\lim_{x \to 5} 2x^2 - 3x + 4$, $\lim_{x \to -2} \frac{x^3+2x^2-1}{5-3x}$, and $\lim_{x \to 1} \frac{x^2-1}{x-1}$ using graphical, numerical, and algebraic approaches. Homework consisted of tasks that required students to graphically, numerically, and algebraically estimate the value of limits similar to those demonstrated in class; to explain in their own words what was meant by expressions such as $\lim_{x \to 2} f(x) = 5$; and to evaluate limits of the type demonstrated in class, using strictly algebraic methods.

2.6.2. *Limit in the traditional learning environment.* The traditional instructors also first introduced the definition of the limit of a function as $x$ approached $a$ as the value that the function approached as $x$ approached $a$. Each instructor then drew a function (by hand on the board) that had removable, jump, and infinite discontinuities and asked students to state the values of given limits. In addition, instructors demonstrated how to make a table of values of $f(x)$ for values of $x$ close to, but not equal to $a$, and how to use this table to approximate the limit of the function as $x$ approached $a$. These graphical and numerical examples took approximately 10 minutes of a 50-minute class period.

In the remaining 40 minutes of the class period, the instructor introduced the limit laws and evaluated multiple limits using the limit laws. In order to evaluate the limit of a function as $x$ approached $a$, students were taught to evaluate the function at $a$. If the function was undefined at $a$, students were taught to algebraically manipulate the function using methods such as factoring, multiplying by the conjugate, and common factor cancellation, until they could evaluate the function at $a$ and arrive at the limit. Once algebraic methods of finding limits had been established, students were not required or encouraged to employ numerical or graphical methods of finding limits. The instructor evaluated limits such as $\lim_{x \to 5} 2x^2 - 3x + 4$, $\lim_{x \to -2} \frac{x^3+2x^2-1}{5-3x}$, and $\lim_{x \to 1} \frac{x^2-1}{x-1}$. Algebraically rigorous methods were presented as the instructor evaluated the limits of functions such as $\lim_{x \to 0} \frac{(3+x)^3-27}{x}$ and $\lim_{x \to 0} \frac{\sqrt{x^2+9}-3}{x^2}$ . For homework, students solved textbook problems that required an execution of the skills and techniques demonstrated in class.

**2.7. Instrumentation.** Common written midterm and common final examinations were designed to measure the degree to which students had mastered not only the procedural elements of calculus, but also their ability to extend knowledge to new situations and explain in words or graphically why standard procedures made sense (see Appendices A and B for the examinations).

Two weeks before each examination, instructors individually submitted potential examination items to the calculus coordinator. In order to ensure that the test items would be comparable to items on examinations given in previous years and to ensure that the rigor of the test items would not be compromised by instructor bias, the calculus coordinator (a tenured full professor who was not one of the four instructors who participated in the study) designed the common examinations. The calculus coordinator would place himself in the traditional camp, especially in the area of test design. However, for the purposes of the study, he was committed to designing balanced exams. The examinations were presented to the four instructors one-hour prior to the administration of the examinations.

2.7.1. *Subscale Development.* Six faculty members independently reviewed the items of the midterm and final examinations. These faculty members assigned each item to either the *Procedural Skill Subscale* or the *Conceptual Understanding Subscale*. To facilitate this item analysis, general definitions of the two subscales were provided by the author. These definitions built upon the descriptions of procedural and conceptual items in the earlier work of Chappell and Killpatrick (2003). A brief description of each follows.

The *Procedural Skill Subscale* was comprised of items that required solution processes similar to those demonstrated in class and practiced in the homework. The successful completion of a *Procedural Skill Subscale* item required the student to recall and use an appropriate procedure. Such procedures included differentiating using the power rule, chain rule, product rule, and quotient rule (or combinations of such rules); finding the equation of a tangent line; computing higher derivatives; implicitly differentiating; using the limit definition of the derivative to find the derivative; evaluating integrals by $u$-substitution; solving related rates problems; analytically finding the maximum and minimum values of a function; and computing average velocities. *Procedural Skill Subscale* items did not require students to explain their answers or to describe the meaning behind a procedure.

The *Conceptual Understanding Subscale* was comprised of two types of items: nonstandard problems that required students to extend knowledge to new situations and items that required students to explain verbally or graphically why a particular procedure made sense. For example, Item 2 on the midterm examination required taking a limit. The limit could be evaluated by recalling and implementing the algebraic procedure for calculating limits. However, this item also asked students to explain why their result was true. They had to defend their answers, in words or graphically, with the meaning of limit.

Midterm Item 4(b) is an example of a "nonstandard" problem, one requiring the extension of knowledge to a new situation. A problem that is nonstandard to one student may be standard to another (Darken et al., 2000). However, neither instructional environment taught a procedure for finding the derivative of an absolute value function and midterm examination Item 4(b) called for the extension of knowledge to this new situation. Therefore, it was considered to be a nonstandard problem.

Inter-rater agreement was 100% on all midterm and final examination items with the exception of midterm Item 10. Three of the six faculty members assigned Item 10 to the *Procedural Skill Subscale*, asserting that a procedure involving factoring, canceling of factors, and setting the denominator equal to 0 could be followed. The other three faculty members assigned Item 10 to the *Conceptual Understanding Subscale*, insisting that this item required a solid understanding of the types of discontinuity. As a result, Item 10 was not assigned to either subscale. Descriptive statistics and univariate test results for midterm examination Item 10 are reported separately.

The six independent reviewers agreed that midterm examination Items 1, 3, 5, and 6 would comprise the *Procedural Skill Subscale*. The point values of these four items summed to 47. The six independent reviewers agreed that midterm examination Items 2, 4, 7, 8, and 9 would comprise the *Conceptual Understanding Subscale*. The point values of these five items summed to 46.

The six independent reviewers agreed that final examination Items 1, 2, 4, 5, and 7 would comprise the *Procedural Skill Subscale*. The point values of these five items summed to 115. The six independent reviewers agreed that final examination Items 3, 6, and 8 would comprise the *Conceptual Understanding Subscale*. The point values of these three items summed to 60.

**2.8. Procedure.** Of the 79 students enrolled in the concept-based learning environment, 37 were enrolled in the section taught by the graduate student instructor and 42 were enrolled in the section taught by the faculty member. Of the 65 students enrolled in the traditional learning environment, 34 were enrolled in the section taught by the graduate student instructor and 31 were enrolled in the section taught by the faculty member. All four sections met for 50 minutes four times a week for 15 weeks.

Five laboratory sessions that utilized MATLAB were graded components of all four sections. In all four sections, students downloaded the lab assignments from the course website. Each lab led students through a series of MATLAB codes. All four instructors viewed the labs as providing very detailed recipes that, when followed correctly, would enable students to get the expected results. All four instructors facilitated the lab sessions by answering technical questions such as "Why did this line of code not produce this output?" None felt that they had facilitated conceptual understanding during these lab sessions. Each laboratory session lasted 50 minutes.

The course syllabus strongly encouraged students to purchase a TI-83 calculator. The differential uses of the graphing calculator by instructors and students in the two learning environments have been discussed.

None of the instructors provided a review sheet before examinations. The 90-minute common midterm examination was administered midway through the semester, and the 120-minute common final examination was administered during examination week. Scratch paper was allowed on both the midterm and the final examinations, but no class notes or textbooks were allowed. Students were not allowed to use a computer algebra system, but were allowed to use the capabilities of the TI-82 or TI-83 graphing calculator. Graphing calculators with symbolic algebra capabilities, such as the TI-89, were not allowed. No information, such as formulas or programs, was allowed to be stored in the calculator. Random calculator checks were conducted during the examinations to ensure that no programs or formulas were stored in them. Approximately 95% of the students owned a TI-82 or TI-83 graphing calculator and the other 5% borrowed a graphing calculator from the department's loan program. Student responses on all examinations were independent, in that each student worked alone.

The midterm examination was scored out of 100 possible points. The final examination was scored out of 175 possible points. To provide assurances that the data used to generate the statistical tests were reliable, to assure objective grading results, and to maintain consistency in scoring across the four sections, the instructors collaboratively designed, agreed upon, and followed a grading rubric for each item on the midterm and final examinations. The same instructor graded the same item on all student papers, adhering to the appropriate rubric. Each instructor then double-checked the scores for their individual classes against the scoring rubrics. Inter-rater reliability was 95% or higher for each item. Discrepancies were resolved by the four instructors in the regularly scheduled weekly meetings. Scores on individual questions were summed to arrive at subscale scores.

**2.9. Design.** The independent variable was learning environment with two levels, concept-based and traditional. The dependent variables were scores on the midterm and final examinations' two subscales: the *Conceptual Understanding Subscale* and the *Procedural Skill Subscale*. To test Hypotheses 1 and 2, students' scores on the two subscales of the midterm and final examinations were subjected to a two-tailed $t$-test for independent samples to analyze the effects of learning environment. The alpha level for the demonstration of significance was $\alpha < .0125$. To test Hypothesis 3, student responses to those *Conceptual Understanding Subscale* tasks requiring students to extend knowledge to unfamiliar situations were analyzed and coded to identify common patterns of response.

## 3. Quantitative Results

Students completed a calculus readiness examination during the first week of class. The students' scores on the calculus readiness examination were subjected to a two-tailed $t$-test for independent samples in order to test for differences between environment means. The overall $t$ was not significant, $t(137) = .771$, $p = .442$, suggesting that there were no significant differences between conditions on the pretest measure.

*Student Achievement on Subscales of the Midterm Examination.* All students, 79 enrolled in the concept-based environment and 65 enrolled in the traditional environment, independently completed the midterm examination. Descriptive statistics and univariate test results are presented in Table 1.

The students enrolled in the concept-based learning environment were expected to score higher than the students enrolled in the traditional learning environment on the 46 point *Conceptual Understanding Subscale* of the midterm examination. As predicted, these students scored significantly higher ($M = 43.29$) than the students enrolled in the traditional learning environment ($M = 37.37$) on the *Conceptual Understanding Subscale*, $t(142) = -3.856$, $p < .001$. The magnitude of the effect was estimated using Hedges' $g$ ($g = .65$).

The scores of the students enrolled in the concept-based learning environment were not expected to differ from the scores of the students enrolled in the traditional learning environment on the 47 point *Procedural Skill Subscale* of the midterm examination. Surprisingly, their *Procedural Skill Subscale* scores were significantly

TABLE 1. Descriptive Statistics and Univariate Test Results for All Subscales

| Measure | Traditional | | | Concept-based | | | t- value | Effect |
|---|---|---|---|---|---|---|---|---|
| | n | M | SD | n | M | SD | | |
| CUS (Midterm) | 65 | 37.37 | 9.89 | 79 | 43.29 | 8.54 | -3.856* | .65 |
| PSS (Midterm) | 65 | 36.62 | 7.61 | 79 | 40.06 | 6.82 | -2.863* | .48 |
| CUS (Final) | 59 | 45.59 | 12.18 | 73 | 50.16 | 7.79 | -2.614* | .47 |
| PSS (Final) | 59 | 72.59 | 24.63 | 73 | 80.30 | 20.60 | -1.958 | |

$*p < .0125$

higher ($M = 40.06$) than those of the students enrolled in the traditional learning environment, ($M = 36.62$), $t(142) = -2.863$, $p = .005$. The magnitude of the effect was estimated using Hedges' $g$ ($g = .48$).

On Item 10 (scored out of 7 points) on the midterm examination, the scores of the students enrolled in the concept-based learning environment ($M = 5.84$) were not significantly different from the scores of the students enrolled in the traditional learning environment ($M = 5.4$), $t(142) = -1.660$, $p = .099$.

**3.1. Responses to the Midterm Item Requiring the Extension of Knowledge.** Since a procedure for finding the derivative of an absolute value function had not been taught in either instructional environment, midterm Item 4(b) required the extension of knowledge to a new situation. Correct student responses were analyzed separately from incorrect student responses. Patterns identified among the student explanations used to support correct answers to midterm examination Item 4(b) are summarized in Table 2. Percentages of students who correctly answered midterm examination Item 4(b) are presented along with percentages for final exam Item 8 in Table 3.

3.1.1. *Concept-based learning environment.* Of the students enrolled in the concept-based learning environment, 95% (75 of 79) correctly answered that the function $f(x) = |x - 3| + 4$ was not differentiable (or had no derivative) at $x = 3$. Of those, 80% (60 of 75) argued that since the graph's slope changed abruptly at $x = 3$ (the slope to the left of $x = 3$ was -1 and the slope to the right of $x = 3$ was +1), no slope existed at $x = 3$. These students had strongly connected slope and derivative. An additional 11% (8 of 75) stated that the function had no derivative at $x = 3$ because of the corner at $x = 3$. Another 9% (7 of 75) stated that the function was not differentiable at $x = 3$ but did not provide any explanation.

Of the students enrolled in the concept-based learning environment, 5% (4 of 79) incorrectly answered that the function $f(x) = |x - 3| + 4$ was differentiable at $x = 3$. Prevailing patterns among the explanations used to support the incorrect answer could not be discerned.

3.1.2. *Traditional learning environment.* Of the students enrolled in the traditional learning environment, 65% (42 of 65) correctly answered that the function $f(x) = |x - 3| + 4$ was not differentiable (or had no derivative) at $x = 3$. Of those, 55% (23 of 42) argued that a function did not have a derivative at a corner. Another 31% (13 of 42) stated that the function was not differentiable at $x = 3$ but did not provide any explanation. An additional 14% (6 of 42) correctly provided slope arguments similar to those provided by conceptually taught students.

TABLE 2. Patterns Among Student Explanations Used to Support Correct Answers to Midterm Examination Item 4(b)

| Percentage of students who: | *Traditional* | *Concept-based* |
|---|---|---|
| Connected slope and derivative in their explanation | 14% | 80% |
| Argued that the function does not have a derivative at a corner | 55% | 11% |
| Provided no explanation | 31% | 9% |

TABLE 3. Percentages of Students Who Correctly Answered Examination Items Requiring the Extension of Knowledge

| Examination Item | *Traditional* | *Concept-based* |
|---|---|---|
| Midterm Examination Item 4b | 65% | 95% |
| Final Examination Item 8 | 54% | 88% |

Of the students enrolled in the traditional learning environment, 35% (23 of 65) incorrectly answered that the function $f(x) = |x - 3| + 4$ was differentiable at $x = 3$. Of these, 74% (17 of 23) stated that the derivative at $x = 3$ was 0 due to the change in direction of the graph. Prevailing patterns among the other 26% (6 of 23) of the explanations could not be discerned.

**3.2. Student Achievement on Subscales of the Final Examination.** Almost all students, 73 enrolled in the concept-based environment and 59 enrolled in the traditional environment, independently completed the final examination. The six students in the concept-based environment and the six students in the traditional environment who did not take the final examination had withdrawn from the course midsemester. Descriptive statistics and univariate test results are summarized in Table 1.

The students enrolled in the concept-based learning environment were expected to score higher than the students enrolled in the traditional learning environment on the 60 point *Conceptual Understanding Subscale* (CUS) of the final examination. As predicted, these students scored significantly higher ($M = 50.16$) than the students enrolled in the traditional learning environment ($M = 45.59$) on this subscale, $t(130) = -2.614$, $p = .01$. The magnitude of the effect was estimated using Hedges' $g$ ($g = .47$).

The scores of the students enrolled in the concept-based learning environment were not expected to differ from the scores of the students enrolled in the traditional learning environment on the 115 point *Procedural Skill Subscale* (PSS) of the final examination. Indeed, their scores on this subscale ($M = 80.30$) were not significantly different from those of the students enrolled in the traditional learning environment ($M = 72.59$), $t(130) = -1.958$, $p = .052$.

TABLE 4. Patterns Among Student Explanations Used to Support Correct Answers to Final Examination Item 8

| Percentage of students who: | *Traditional* | *Concept-based* |
|---|---|---|
| Primarily supported integral expressions by connecting area and integration | 25% | 77% |
| Supported integral expressions with textbook definitions and references to the car's movement | 59% | 20% |
| Supported integral expressions with references to the car's movement | 16% | 0% |
| Provided no explanation | 0% | 3% |

## 3.3. Responses to the Final Exam item Requiring the Extension of Knowledge.

Applications that involved integration to solve displacement and total distance traveled problems had not been presented in either instructional environment. Therefore, final examination Item 8 required the extension of knowledge to a new situation.

Student responses that included the correct integral expressions for both parts 8(a) and 8(b) were analyzed separately from student responses that did not. For each instructional group, the percentages of students who correctly answered final examination Item 8 are presented in Table 3. Patterns identified among the student explanations used to support correct answers to this item are summarized in Table 4. A detailed analysis of student responses follows.

3.3.1. *Concept-based learning environment.* Two patterns emerged among the explanations offered by students enrolled in the concept-based learning environment to support a correct integral expressions. First, these students strongly connected integration and area in their explanations. Second, these students explicitly determined the intervals over which the velocity was positive and negative. The student response shown in Figure 1 is representative of the responses from these students.

The correct integral expressions for both parts 8(a) and 8(b) were provided by 88% (64 of 73) of the students from the concept-based learning environment. Of these, 77% (49 of 64) strongly connected integration and area in their explanations; they related net area to displacement and total area to total distance traveled. They interpreted the displacement as the area of the region above the $x$-axis and below the graph of $f$ minus the area of the region below the $x$-axis and above the graph of $f$. Although textbook definitions and references to the movement of the car often supplemented these area arguments, the connection between integration and area was of primary emphasis. Of the students who provided the correct integral expressions, 20% (13 of 64) did not connect integration and area in their explanations. Instead, these students supported their integral expressions with the textbook definitions of displacement and total distance traveled and made references to the movement of the car. Another 3% (2 of 64) provided the correct integral expressions but did not support these with explanations.

In setting up the integral expression that represented the total distance traveled, 83% (53 of 64) of the students enrolled in the concept-based learning environment who provided correct integral expressions explicitly determined the intervals over

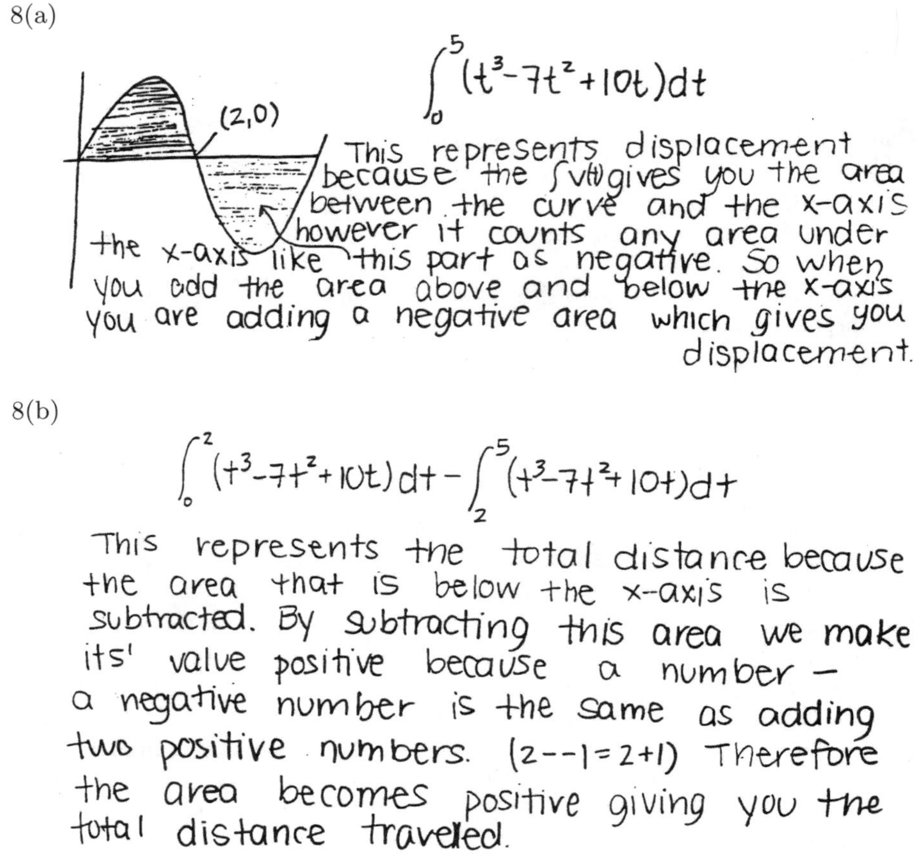

8(a)

$$\int_0^5 (t^3 - 7t^2 + 10t)\,dt$$

(2,0)

This represents displacement because the $\int v(t)$ gives you the area between the curve and the x-axis however it counts any area under the x-axis like this part as negative. So when you add the area above and below the x-axis you are adding a negative area which gives you displacement.

8(b)

$$\int_0^2 (t^3 - 7t^2 + 10t)\,dt - \int_2^5 (t^3 - 7t^2 + 10t)\,dt$$

This represents the total distance because the area that is below the x-axis is subtracted. By subtracting this area we make its' value positive because a number − a negative number is the same as adding two positive numbers. (2--1=2+1) Therefore the area becomes positive giving you the total distance traveled.

FIGURE 1. A representative correct student response from those enrolled in the concept-based learning environment.

which the velocity was positive and negative. These students then adjusted the integral expression so that each piece contributed a positive value to the sum. The remaining 17% (11 of 64) did not explicitly determine these intervals. Instead, these students provided the integral of the absolute value of the velocity function.

Of the students enrolled in the concept-based learning environment, 12% (9 of 73) did not provide the correct integral expressions for parts 8(a) and/or 8(b). Of these, 56% (5 of 9) provided integral expressions of functions other than velocity. Prevailing patterns among the other 44% (4 of 9) of explanations could not be discerned.

3.3.2. *Traditional learning environment.* Two patterns emerged among the explanations that students enrolled in the traditional learning environment used to support a correct integral expressions. First, these students most often defended their integral expressions with textbook definitions and references to the car. Second, these students did not explicitly determine the intervals over which the velocity was positive and negative. The student response shown in Figure 2 is representative of the correct responses received from these students.

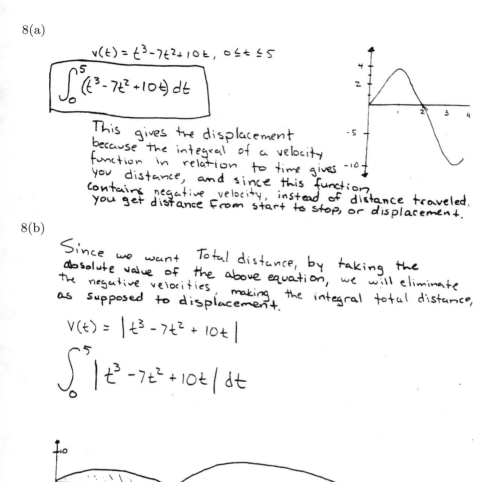

8(a)

$v(t) = t^3 - 7t^2 + 10t, \; 0 \le t \le 5$

$\int_0^5 (t^3 - 7t^2 + 10t)\, dt$

This gives the displacement because the integral of a velocity function in relation to time gives you distance, and since this function contains negative velocity, instead of distance traveled you get distance from start to stop, or displacement.

8(b)

Since we want Total distance, by taking the absolute value of the above equation, we will eliminate the negative velocities, making the integral total distance, as supposed to displacement.

$V(t) = |t^3 - 7t^2 + 10t|$

$\int_0^5 |t^3 - 7t^2 + 10t|\, dt$

FIGURE 2. A representative correct student response from those enrolled in the traditional learning environment.

Correct integral expressions to both parts 8(a) and 8(b) were provided by 54% (32 of 59) of the students enrolled in the traditional learning environment. Of these, 59% (19 of 32) presented the textbook definitions of displacement and total distance traveled. Specifically, they argued that the integral of the velocity function was the displacement and that the integral of the absolute value of the velocity function was total distance traveled. The students supported these textbook definitions well, referencing the positive and negative distances traveled by the car with phrases such as "backward and forward movement" and "traveling backward and forward." Of these, 25% (8 of 32) connected integrals to area. These students did not reference textbook definitions or the movement of the car. Another 16% (5 of 32) referenced

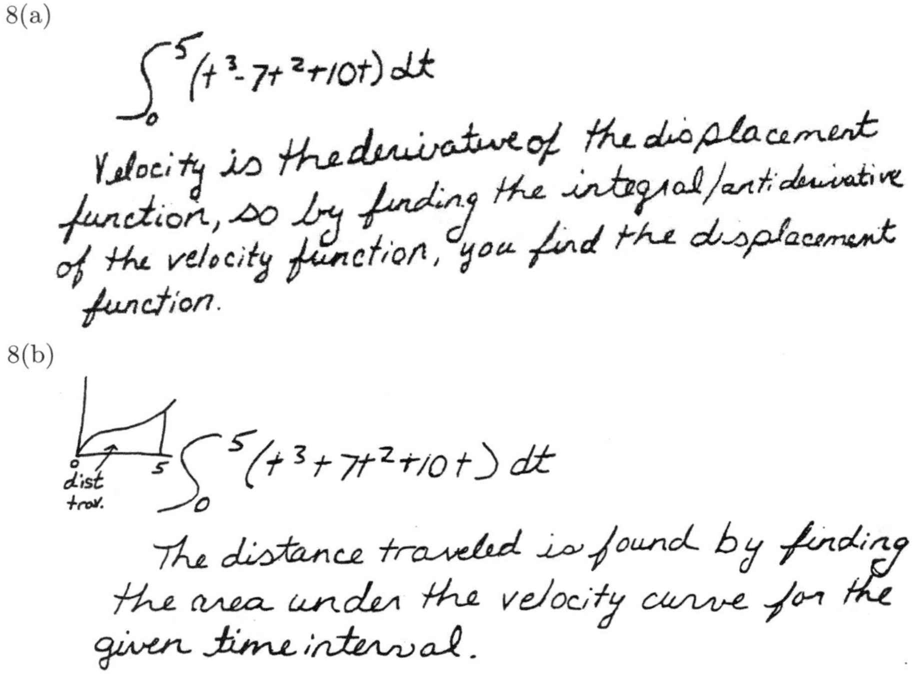

8(a)

$$\int_0^5 (t^3 - 7t^2 + 10t)\, dt$$

*Velocity is the derivative of the displacement function, so by finding the integral/antiderivative of the velocity function, you find the displacement function.*

8(b)

$$\int_0^5 (t^3 + 7t^2 + 10t)\, dt$$

dist
trav.

*The distance traveled is found by finding the area under the velocity curve for the given time interval.*

FIGURE 3. A representative incorrect student response from those enrolled in the traditional learning environment.

the positive and negative distance traveled by the car without referencing area or textbook definitions.

To set up the integral expression that represents the total distance traveled, the students enrolled in the traditional learning environment did not explicitly determine the intervals over which the velocity was positive and negative. Instead, they provided the integral of the absolute value of the velocity function.

Of the students enrolled in the traditional learning environment, 46% (27 of 59) did not provide the correct integral expression(s) for parts 8(a) and/or 8(b). Of these, 59% (16 of 27) provided the same integral expression for both displacement and total distance traveled. Their explanations supported the interpretation that these students saw no difference between displacement and total distance traveled. Another 30% (8 of 27) provided integral expressions of functions other than velocity. The remaining 11% (3 of 27) reversed the integral expressions for parts 8(a) and 8(b). The student response shown in Figure 3 is representative of the incorrect responses received from the students enrolled in the traditional learning environment.

The quantitative data provided evidence that concept-based instruction was more effective than traditional instruction, especially in the areas of conceptual understanding and the ability to solve unfamiliar problems. The qualitative component of the study, presented next, was based on classroom interviews and independent written questionnaires.

## 4. Qualitative Method

**4.1. Participants.** Interview participants were 63 of the 73 concept-based students who participated in the quantitative components of the study. Ten students were absent on the day of the interviews. The traditional students were not interviewed.

**4.2. Interview Questions.** The interview questions were constructed by the instructors involved in the study and the department head. The Department found it important to limit the number of questions in order to get detailed and quality responses.

Question 1: Is the instruction in this class different from the instruction in the mathematics classes that you have taken in the past? Please thoroughly explain the reasoning behind your response.

Question 2: Critics charge that the calculus reform movement, in its efforts to enhance the conceptual understanding of students, has "watered down" secondary and post-secondary calculus courses by teaching only a superficial use of skills. In contrast, other researchers assert that even though procedural skills are not the emphasis of concept-based classes, conceptually taught students are not outperformed by their traditionally taught peers on procedural questions. Reflecting on your experience in first semester calculus, please comment on the merits of such statements.

Question 3: The Mathematics Department recognizes that it is important to prepare students both to understand mathematical ideas and to efficiently execute mathematical procedures. The Mathematics Department struggles with whether the primary instructional focus should be on the understanding of central mathematical concepts or on the practicing of procedures. Would you prefer to take a mathematics course that primarily emphasizes procedures and the practicing of procedures or would you prefer to take a mathematics course that primarily emphasizes the understanding of mathematical concepts? Please thoroughly explain the reasoning behind your response.

Question 1 was asked to empirically document that the students' perceptions of their learning environment were consistent with the designated learning environment. Question 2 directly addressed the research question: From the students' perspective, why might conceptually taught students be able to perform equally as well as traditionally taught students on procedural problems when procedures were not the emphasis of concept-based classes? Question 3 was purposefully more general than Question 2 and was asked in the hope that the ensuing whole-class discussion would provide additional insight into the response themes established regarding Question 2.

**4.3. Interview Procedure.** The two concept-based classes were interviewed separately as whole classes. The interview procedure was the same for both classes.

A written questionnaire consisting of the three interview questions was given to the students two days before the whole-class interview. Students were asked to independently respond to each of the questions on their own notebook paper, thoroughly explaining the reasoning behind each response. The students were assured that their instructor would not be permitted to read any of the responses until grades had been submitted. The completed questionnaires were submitted to the department head immediately before the interview began.

The whole-class interviews were conducted during the last week of classes, during the regularly scheduled class sections, in the normal classroom, and each interview lasted 50 minutes. The interview questions were identical to those on the written questionnaire. The department head officially conducted the interviews and two other faculty members, who were not involved in the study, were present to ask students to expand or clarify answers they gave. Students volunteered responses but votes were taken for each question and recorded by two department secretaries in order to measure the perspectives of the class at-large. Both interviews were audiotaped. Neither the concept-based instructors nor the traditional instructors were present at either interview. The students were assured that the interview was to be strictly confidential and that their instructor would not be permitted to listen to the audiotapes until grades had been submitted. The department head asked the students to please say anything that they wanted and underscored that the department was really interested in their opinions.

**4.4. Data Analysis.** The audiotapes were transcribed. The results from the classroom votes were cross-referenced with the written responses to check that classroom votes were consistent with the independent written responses. This was done to ensure that students were not unduly influenced by majority opinion in the classroom setting. For each question, student responses were analyzed, categorized, and coded to identify prevailing themes.

## 5. Qualitative Results

The general student perspectives that emerged are presented below. Student quotations (extracted from written questionnaires and audiotapes of classroom interviews) serve to illustrate and highlight major perspectives. No student is quoted more than once and quotations come from both classes. The results of the second interview were consistent with the results of the first interview, increasing the generalizability of the results.

**5.1. Question 1.** In the first classroom interview, 28 students agreed that the instruction in their class was different from their past mathematics instruction and 4 students said that the instruction was no different from past mathematics instruction. In the second classroom interview, 30 students agreed that the instruction in the class was different from their past mathematics instruction and 1 student asserted that the instruction was no different from past mathematics instruction. These votes accurately reflected their independent written responses.

As the votes indicated, the overwhelming majority of students agreed that the instruction in their concept-based class was different from mathematics classes that they had taken in the past. They commented that the class focused far more on understanding than their previous mathematics classes had. The student responses established that the students did indeed view the course as concept-based. The 5 students who indicated that the instruction was no different from past mathematics instruction, said that their past mathematics classes had also focused on understanding. The following student quotation summed up majority opinion.

> In this class, there is a lot more explanation of the math concepts.
> Instead of just memorizing how to do a problem, we learn why
> we do a problem. I am taught more about where equations come

from and what equations actually mean rather that just being given the equation and told to do the procedures.

The following student quotations represent the diversity of student responses but they all have the same gist as the previous quote.

> The teacher concentrates on why does this or that makes sense. We have to know the reasons behind the procedures that we are doing rather than learning the formulas and practicing them until we are proficient.

> There's more focus on understanding than just plug and chug. All previous math classes [were] taught by practice and repetition methods. This course has forced me to think about what I am doing instead of just being a trained monkey.

> The teacher gives a lot more explanation of how things work instead of the usual "here the equation is." The teacher takes the time to explain the meaning of what we are doing instead of just telling us how to do it. Previous classes just tell you the way it is and this class explains why.

> Our teacher makes sure we understand why the equations work. Most math classes I have taken in the past have been straight memorization of a lot of formulas.

> The teacher spends a tremendous amount of time making sure that we understand why we do a certain procedure this way and what each portion of it means. In general, it's our responsibility to determine how much practice we actually need.

**5.2. Question 2.** In the first classroom interview, 30 students agreed on reasons why conceptually taught students may not be outperformed by their traditionally taught peers on procedural questions. Two students felt that the emphasis on conceptual understanding did indeed "water down" the use of procedural skills. In the second classroom interview, 30 students also agreed on why conceptually taught students may not be outperformed by their traditionally taught peers on procedural questions. One student felt that the emphasis on conceptual understanding did indeed "water down" the use of procedural skills. These votes accurately reflected the written responses.

Three major themes emerged from the student discussion of Question 2. Each of the three themes is supported by student quotations. The three themes were:

> *Theme 1:* If a student understands a procedure, he or she can remember, recall, and retain it more easily than a procedure memorized with little or no understanding.
> *Theme 2:* Conceptual frameworks make it easier to learn and apply new procedures.
> *Theme 3:* Information acquired with understanding makes students better problem solvers in that they can extend information that they understand much more easily to new situations than information memorized with little or no understanding.

Regarding Theme 1, students emphasized that if procedures were memorized with little or no understanding, they would "exist in the student's brain as disconnected bits of knowledge." Students felt that conceptual understanding provided "natural links and connections between procedures" that made remembering, recalling, and retaining procedures more manageable:

> In my past math classes, I would forget procedures that I had practiced repeatedly in the homework. I find things easier to recall in this class because I understand how and why.

> If I learn the full mathematical concept, I can retain what was taught. If I just program my mind to do a procedure, I often forget some of the procedure when the test comes at a later date.

> By just practicing and practicing procedures, one will eventually be able to do the process but after there is no more practice a person will lose the information. Procedures and practice will only get a student through the next test.

> If you go more in depth on a concept and make students think about how and why the procedure can solve the problem, it sticks in the student's mind and is retained.

> I took calculus in high school. I don't remember the definition of the derivative formula at all from high school. In here, the teacher brought up that it is just slope. She showed us how it is just a point here minus a point here just like you calculated slope in algebra. It will always stick in my mind. The one I learned in high school I don't remember at all because it was just some formula and all I used it for was for the test. In here, she taught us to use the concept of how it works so it stuck in my mind.

In Theme 2, students argued that a conceptual framework made it easier to learn and apply new procedures. In their opinion, with conceptual learning, each new topic "built naturally" on the previous topic. However, they reported that when they memorized information, they felt like each procedure presented was entirely new.

> I believe that if one understands a concept, all variation on notation or format would only need explanation once or twice, just to show what they mean. The result is that less time needs to be spent on the new material. However, in a practice based teaching environment, a teacher must explain each new notation as though it was a new concept.

> This approach gives me a chance to really think about and grasp the understanding rather than just going through the motions of doing the procedure and not really understanding what I am doing.

> When I took calculus in high school, it was all procedure. I found it hard to remember all of the things taught. Now, if I know where the formula comes from, I can always figure out the

equation to be used for the problem. If I understand something, I can go back and figure out the procedure if I need to. If I only understand the procedure, I am usually only able to solve problems that look just like problems I have seen and practiced before.

Once a student understands a concept in full, they are able to grasp any related material more quickly due to their greater understanding. When I understand the concepts, it is easier for me to understand the procedures and I can spend less time rehearsing the procedure.

In Theme 3, students agreed that information acquired with understanding made them better problem solvers. They underscored that knowledge grounded by conceptual understanding was much easier to extend to new situations than knowledge that was acquired with little or no understanding. Reflecting on previous mathematics courses, the students agreed that their primary problem solving strategy had been to identify the problem type, through the wording or context of the problem, connect the new problem to a similar problem that they had seen before, and then apply an associated procedure. The students observed that they were "out of luck" if the problem at hand was not like any problem that had been directly taught in class or completed for homework. They further recounted that prior to the concept-based calculus class they often had been unable to recall and apply the definitions and procedures that appropriately related to the unfamiliar problem, remembered procedures incorrectly, or interchanged procedures.

As they reflected on their experiences in concept-based calculus, students reported feeling that they were much less inclined to employ the "conjure up a similar problem and its procedure strategy." If they happened to forget a procedure, they felt that their understanding of the meaning behind the procedure provided the necessary framework to reconstruct it. If a problem was unfamiliar, they said that they now resorted to basic concepts that they knew and worked from there:

When I only know the procedure, I feel locked into doing the problem in a set way. When I know the concept, I can figure out an answer in many different ways.

By understanding the concepts inside and out, it is easier to apply them to other related problems. You know why you are doing a problem that way instead of just completing the action.

Understanding the concept helps you solve more complex problems. If I understand the concept, I am usually able to figure out how to do the problem even if I haven't seen it before. If I only know the procedure and can't recognize the problem as one I've seen before, I am stuck.

If I learn the concept, it doesn't matter what you put in front of me. I can probably do it and understand what I am doing. If I just know a procedure, then I am gonna keep saying "what about this situation?" and "what about that situation?"

**5.3. Question 3.** In the first classroom interview, 30 students voted that they preferred to take a mathematics course that primarily emphasized concepts and 2 students voted that they preferred to take a mathematics course that primarily emphasized procedures. In the second classroom interview, 30 students voted that they preferred to take a mathematics course that primarily emphasized concepts and 1 student voiced a preference for a mathematics course that primarily emphasized procedures. These votes accurately reflected the independent written responses.

The reasons given by students were the same as those for Question 2. This came as no surprise. If students felt that conceptual understanding helped them recall material better and use procedures more easily, it would seem natural they would value concept-based instruction. However, several new and interesting themes emerged. Although the overwhelming majority of students agreed that the primary emphasis should be on conceptual understanding, they insisted that:

*Theme 4:* There must be a balance between understanding and practice of procedures.

*Theme 5:* The teacher must be able to implement reform agendas well by effectively balancing conceptual understanding and procedural skill.

*Theme 6:* Adapting to a concept-based learning environment is frustrating, difficult, and takes time.

Theme 4 arose from student assertions that the debate between "traditional" approaches and "experimental" concept-based approaches to mathematics instruction was counter-productive. The students agreed that they preferred to take a mathematics course that primarily emphasized concepts but insisted that it was important not to avoid procedures altogether. They stated that their instructor had developed concepts but had then linked these concepts to the appropriate procedures. They stressed that this linkage was prerequisite for student success.

> Our instructor spent the time in class helping us understand the math concepts, connected the concepts with the procedures, and then went over a few examples in class that would help us with the procedure. She then assigned homework that let us practice the procedures. This was a wonderful way of teaching the material because we learned to understand the math concepts in class and practiced the procedures at home.

> Procedures are an important component of mathematics. But without grasping the understanding of why these procedures are commonplace through grasping the underlying concepts, we learn only how to regurgitate absorbed procedural steps.

> I feel you need a balance between understanding and practice because the whole idea is to put the understanding into practice.

> Understanding concepts is equally important as the ability to perform the problems. Concepts do nothing for you if you can't procedurally work through it.

> Sometimes you are given a problem on a test and you just have to grind it out. I would sit and stare at the exam and I could only remember why we were supposed to do the problem. It

can't be one way. There needs to be a mix of the two, where you
can say where it comes from but you can also know how to do
it.

Under Theme 5, students unanimously agreed that teacher effectiveness influenced their answers to Question 3. The following quote is how one student
summarized the discussion regarding the difficulty of separating teaching method
from teacher. Another classroom vote was taken after the statement was made. It
revealed that all students agreed with this student.

As far as conceptual teaching goes, I think it has a lot to do with
the teacher. I know that you can have a class that is conceptual
and the teacher can just move on and you don't get the concept
and you don't get the procedure. You are kind of in a rut there.
We are lucky with our teacher because she is effective at what
she does. If we didn't have an effective teacher, this might be a
different discussion.

In Theme 6, students insisted that adapting to a concept-based learning environment was frustrating, difficult, and time consuming. They agreed that at first
they were quite resistant to a teaching method that was counter to their past experiences with mathematics instruction. However, they stated that as they began
to reap the rewards associated with conceptual understanding, their viewpoints
shifted and resistance diminished:

Initially everybody was afraid of this new way of learning but
we accepted it more and more as the class went on.

I kind of noticed a shift in the class in general. When she first
presented this method of teaching, I could just see the looks on
everybody's faces like, "Oh God, I have to understand this now."
People changed. We can explain things now and know why we
are doing them.

If the department decides to try this new method of teaching,
just know that you are going to get mixed messages for the first
couple of weeks. It was really hard going from the basic way of
learning to learning to understand the procedures. On the tests
we had to explain what we were doing and I had never had to
do this before. A lot of us struggled with putting our ideas into
words. But in the end, it was the best.

5.3.1. *Opposing Viewpoints.* The minority opinions regarding Questions 2 and
3 came from the same students and were primarily driven by a concern that conceptually taught students would not be able to keep up in traditionally taught second semester calculus where they knew procedures were likely to be of paramount
importance. These students argued that, due to the emphasis on conceptual understanding, they were only exposed to the most basic of procedures and were thus
unprepared to solve problems that might involve procedures of greater complexity.
The following two quotations nicely encapsulate the concerns of these students.

When we move on to calculus two, I am not sure that we will
be able to hang with the students that were taught more traditionally. Only basic procedures were demonstrated in class and

assigned for homework. What will happen when the procedures get much more complex?

The reason why I said that a calculus class needs to be more procedure than concepts is because I know that a lot people are not going to take just this one class. We will go on to second semester calculus. I feel that right now it is more acceptable to be more procedure than concepts. If we get hooked on trying to understand why it works instead of how it works then it will be harder later on.

## 6. Conclusion and Discussion

The quantitative component of this study, involving 144 calculus students and four instructors, investigated the effects of instructional environment (concept-based or traditional) on the performance of calculus students on common assessments that measured conceptual understanding, procedural skill, and ability to extend knowledge to unfamiliar problems. The initial goal of the study was to answer two questions: If an instructor chooses to emphasize the learning and practice of procedures and de-emphasize concept development, will students be able to perform well on examinations that require the extension of knowledge to new situations and that require students to explain verbally or graphically why a particular procedure makes sense? If an instructor chooses to emphasize concept development and de-emphasize the learning and practice of procedures, will students be able to perform well on examinations that require the recall and implementation of procedures and that require the extension of knowledge to new situations?

The two traditional instructors emphasized the learning and practice of procedures and de-emphasized concept development. The two concept-based instructors emphasized concept development and de-emphasized the learning and practice of procedures. To safeguard that the effects would be due to the treatment and not to teaching effectiveness, all four instructors were selected for the study based on documented teaching accomplishments.

Results from common midterm and final examinations, designed to determine differences in both student understanding and procedural competency, provide post-secondary level empirical evidence that concept-based instruction has more reach than traditional instruction, especially in the areas of conceptual understanding and the ability to solve unfamiliar problems. Specifically:

- Concept-based instructional programs can effectively facilitate the development of student understanding without sacrificing skill proficiency.
- Conceptual understanding supports the acquisition of procedural skills.
- Knowledge acquired with understanding can be extended to solve unfamiliar problems more readily than strictly procedural knowledge.

Students enrolled in the concept-based classes outperformed those students enrolled in traditional classes on examinations designed to test conceptual understanding. The conceptually taught students were better able to explain in words or graphically why a particular procedure made sense or why they had applied a certain mathematical procedure to solve a problem. Very often, traditionally taught students were able to successfully complete tasks on a computational level but could not explain in words or graphically why a particular procedure made sense or why they had applied a certain mathematical procedure to solve a problem.

The differential performance of the two groups on conceptual questions was to be expected given the emphases of the two learning environments. However, conceptually taught students performed as well on measures of procedural skill as the traditionally taught students. These results suggest that the same degree of procedural proficiency can be achieved by repeatedly demonstrating procedures or by developing a conceptual context first and then extending that understanding to the appropriate procedures. Although the concept-based instructors had demonstrated only basic examples of each procedure, they showed how procedures arise naturally from the conceptual context. Conceptual understanding seemed to provide a cognitive structure on which later skill development could build. Even though the conceptually taught students had only seen basic procedures demonstrated in class, they were able to perform as well on complex procedures as students who had seen such procedures demonstrated in class.

The item analysis supported the conjecture that knowledge acquired with understanding can be extended to solve unfamiliar problems more readily than strictly procedural knowledge. Conceptually taught students were better able to extend knowledge to new situations than traditionally taught students. On items that required the extension of knowledge, the incidence of correct answers was higher among students enrolled in the concept-based learning environment than among students enrolled in the traditional learning environment (95% vs. 65% on midterm examination Item 4(b) and 88% vs. 54% on final examination Item 8). The concept-based students displayed a strong understanding of two of the most fundamental calculus concepts: integrals represent area and derivatives represent slope. They were generally able to accurately recall and extend these basic concepts to solve unfamiliar problems. The traditionally taught students were more reliant on specific definitions and formulas related to unfamiliar problems, but they were often unable to recall and apply them. These students also often misinterpreted definitions and interchanged procedures.

The results of the item analysis underscore that knowledge grounded in conceptual understanding is much easier to recall and extend to new situations than knowledge isolated from conceptual understanding. Through classroom interviews and independent written questionnaires, the qualitative component of the study examined the relationship between conceptual understanding and procedural skill by exploring students' perspectives regarding the effects of concept-based instruction on the acquisition of procedural skills and the ability to extend knowledge to unfamiliar problem situations.

Major themes that emerged from whole-class interviews included the importance of conceptual foundations for learning and understanding procedures, the value of concept-based instruction for extending understanding to new problem solving situations, the need for a balance between effective concept-based and procedural teaching, and the challenge students felt they faced in adjusting to a concept-based learning environment. Each theme addresses why conceptually taught students are able to perform as well as traditionally taught students on procedural problems, even when procedures are not the emphasis of concept-based classes. Student interview responses also proposed reasons why conceptual understanding transfers more easily to unfamiliar problems. However, the question of how conceptual understanding and procedural skill interact remains a significant and unresolved issue, especially at the secondary and post-secondary levels.

## 7. Future Research

The results presented here raise several important questions and possibilities for further research. The present study involved students intending to major in engineering, physical science, or mathematics. Traditionally, such students enter their collegiate mathematics courses with strong procedural skills. The study should be replicated to investigate whether the connection between conceptual understanding and procedural skill is dependent upon the procedural skill level of entering students. More specifically, do concept-based instructional programs effectively facilitate student understanding without sacrificing skill proficiency among student populations entering into university-level mathematics courses with weak procedural skills, such as business calculus students or college algebra students?

One of the most important measure of an instructional method's success is student performance in subsequent courses. Critics fear that reform calculus does not equip students for traditionally taught higher-level mathematics courses (Baxter et al., 1998). Some longitudinal studies have investigated the long-term effects of emphasizing concepts over procedures within a calculus curriculum (Baxter et al., 1998). Also, several studies (Armstrong, Garner, & Wynn, 1994; Baxter, Majumdar, & Smith, 1998; Bookman & Friedman, 1994) suggest that students who complete a reform first semester calculus course are not as successful as their traditionally taught peers in traditionally taught second semester calculus. These studies allude to, but do not empirically identify, transition problems that students face as they change learning environments. Further studies are needed to determine transition issues and long-term effects of emphasizing concepts over procedures within a calculus curriculum.

## References

Alarcon, F. E., & Stoudt, R. A. (1997). The rise and fall of a Mathematica-based calculus curriculum reform movement. *PRIMUS, 7*(1), 73-88.

Armstrong, G., Garner, L., & Wynn, J. (1994). Our experience with two reformed calculus programs. *PRIMUS, 4*(4), 301-311.

Armstrong, G. M., & Hendrix, L. J. (1999). Does traditional or reform calculus prepare students better for subsequent courses? A preliminary study. *Journal of Computers in Mathematics and Science Teaching, 18*(2), 95-103.

Baxter, J. L., Majumdar, D., Smith, S. D. (1998). Subsequent-grades assessment of traditional and reform calculus. *PRIMUS, 8*(4), 317-329.

Bookman, J., & Friedman, C. P., (1994). A comparison of the problem solving performance of students in lab based and traditional calculus. In E. Dubinsky, A. H. Schoenfeld, & J. Kaput (Eds.), *Research in Collegiate Mathematics Education. I.* (pp.#101-116). Providence, RI: American Mathematical Society.

Brownell, W. A. (1935). Psychological considerations in the learning and teaching of arithmetic. In W. D. Reeve (Ed.), *The teaching of arithmetic: Tenth yearbook of the National Council of Teachers of Mathematics* (pp. 1-31). New York: Columbia University, Teachers College.

Brunett, M. R. (1995). A comparison of problem-solving abilities between reform calculus students and traditional calculus students (Doctoral dissertation, American University, 1995). *Dissertation Abstracts International, DAI-A57/01.*

Chappell, K. K. (2003). Transition issues that reform calculus students experience in traditional second semester calculus. *PRIMUS, 13*(2), 129-151.

Chappell, K. K., & Killpatrick, K. (2003). Effects of concept-based instruction on students' conceptual understanding and procedural knowledge of calculus. *PRIMUS, 13*(1), 17-37.

Darken, B., Wynegar, R., & Kuhn, S. (2000). Evaluating calculus reform: A review and a longitudinal study. In E. Dubinsky, A. H. Schoenfeld, & J. Kaput (Eds.), *Research in Collegiate Mathematics Education. IV.* (pp. 16-41). Providence, RI: American Mathematical Society.

Davis, R. B. (1984). *Learning mathematics: The cognitive science approach to mathematics education.* Norwood, NJ: Ablex.

Dreyfus, T., & Eisenberg, T. (1990). Conceptual calculus: Fact or fiction. *Teaching Mathematics and Its Applications, 9*(2), 63-65.

Ganter, S. (1997). Impact of calculus reform on student learning. *AWIS Magazine, 26*(6), 10-15.

Heid, M. K. (1988). Resequencing skills and concepts in applied calculus using the computer as a tool. *Journal for Research in Mathematics Education, 19*(1), 3-25.

Heid, M. K., Blume, G., Flanagan, K., Iseri, L., Deckert, W., & Piez, C. (1998). Research on mathematics learning in CAS environments. In G. Goodell (Ed.), *Proceedings of the Eleventh Annual International Conference on Technology in Collegiate Mathematics,* (pp. 156-160). Reading, MA: Addison Wesley.

Hershberger, L. D., & Plantholt, M. (1994). Assessing the Harvard Consortium Calculus at Illinois State University. *Focus on Calculus, 7,* 6-7.

Hiebert, J. (1999). Relationships between research and the NCTM standards. *Journal for Research in Mathematics Education, 30*(1), 3-19.

Hiebert, J., & Carpenter, T. P. (1992). Learning and teaching with understanding. In D. A. Grouws (Ed.), *Handbook of research on mathematics teaching and learning* (pp. 65-97). New York: MacMillan.

Holdener, J. (1997). *Calculus and Mathematica* at the U.S. Air Force Academy: Results of an anchored final. *PRIMUS, 7,* 62-72.

Hollar, J. C., & Norwood, K. (1999). The effects of a graphing-approach intermediate algebra curriculum on students' understanding of function. *Journal for Research in Mathematics Education, 30*(2), 220-226.

Judson, P. (1994). Calculus reform blooms in Texas. *In General Terms, 2,* 3.

Lefton, L. E., & Steinbart, E. M. (1995). *Calculus & Mathematica*: An end-user's point of view. *PRIMUS, 5,* 80-96.

Mathematical Association of America. (1988). *Calculus for a new century: A pump, not a filter.* (L. Steen, Ed.). Washington, DC: Author.

Meel, D. E., (1998). Honors students' calculus understandings: Comparing Calculus & Mathematica and traditional calculus students. In E. Dubinsky, A. H. Schoenfeld, & J. Kaput (Eds.), *Research in Collegiate Mathematics Education. III.* (pp. 163-215). Providence, RI: American Mathematical Society.

National Council of Teachers of Mathematics. (1989). *Curriculum and evaluation standards for school mathematics.* Reston, VA: Author.

O'Callaghan, B. R. (1998). Computer-intensive algebra and students' conceptual knowledge of functions. *Journal for Research in Mathematics Education, 29*(1), 21-40.

Palmiter, J. R. (1991). Effects of computer algebra systems on concept and skill acquisition in calculus. *Journal for Research in Mathematics Education, 22* (2), 151-156.

Park, K., & Travers, K. J. (1996). A comparative study of a computer-based and a standard college first-year calculus course. In E. Dubinsky, A. H. Schoenfeld, & J. Kaput (Eds.), *Research in Collegiate Mathematics Education. II.* (pp.155-176). Providence, RI: American Mathematical Society.

Penn, H. L. (1994). Comparison of test scores in Calculus I at the Naval Academy. *Focus on Calculus, 6*, 6-7.

Schoenfeld, A. H. (1994). What do we know about mathematics curricula? *Journal of Mathematical Behavior, 13*, 55-80.

Schwarz, B. B., & Hershkowitz, R. (1999). Prototypes: Brakes or levers in learning the function concept? The role of computer tools. *Journal for Research in Mathematics Education, 30*(4), 362-389.

Stewart, J. (1999). *Single variable calculus* (4th ed.). New York: Brooks/Cole.

Stiff, L. V., McCollum, M., & Johnson, J. (1992). Using symbolic calculators in a constructivist approach to teaching mathematics of finance. *Journal of Computers in Mathematics Education, 11*(1), 75-84.

Tall, D., & Thomas, M. (1991). Encouraging versatile thinking in algebra using the computer. *Educational Studies in Mathematics, 22*(2), 125-147.

Tidmore, E. (1994). A comparison of calculus materials used at Baylor University. *Focus on Calculus, 7*, 5-6.

Wilson, R. (1997). A decade of teaching "reform calculus" has been a disaster, critics charge. *Chronicle of Higher Education, 43*(22), A12-A13.

## Appendix A

### Midterm Examination

1. (21 points) Calculate the following derivatives:

    (a) $\frac{d}{dx}(-2x+15)^{3/2}$      (b) $\frac{d}{dx}\frac{\sqrt{x+2}}{\tan x}$      (c) $\frac{d}{dx}x\sqrt{\sin x}$

2. (7 points) Evaluate the following limit. Explain why your results are true.

    $\lim_{x \to -1}\frac{(x+1)^2}{2x^2+4x+2}$

3. (7 points) The volume of a right circular cone is given by the equation $V = \frac{\pi r^2 h}{3}$ where $r$ is the radius of the base and $h$ is the height of the cone. Find the rate of change of the volume with respect to time if the radius is constant and the height of the cone is changing at a rate of 10 $m/sec$.

4. (12 points) The graph of $f(x) = |x-3| + 4$ is given below.

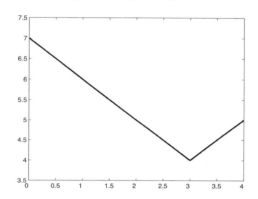

    (a) What is the derivative of $f$ at $x = 2$? Explain.

    (b) What is the derivative of $f$ at $x = 3$? Explain.

5. (7 points) Find the slope of the tangent line to the curve $y^2 + 3xy + 2x^2 = 4$ at the point $(3, -2)$.

6. (12 points) Find the absolute maximum and minimum values of the function $f(x) = x^3 - 3x^2 + 2$ on the closed interval $[-2, 1]$.

7. (7 points) Given the Mean Value Theorem [Let $f$ be a function that satisfies the following hypotheses: (1) $f$ is continuous on the closed interval $[a, b]$. (2) $f$ is differentiable on the open interval $(a, b)$. Then there is a number $c$ in $(a, b)$ such that $f(b) - f(a) = f'(c)(b - a)$], find the number $c$ that satisfies the conclusion on the Mean Value Theorem where $f(x) = 5x^2 + 4x + 3$ and the interval is $[-1, 2]$. Explain why the hypotheses of the Mean Value Theorem are satisfied.

8. (12 points) An arrow is shot upward on Venus with an initial velocity of 8 $m/sec^2$. Its height given in meters after $t$ seconds is $h(t) = 8t - t^2$.

    (a) Find the average velocity over the time interval from $t = 1$ to $t = 2$.

    (b) Use the limit definition of the derivative to find the velocity of the arrow at time $t = 2$.

    (c) Using derivative rules give formulas for the velocity and acceleration of the arrow at the time $t$.

    (d) At what time is the arrow at its highest and how high is it?

9. (8 points) At which point(s) does the graph of the equation $y = \frac{1}{3}x^3 - \frac{3}{2}x^2 + 2x$ have a horizontal tangent line? Find these points analytically (using equations, etc.) and show why your results are true graphically.

10. (7 points) Find the point(s) of discontinuity of the function $f(x) = \frac{x^2-9}{x^2+x-6}$. Label the discontinuities as removable, infinite or jump.

## Appendix B

### Final Examination

1. (28 points) Calculate the following derivatives. You do not need to simplify.
   (a) $\frac{d}{dx}\frac{x^2}{\sin^2(3x)}$
   (b) $\frac{d}{dx}e^{\tan 3x}$

2. (42 points) Evaluate the following integrals. Show all of your work. Calculator answers are not sufficient. If you use a $u$-substitution, show what $u$ and $du$ are.
   (a) $\int \frac{x^2+1}{\sqrt{x}}\, dx$
   (b) $\int_0^{\pi/4} \frac{1+\cos^2\theta}{\cos^2\theta}\, d\theta$
   (c) $\int \frac{4x}{(x^2+1)^{3/2}}\, dx$

3. (28 points) Calculate the following limits. Show work. A calculator or numerical table is not sufficient (even though they are helpful). If the limit does not exist, explain why.
   (a) $\lim_{x\to\infty} \frac{3x^4+2x^3+x^2+1}{2x^4+3x^2+7}$
   (b) $\lim_{x\to\infty} \sin x$

4. (14 points) Use the definition of the derivative to calculate $f'(2)$ where $f(x) = 1/x$.

5. (15 points) Find the equation of the tangent line to the curve $\ln y = 3x + 1$ at the point $(0, e)$.

6. (16 points) Which of the graphs given below is the graph of the function $f$ that satisfies $\frac{df}{dx} = -x$ and $f(-1) = 1$? Explain your answer.

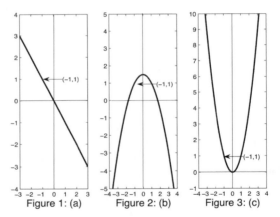

Figure 1: (a)        Figure 2: (b)        Figure 3: (c)

7. (16 points) Find the inverse of the function $f(x) = \sqrt{2 + 5x}$. Give the domain and range of both $f$ and $f^{-1}$.

8. (16 points) The velocity function (in miles per hour) for a car moving along a straight highway is given by $v(t) = t^3 - 7t^2 + 10t, 0 \le t \le 5$.
   (a) Set up an integral expression that represents the displacement of the car during the given time interval. Explain why the integral expression that you give represents displacement.
   (b) Set up an integral expression that represents the total distance traveled by the car during the given time interval. Explain why the integral expression you set up represents the total distance traveled.

DEPARTMENT OF MATHEMATICS, COLORADO STATE UNIVERSITY, 111 WEBER BUILDING, FORT COLLINS CO 80523
*E-mail address*: `chappell@math.colostate.edu`

CBMS Issues in Mathematics Education
Volume **13**, 2006

# Constructing a Concept Image of Convergence of Sequences in the van Hiele Framework

Maria Ángeles Navarro and Pedro Pérez Carreras

ABSTRACT. The goal of this paper is to describe the use of the van Hiele model as the theoretical basis for a new teaching methodology; the method provides the student with a suitable concept image for the convergence process of a sequence of real numbers. To this end, we analyzed the reasoning followed by students in understanding this concept and have provided the gradation of levels proposed by the model as well as their descriptors. We describe the experimental part of a study that was carried out by means of semi-structured clinical interviews and that used an interactive computer screen with a suitable visualization of convergence. The conclusions reached in the experimental part were used to design a teaching methodology that allows the early introduction of the convergence process for a sequence of real numbers. Existing research on convergence is mentioned and we have indicated which of the observed obstacles are overcome with the proposed methodology.

## 1. Introduction

The concept of limit of a sequence of real numbers as defined by Cauchy-Weierstrass occupies a central position in the construction of calculus. Understanding of basic notions such as continuity, differentiation, and integration of functions are based on a mastery of the concept of the limit of a sequence. In our research we have studied the structure of the stages through which the reasoning of students must pass in order to construct the notion of convergence of an ordered infinite process. To this end, we framed our work within the van Hiele model (1986). This research was carried out during 2000 and 2001 at the University of Sevilla in Spain (Navarro, 2002).

Our main goal was to study visually and verbally an ordered infinite process that may, or may not, stabilize. The chosen mathematical representation of such a process was the concept of convergent or divergent sequence disguised as a "cloud of points," a bi-dimensional set of points in the plane that is progressive and infinite. Our aim was to develop an educational experience that:

(1) Produces conjectures as to whether the process stabilizes or not.

The authors would like to thank E. Dubinsky, G. Harel, A. Selden, and several anonymous reviewers for their detailed and helpful comments on previous drafts of this article.

(2) Provides a computer-oriented tool to secure, modify, or produce a new conjecture.

(3) Encourages the verbal use of logical quantifiers by repeated use of the tool.

(4) Produces a concept image essentially coincident with the notion of convergence of sequences, once we move to this context.

(5) Forces the need to proceed further with a concept definition of convergent sequence, once the main goal is reached.

(6) Guarantees the helpfulness of the experience by testing the ease of transition from concept image to concept definition.

(7) Produces a reliable methodological proposal to construct the desired concept image in students who do not have the logical-algebraic maturity necessary for understanding the concept definition of convergence. Our aim was not to develop a substitute for the concept definition, but rather a battery of actions that can be used prior to mathematical instruction in this topic.

To attain our goal we imposed upon ourselves the following six conditions. First, see how far one could proceed verbally (in the Socratic tradition) and visually (by means of a mathematical assistant described in Section 3.1) in the absence of automated reasoning stirred up by students' previous knowledge of convergence as well as the use of logical quantifiers that clearly followed linguistic cues.

Second, deliberately use the "black box" effect when dealing with computers. That is, focus a student's attention on the object of study as well as its behavior when data are changed. At the end of the experience, the black box content was revealed to students. Our decision to proceed in this way was dictated by former experiences dealing with computers in the classroom. Where concepts are dealt with, it is more effective to treat the first concept visually with the right image, or a sequence of them, and then proceed by translating images to symbols, and not the other way around. This approach can be time consuming. Though coming up with the right image is not always easy, it pays off.

Third, avoid the cognitive obstacles that researchers have singled out in the treatment of convergence, many of them due to symbolic representation. Also, not introduce other obstacles, such as those derived from the representing of sequences as functions.

Fourth, from all possible associated images of the concept, choose the simplest in which only points and order matter. Choose a bi-dimensional representation (computer screen) for better visualization of changes that occur when manipulating data and prepare students for a subsequent interpretation of a sequence as a function. Allow the image to be dynamic, rather than static, to create the illusion of going to infinity.

Fifth, given the intimidating nature of face-to-face mathematical experiences with professors, proceed by means of a metaphor. We hoped that as the experience developed, students would progressively concentrate on essentials, and that is what happened. Finally, introduce the loss of certainty as a key ingredient in progress to understanding.

Given these constraints, a clinical interview was designed, aimed at making students aware of, and reflect on, the relations and properties used in their reasoning. Our intention was that the product of this reflection would be a suitable concept image of convergent sequence, one that favored verbal expression of a correct definition, even if not of the formal kind. We think that concept formalization should be carried out in a subsequent stage, and because of this, formalization was not one of our research objectives.

To attain our goal we created a model that postulated the existence of three well-differentiated levels of reasoning prior to formalization. These levels establish the boundaries between two phases of an acceptance-confidence-crisis process, reaching the goal in its optimal level (Level 3). The van Hiele model is of this type. Its selection was due to our previous positive experiences with it in dealing with other delicate concepts of real analysis.

Below, we describe the theoretical framework in which we developed our investigation, followed by a brief survey of existing literature. Next we outline the research methodology employed and the results obtained. Finally, we propose a teaching methodology that leads students to an intuitive construction of the convergence process and later gives rise to understanding the terminology and notation used in the classic formal definition of limit of a convergent sequence.

## 2. Theoretical Framework

**2.1. The van Hiele Model.** The van Hiele model (1957, 1986) provides a description of the learning process by postulating the existence of levels of reasoning. These levels are not identified with computational skills. In our study, they are classified as Level 0 (Predescriptive), Level 1 (Visual Recognition), Level 2 (Analysis), Level 3 (Classification and Relation) and Level 4 (Formal Deduction). However, we did not study Level 4; the van Hieles themselves asserted that its detection was difficult and only of theoretical interest. Van Hiele (1986) and other researchers (Clements & Battista, 1992; Jaime & Gutiérrez, 1990) who have studied and applied this model, agree that in Level 1 students are guided by a series of visual characteristics and they are lead by their intuition.

In Level 2, individuals notice the existence of a network of relationships. Students recognize and can characterize shapes by their properties. This is the first level of reasoning that we can call "mathematical" because students are able to describe and generalize (through observation and manipulation) properties that they still do not know. Reasoning in Level 3 is related to the structure of the second level. Conclusions are no longer based on the existence or non-existence of links in the network of relationships of the second level, but rather on existing connections between those links.

Some studies (Campillo, 1998; de la Torre, 2000; Esteban, 2000; Jaramillo, 2000), using the terminology of Llorens (1994), take Level 0 into account. This level is concerned with the degree to which students recognize the basic objects of study. It is the starting point for an individual who is beginning to work on the construction of a concept.

The application of this model to a specific subject requires the establishment of a series of descriptors for each level, to enable their detection. To be considered within the van Hiele model:

(1) These levels must be hierarchical, recursive, and sequential.
(2) The levels must be formulated so that they include a progression in the level of reasoning as a result of a gradual process, resulting from learning experiences.

(3) Tests designed for the detection of levels should take into account the existing relationships among levels and the language used by apprentices.

(4) The fundamental objective of the design must be the detection of levels of reasoning, without confusing them with levels of computational skill or previous knowledge.

The van Hiele model was initially applied to geometric concepts in elementary levels of primary education. Several doctoral theses (Campillo 1998; Campillo & Pérez Carreras, 1998; de la Torre 2000; Esteban, 2000; Jaramillo, 2000; Llorens, 1994; Llorens & Pérez Carreras, 1997) have shown the possibility of extending the model to concepts of real analysis studied in the final years of secondary education and in the first university year. There is a common strategy in all if these works: (a) study concepts (not results, theories, or techniques); (b) provide understanding of the actual definitions of concepts through the construction of an appropriate concept image that makes use of the computer as a cognitive tool; (c) frame the study into an educational model; (d) design a methodology that, on one hand, confirms that the model works, and on the other hand, provides evidence that progress in understanding has been achieved; and (e) propose a teaching methodology, independent of the model, that can be implemented in a reasonable amount of time.

The core of these works was an appropriate Socratic interview design that, within the context of the van Hiele model, allowed the detection of students' levels of thinking with respect to the specific mathematical concept treated there. The van Hiele model (1957, 1986) was chosen because it mimics the genesis of some mathematical concepts. That is, first, the discovery of isolated phenomena; second, the acknowledgement of certain characteristics common to all of them; third, the search for new objects, their study and classification and, fourth, through consideration of examples and counterexamples to proposed definitions, the emergence of definitive formulations.

**2.2. Comments about the Literature.** Various studies have shown that one of the main difficulties students find in the study of calculus is associated with the conceptualization and the formalization of the notion of convergence. Most research articles agree on several aspects that block the acquisition of this notion. We discuss these.

**The change in the lines of reasoning that students must follow.** For instance, the necessity of new rules for the treatment of equality frequently constitutes an obstacle for students. Calculus supposes an understanding that $\forall \varepsilon > 0$, $|a - b| < \varepsilon$ indicates the equality of $a$ and $b$. Understanding this is not easy as students are accustomed to managing chains of equalities (Artigue, 1995).

**Spontaneous conceptions.** Monaghan (1991) analyzed the effects of the language used in teaching and learning the concept of limit. He focused on the inherent ambiguities of the four expressions: "to tend to," "to approach," "to converge," and "the limit." He indicated that there are terms that are synonymous in mathematical language but have different connotations in natural language, and this can evoke different conceptual images for the same mathematical concept, depending on the student. Tall and Schwarzenberger (1978) remarked that informal translation of the definition of limit of a sequence contains two opposing dangers. On one hand, taking a high level delicate concept and speaking about it at lower level can mean a loss of precision that increases the conceptual difficulty. On the other hand, the informal language of translation can contain other unwanted colloquial connotations. Vinner and Tall (1981) indicated that if students have a concept image that does not allow $s_n$ to be equal to $s$, then even on showing them an example where this occurs, they continue to conjecture that the terms of the sequence cannot take that value. This situation is exemplified through the following sequence:

$$ s_n = \left\{ \begin{array}{ll} 0 & (n \text{ odd}) \\ 1/2n & (n \text{ even}) \end{array} \right. $$

When viewing this example, students insisted that it was not one sequence, but two. Although they were given a formal definition, the image of the concept definition that had been formed in their cognitive structure was very weak. According to Tall and Vinner, problems in understanding a concept caused by the co-existence of a strong concept image with a weak concept definition permeate university studies of calculus, especially when there are potentially conflicting factors among them. Furthermore, they observed the great initial difficulties that students have when using the quantifiers "all" and "some" in the usual formal definition of limit; these often caused further difficulties in the proof of some properties.

Mathews, McDonald, and K. Strobel (2001), members of RUMEC group, used APOS theory (Asiala et al., 1996) to examine students' cognitive construction of the concept of sequence. They showed that students tend to construct two distinct cognitive objects and refer to both as a sequence. One construction, that the authors called SEQLIST, is what one might understand as a listing representation of a sequence. The other, that they called SEQFUNC, is what one might interpret as a functional representation of a sequence. As the connections between these two entities became stronger and the students reflected on these connections, they began to understand sequence as a single cognitive entity, with SEQLIST and SEQFUNC both being mathematical representations of this entity. In that paper, Mathews, McDonald and Strobel showed that all interviewed students had an object conception of the construct SEQLIST, while in contrast, only half of the students clearly demonstrated an object conception of SEQFUNC. They indicated that the development of an object conception of SEQFUNC is tied to students' function schema, and depending upon its richness, this function schema may be quite helpful in the development of SEQFUNC as an object, or it may be a hindrance.

**Mental models.** Robert (1982, 1983) studied the different models by which students retain the notion of limit of a sequence. He reached the conclusion that, despite knowing the formal definition, when students were asked to describe this concept, they had a tendency to recall different aspects of their previous conceptions. Four main categories of model were postulated by Robert.

(1) *Primitive models* consist of spontaneous conceptions corresponding to erroneous representations that are incomplete.

(2) *Dynamic models*, in which "to converge" is described as "to approach"; these are representations in terms of "action" that are described by students through expressions such as, "The values approach a number more and more closely." This category also includes dynamic models of a monotonic kind, where the idea is: "A convergent sequence is an increasing (or decreasing) sequence that approaches a limit."

(3) *Static models* are divided into *prestatic models* and *static models*. In *prestatic models* for $u_n$ approaching $L$, students express their idea of convergence through statements like: "The elements of the sequence end up being sited in a neighborhood of $L$." *Static models* are natural language formulations and express the formal definition of the convergence of a numeric sequence. These formulations are always geometric and of the kind: "Every interval around $L$ contains all $u_n$ except a finite number of $u_n$."

(4) *Mixed models* are a combination of static and dynamic formulations.

In studying the relationship between students' mental models and their procedures for resolution of problems, Robert observed a strong association between primitive models and incorrect procedures, regardless of whether the proposed exercises needed use of the formal definition. Also, a positive relationship between the static and mixed models and the correct procedures was detected; students who had static mental models solved problems involving the notion of convergence better, regardless of whether they used the formal definition.

**Cognitive obstacles.** Cornu (1991), after analyzing the history of the evolution of the limit concept, described the following epistemological obstacles.

(1) *The failure to link geometry with numbers.* He referred to the difficulty the Ancient Greeks had in passing from the geometric context to a purely numerical interpretation of the method of exhaustion. Cornu concluded: "The geometrical interpretation, and its success in resolving pertinent problems, is therefore seen to cause an obstacle which prevents the passage to the notion of a numerical limit" (p. 160).

(2) *The notion of the infinitely large and infinitely small.* Contemporary students view the symbol $\varepsilon$ as representing a number that is not zero, yet is smaller than any positive real number. There is a corresponding belief in the existence of an integer bigger than all the others, yet which is not infinite.

(3) *The metaphysical aspect of the notion of limit.* According to Cornu, "The notion of limit is difficult to introduce in mathematics because it seems to have more to do with metaphysics or philosophy" (p. 161). The concept of limit is linked to the existence and handling of infinity, and students may have difficulties using this concept. Cornu exhibited these difficulties using students' answers in an interview: "It isn't rigourous, but it works," "it doesn't exist," "it is very abstract," "the method is all right, provided you are content with an approximate value."

(4) *Is the limit attained or not?* This is a debate that has persisted throughout history and still arises when students talk about limits. Generally, students have difficulty in clearly separating the limit as an object from the process that allows its construction.

*Conceptions of the infinite.* Sierpinska (1985, 1987) analyzed different ideas of infinity that students had in connection with the concept of the limit of a sequence. She classified students into groups. There are *unconscious infinitists* who may say "infinite," but think "very big." They consider the limit to be the last value of the terms of a sequence; for some, this last value is either plus infinity (a very big positive value) or minus infinity. There are *conscious infinitists* who believe that infinity is about something metaphysical that is difficult to grasp with precise definitions. There are *kinetic infinitists* who connect the idea of infinity with the idea of time. There are *potentialists* for whom infinity exists only potentially. Still others are *potential actualists* who think that infinity can ultimately be actualized, that it is possible to make a "jump to infinity."

*The obstacle of the symbol.* The notation used nowadays in the definition of limit of a convergent sequence is particularly complex for our students. Some of the difficulties associated with the notation are related to the epistemological obstacles described by Cornu (1991). When introducing the notion of the limit of a numeric sequence, we start by introducing a new nomenclature, specifying the terms of a sequence using subscripts, in order to later introduce the general term in which the concept of function appears either explicitly or implicitly. Also, the use of a potentially infinite process and the introduction of a quantity $\varepsilon > 0$ brings in some of the aforementioned epistemological obstacles.

## 3. Research Methodology

In order to carry out the study, individual clinical interviews were used and certain characteristics were taken into account in their design. On one hand, students must take an active part so that through experience the concept is acquired using elements that take part in the process of convergence of a numeric sequence. On the other hand, the method used must allow the gradual construction of the concept, in such a way that we can observe the external signs of each student's reasoning process. Furthermore, students should be able to carry out the reasoning at their own speed.

Putting these two characteristics together, we used a method of visualization that permitted the establishment of an individual relationship between teacher and student with the help of a computer and a mathematical assistant. Through questions and responses, we progressively helped students construct the concept and analyzed the reasoning that occurred in their thinking processes.

**3.1. Visualization.** We introduced the concept of sequence and limit of a numeric sequence intuitively using available computer technology (Pérez Carreras, 2000). The capability of modern computers allows students' geometric intuitions to be connected to some particularly problematic ideas related to the concept of infinity. In order to reflect on the chosen visualization, an interactive computer screen was used for visualization (see Figure 1). We will refer to it as the mathematical assistant, or simply, assistant. Its use does not depend on the program with which it was designed (*Matlab* 5.3) so students did not have to learn about new software.

An image of each studied sequence, using points, was displayed in the area on the left. The screen was interactive so students could use it as a tool. That is, they could click on different functions on the right. Points could be added to the image one after another, or by specifying a certain number of points. By drawing a horizontal band that limited the size of the box, a region could be enlarged (as shown in Figures 2 and 3). Points could be marked on a blue line at the left to indicate a student's conjecture about the limit of the studied sequence. The original screen was colored and the blue line was the vertical line on the left. Because figures are in black and white in this paper, the blue line will be referred to as "the grey line (or the grey segment) on the left."

At no time did we use the nomenclature and notation typical of sequences of real numbers. Students did not see any formulas, and the language used was the most colloquial possible. We referred to the images as "clouds of points" in the plane

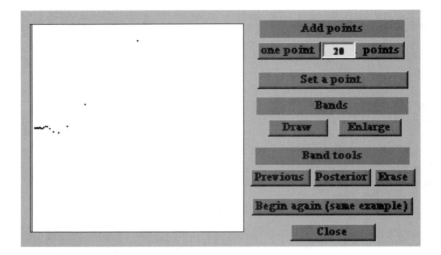

FIGURE 1. User interface for the interactive computer software
used for visualization.

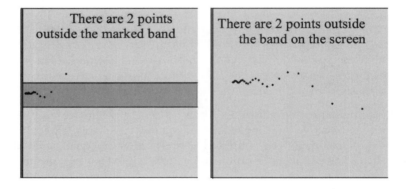

FIGURE 2.                              FIGURE 3.

whose abscissas tended to a limit and whose ordinates represented the sequence,
using the parametric representation $(1/n, s_n)$. In order to avoid preconceived ideas,
a cloud of points was introduced as the footprints of an invisible animal. The limit
of the sequence, if it existed, was to be given by a point on the grey line on the left
whose abscissa was the limit of the abscissas of the sequence. Students understood
this to be the point towards which the animal walked and identified it as the "point
of stabilization of the cloud." The epsilon variable, in the definition of limit, was
portrayed in terms of the width of a horizontal band centered on the limit in such
a way that the final inequality could be expressed by saying that "all cloud points
are in the band starting from a certain place," or better still, that "the number
of points outside any band is finite." At no time in this presentation were the
mathematical terms and variables mentioned.

Using this visualization, students were inclined to handle, by themselves in a intuitive way, those notions and lines of reasoning of an infinite kind where an obstacle has traditionally been found. We think such visualization contributes to the formation of a suitable concept image for the notion of the limit of a convergent sequence. The general terms used to generate the figures shown in this and the next section are provided in Appendix B.

**3.2. The Clinical Interview.** Using this visualization, we designed a script for a clinical interview consisting of 35 questions, divided into three blocks (for further detail, see Navarro, 2002, pp. 38–61). The first block of questions permits students to construct a valid image of the concept of sequence. In the second block, a tool intervenes that allows students to discern different behaviors of the clouds of points. The last block is devoted to the search for a verbal definition that will be made ever more precise until the concept definition is reached.

During our research we carried out 25 clinical interviews, of which 20 were with this version of the interview protocol. Each interview lasted approximately one hour and fifteen minutes. The students came from different levels of education. Ten students were from the first year of university and ten students were from the three last years of secondary education. Of those, three came from the fourth year of E.S.O.(15 to 16 years old), five from the first year of bachillerato (17 to 18 years old), and two from the second year of bachillerato (18 to 19 years old). In the selection of students, only the school or university year was considered. We did not know their marks or other information about them. The interview experience was carried out as an activity independent from the courses studied by interviewees. The details of the visualization and interview process are presented below in the context of the results.

## 4. Results

In the first questions we examined what concept of point a student had. If it was not correct, we did not think it possible for the student to understand what a numeric sequence with all its characteristics entailed. To this end, the student was shown the bisection process for a segment, that can be repeated mentally as many times as required. A student's point conception was indicated by the answer to the question: "How many times can this division process of the segment be repeated?" Most students told us that this process never ends. However, some students thought that the process would end after finitely many divisions, obtaining a point in the final division.

In this introductory part of the interview, students had to understand that there are mental processes that are finite, but that can be executed as many times as desired. They also needed to accept the existence of unlimited processes that give rise to finite and precise conclusions.

**4.1. Construction of Clouds of Points.** After this introduction, the construction of the sequence concept began with an image of footprints of an invisible animal. We constructed the concept of a "cloud of points" as an endless process, in which the points appeared one after another, in an ordered way. Also, each new point that appeared was nearer the grey line on the left than the previous point. We showed various examples, so interviewees became familiar with the concept.

They started to conjecture about the existence or non-existence of a point on the grey line on the left towards which the "footprints" were directed.

In this initial part of the interview the key questions were:

(1) Do you think that the animal has the intention of walking towards some specific point on the grey line on the left or that it is walking aimlessly?

(2) What is your conjecture?

These questions were formulated in each of the first examples, in connection with the images shown on the screen. At the same time, the interactive screen was explained and students were encouraged to use it in order to make conjectures. Students' answers showed us that they were starting to understand the dynamic character of clouds of points:

> *Student:* Following its trajectory it appears that the animal is getting closer to the grey line on the left and it would be possible to... determine toward which point [the cloud] is directed but... nobody can guarantee to me that it cannot change direction at a certain moment.

The next step was to show the lack of reliability of conjectures that were made from first observations and the necessity of obtaining a considerable number of points and some enlargements of the image. To this end, in one of the examples, we asked the students to make a conjecture when only six points of the sequence were on the screen (Figure 4). The interviewees who understood the concept indicated to us their need to see more points in order to make a conjecture. That showed us that these students were prepared to continue making progress in their reasoning level. In contrast, others dared to conjecture a chosen point position of the animal. After having added 100 points to the image (Figure 5), we asked, "What do you think of this, now?" This made them see their mistake and led them henceforth to consider the need to have a greater number of points on the screen before making a conjecture about the behavior of a cloud of points.

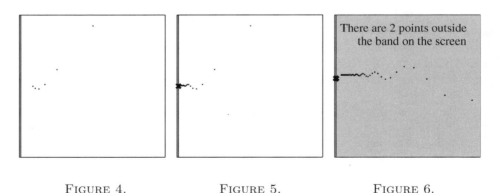

FIGURE 4.                 FIGURE 5.                 FIGURE 6.

In the same example we began to show that, even when a conjecture looked reasonable at first sight, enlargement of the image could lead one to change the conjecture. To this end, after adding a considerable number of points and enlarging the image (Figure 6), we asked, "Do you still maintain your conjecture or do you want to change it?" When students saw the enlarged image, most of them changed their conjectures. During the rest of the interview, we observed students adding large numbers of points to the clouds and enlarging the image one or more times before making a conjecture about the behavior of the clouds of points.

In the first stage of the interview, students should have realized that clouds of points are dynamic processes whose behavior can change at any given moment. This would have helped them understand the necessity of a tool that allowed them to enlarge the images. Students observed that it was necessary to have a considerable number of points on the screen if they wanted their conjectures to be reliable. This was the most delicate part of the experience. It was the one in which the interviewer interacted with students using common language to introduce nomenclature that would be used throughout the remainder of the interview.

Completion of this phase of the interview entailed students' understanding the uncertainty generated by infinite processes (with the necessity of a conjectural jump from finite to infinite) and understanding the need for a mechanism that permits a part of the image to be enlarged.

**4.2. Introduction of the Point of Stabilization of a Cloud.** The next significant intellectual idea that we aimed to help students realize was that, in order to make a conjecture reliable, the important thing was not the beginning of the cloud but its tail. Moreover, they had to observe that they needed to refine their first mechanism. For this, we proposed the idea of enlarging the image by means of bands with an indication of the number of points that were outside these bands. Students had to understand why this last piece of information was important.

We began the second part by asking interviewees if they would like to change their conjectures when the first points of the cloud had disappeared from the image. The answer to this question was important because it showed us how the reasoning of students was evolving. Some interviewees asked which the first points were, and this question showed us that they had not understood the way of generating clouds of points. Reminding them that the animal always walked closer and closer to the grey line on the left, and it could not move back, resolved this problem.

Some students answered that the disappearance of the first points would cause a change in their conjectures. It was observed that these students paid attention to characteristics of a geometric kind, such as the symmetrical positioning of the points. Those who did not overcome this obstacle only used information of a visual kind during the rest of interview and made no progress in their reasoning level.

The rest of the students told us that the disappearance of the first points would not make any change in their conjectures. Their explanations were of the following kind: "If I removed the first points, I would see the rest in exactly the same way and I would continue marking the same point." These students could do without the first points when studying the behavior of the various examples. They did not need to see how the clouds were generated point by point. This represented a qualitative change in their reasoning.

We next introduced the idea that the horizontal bands were something more than a tool for the enlargement of images. At first, we used the drawing of a road that contained the point chosen by the animal. We then transferred this idea to the images of the clouds of points, relating it to the bands that were used as an image enlargement tool. We continued by constructing the concept of a "point of stabilization" of a cloud. To this end, we provided students with a purely intuitive and visual definition:

*Interviewer:* We will say that a cloud of points stabilizes when the animal, whose footprints provide the points, has chosen one specific point on the grey line on the left, with the intention of walking towards it. The name of this point will be the "point of stabilization" of the cloud.

Our intention was that interviewees should reflect on the relation between the increase or non-increase in the number of points outside the drawn bands and the position of the point of stabilization of a cloud. To this end, we asked students to make their own conjectures about the existence of a stabilization point in one of the examples, for which they had drawn one or several bands containing the possible stabilization point. When we asked "Would the number of points outside this band increase?" A typical student's answer was that number would not change "because if the animal is approaching, no point will be situated outside." When the images were of clouds of points with only one lateral "peak," we observed that the conjectures of all students were similar; namely, the cloud of points stabilized, and from a given moment, the new points must be situated inside a band.

The next stage was for students to associate the existence of the point of stabilization with its uniqueness. We used examples and questions that contradicted students' initial intuitions. We started by showing the image of a sequence that had three convergent subsequences and that generated a cloud of points with three lateral peaks (Figure 7). We asked students for a conjecture about its behavior and their first answers were that "the cloud stabilizes at three points" or that the cloud stabilizes at the point corresponding to the central peak. In view of this, we drew a horizontal band around one of the lateral peaks (Figure 8), enlarged the image, and asked, "What would your conjecture have been if you had seen only this part?" (Figure 9). The students usually answered that they would have thought that "the animal has chosen a point and it looks as if it is heading for it."

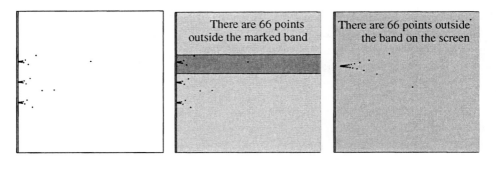

FIGURE 7.                    FIGURE 8.                    FIGURE 9.

We continued by asking: "If we add 100 more points, will the number of points marked outside this band increase?" Here the answers were of two kinds. There were students who forgot the previous images and answered that the number would not increase, but when they saw the increase, they remembered the other two peaks and told us that the number increased because the points accumulated around three different positions. In contrast, other students bore their previous images in mind and had no problem in answering correctly. The response to the next question was usually immediate "What would you have answered if you had not seen the previous image before?" The students told us that they would have thought that the number would not increase.

We continued by asking: "What would you have thought if you saw that the number increased?" To this question, the answers were usually vague and imprecise. Some students told us that they would not be able to interpret this piece of information. Our next question usually made the interviewees reflect. "If the animal had chosen a point inside this band, do you think the points should stop appearing outside this green band, from any given moment?" Usually the answer was affirmative. Then we continued adding points, one after another, observing that the number of exterior points did not stop increasing.

From here, we formulated three questions that were fundamental in the development of the interview's second part:

(1) Let us suppose that the cloud continues behaving this way. Do you think that if we continue adding points, then the number of points outside this band will stop increasing?

(2) Will the possible point of stabilization be inside this band?

(3) Will it have a point of stabilization?

In the last shown example (Figures 7, 8, and 9), students ended up saying that this cloud would not be stabilized because "according to the definition, it only can have one stabilization point." These three questions were formulated for each of the examples considered in the second part, and after students had answered, we always asked them for an explanation of their reasoning.

The next example (Figure 10) showed a sequence with two convergent subsequences that were situated on the screen so that it was only possible to see one of them. Our intention was to check if the increase in the number of points outside the bands had been correctly interpreted by the interviewees.

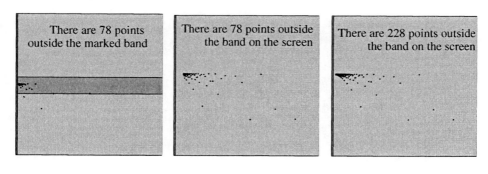

FIGURE 10.  FIGURE 11.  FIGURE 12.

First they were disconcerted by the increase in the number of points outside the first band that they drew (Figures 11 and 12). With all students it was necessary to return to a previous example and remind them what had happened when we enlarged the image and saw only one of the peaks. The students responded that if the cloud's behavior did not change, then it would not stabilize. Their reasoning was always of the same kind:

(a) "It is approaching the grey line on the left at two different places."

    (b) "There are points that are situated outside and do not tend towards that
        point."

    (c) "The animal has not chosen a specific point."

Before telling us what their conjectures were, the students drew one band and added
points to the cloud to observe whether the number of points outside the band had
stopped increasing.

    At this stage, most of the students had grasped the uniqueness of the point of
stabilization, and they considered the increase in the number of points outside the
band as a tool that permitted them to study the existence of this point. The next
questions and examples had the purpose of leading students to observe the need
to draw more than one band and to observe the relation between the halt in the
increase in the number of points outside the bands considered and the existence of
the point of stabilization.

    The interview continued with another cloud of points that had two peaks sit-
uated so closely (Figure 13) that only after making several enlargements was it
possible to see the separation between them (Figures 14 and 15). Most of interviewees, after making various enlargements, observed this and told us that it was a
similar case to the previous one. But with some students it was necessary to draw
a band in one of the peaks, add points, and observe how the number of exterior
points increased.

 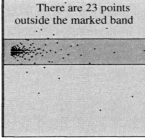 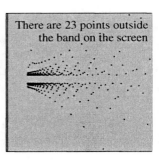

FIGURE 13.             Figure 14.             FIGURE 15.

    In order to finish this example, we asked students questions whose answers
would show us their reasoning:

    (1) Is it important that the number of exterior points to a band stops increas-
        ing?

    (2) What information is provided by this fact?

    (3) What differences do you observe between clouds that stabilize and ones
        that do not stabilize?

A number of students made reference to the images of the previously studied
examples. In their reasoning, they spoke of characteristics of a visual kind, indicating that if the number continued increasing it was because the cloud had more than
one peak. They utilized this increase to decide whether some peak was situated
outside the screen. For example, one student described the differences between
clouds that stabilized and ones that did not stabilize in the following way: "The

clouds that stabilize go to one point... and the others not. This, for example, goes to two points and the other goes to three."

The rest of the students made other interpretations of the halt in the increase in the number of points exterior to the drawn bands. For example, one student expressed it saying: "Well it indicates to me that... that it is coming closer to the grey line on the left with a very narrow range...smaller and smaller until a specific point." He continued his explanation by saying that the number stopped increasing because "As I have said before, the points are going towards a point and their dispersion is smaller."

In order to justify the differences observed between clouds that stabilized and ones that did not, students usually expressed themselves using quite imprecise language. A typical explanation was: "In the clouds that stabilize all points go towards a point and in those that do not stabilize the points come out of the band."

**4.3. Process of Drawing Bands.** The next steps in the students' reasoning were the acceptance of the need for a definition in order to characterize the clouds of points that stabilized, the attempt to find this definition, and its application to specific examples. In this third part of the interview, we tried to direct students to the concept definition using a concept image, constructed in a suitable environment, that favored a verbal expression of this definition. To this end, we assumed that the clouds studied stabilized at a fixed point and that a characterization of this point would be sought. We started by showing an example (Figure 16) that had such slow convergence that it did not let students make a conjecture from a simple visual observation. They had to use the halt in the increase of the number of points outside the bands (Figures 17 and 18). Our intention was that the lack of certainty, and the small amount of visual information provided by this example, would lead students to think about how many bands they would need to draw in order to be able to achieve a reliable conclusion.

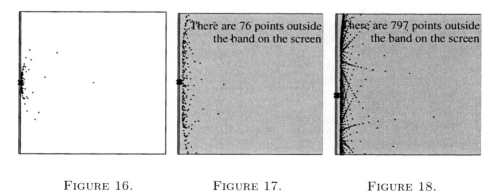

FIGURE 16.              FIGURE 17.              FIGURE 18.

At this time, a red point was marked on the grey line on the left. After the student had drawn various bands and observed the increase, or halt in the increase, in the number of exterior points, we asked:

(1) What can be deduced from this piece of information?

(2) Can we be sure that the red point is the stabilization point of the cloud?

In their answers, some students continued reasoning visually. They utilized the increase or halt in the increase in the number of exterior points to the bands in order to decide whether some part of the cloud was off the screen. When they were sure that this had not happened, they considered the cloud of points as a geometric figure to which they added large numbers of points and repeatedly made enlargements in the image. Interviewees who followed this kind of reasoning were seldom able to achieve a conclusion for this example (Figures 16, 17, and 18).

Other students made progress in their reasoning level. They used the increase or halt in the increase of the exterior points to the bands to obtain conclusions about the existence of the stabilization point. There were students who, after drawing various bands and observing that the number of exterior points to each band had stopped increasing, expressed the view that "the cloud probably ends up stabilizing" and they were reasonably sure of this fact. For others, greater uncertainty was observed when they told us: "The cloud may or may not have a stabilization point." They indicated that they could not guarantee that the cloud stabilized, but if this happened "The stabilization point must be inside the band." When we asked them what would be needed in order to be convinced of this, they said that "It would be necessary to draw an infinite number of bands and... we could never draw them."

From here on, reasoning of a more general kind was sought in an attempt to lose a large part of the visual component that had been present in the concept so far. To this end, we did without images and began to introduce existential and universal quantifiers. Although we initially made reference to the image of the last example studied, we later formulated the questions in a more general context, without images, looking for conditions that characterized the stabilization point of any convergent cloud of points.

**4.4. Students' Definitions.** We started this part of the interview by indicating to the interviewees that we wanted them to think about how any cloud of points behaves when it stabilized at a specific point. We indicated that they should consider the whole cloud, with all its points, and that they should reason within this context. We tried to lead the students to express their own definition of a stabilization point of a cloud by asking them to formulate answers to the following questions:

(1) Consider an arbitrary band that is centered at the stabilization point. Will all points of the cloud be situated inside this band, apart from a finite number?

(2) Is there another point in the grey line on the left that verifies this property?

By means of these questions we made the students reflect on the process under study and the properties of the stabilization point. Their answers to the following questions provided us with a summary of their reasoning characteristics:

(1) How could we characterize the stabilization point of a cloud?
(2) What would your definition be?

We observed three kinds of definitions. The first kind were descriptions of the appearance of the clouds of points, expressed in terms of an approximation to a point on the grey line on the left. Definitions of this kind were:

*Definition 1(a)* "The meeting point of all points."
*Definition 1(b)* "The point towards which the points are approximated."
*Definition 1(c)* "The point towards which the other points approach."

The second kind of definition included all the properties detected by the students. Usually, properties of a visual kind were mixed with properties concerning the number of points inside or outside specific bands. Definitions of this kind were:

*Definition 2(a)* "It is a point such that, if we draw two lines... one higher than this point and the other lower than it, even if they are as close together as we want, there is always going to be an infinite number of points between them."

*Definition 2(b)* "It is a point such that, if we draw a band centered at this point and we continue adding points to the cloud, the number of points outside this band does not increase."

*Definition 2(c)* "If we draw a band around the stabilization point... from a certain point, there is an infinite number of points inside the band and the number of points outside it stops increasing."

Students who expressed the third kind of definition usually used more precise language. Some of them utilized logical quantifiers, and all of them referred to generic bands in their definitions. Definitions of this kind included:

*Definition 3(a)* "It is a point in which... any band that is centered on it, is always going to have a finite number of points outside."

*Definition 3(b)* "In any band around the stabilization point, there is a finite number of points outside this band."

*Definition 3(c)* "It is a point towards which the rest are directed and, no matter how narrow the chosen band is, assuming it is centered at the stabilization point, there will always be an infinite number of points inside this band and a finite number of points outside."

After the interviewees had expressed their definitions, we concluded the experience by asking them to apply these to various examples, some of which were new and others of which had been previously studied. In applying their definitions, students who had expressed definitions of the first kind only utilized image enlargements after adding large numbers of points. In their reasoning they sometimes used characteristics of a geometric kind, such as the symmetric positioning of the points in some clouds. For example, some students told us that they needed to draw a perpendicular bisector to the grey line on the left in order to find the stabilization point.

Students who had expressed definitions of the second kind drew one or more bands and observed the change in the number of points outside the bands. In most cases, their reasoning led them to perceive the mistakes contained in their definitions and they expressed new definitions that were generally more precise than the previous ones. For example, one student first defined the stabilization point in the following way: "The definition of the stabilization point of a cloud, I think that it would be... the point around which the band, that we draw... will always have inside all the... infinite points." After trying to apply this definition to a specific example, he observed his mistake and stated a new definition: "It is the point around which the band drawn by us, always... will always have a finite number of points outside. This band can be made narrower and narrower."

Students who had expressed definitions of the third kind were able to apply them to the shown examples and express the negations of their definitions. For example, one student expressed his negation in the following way: "... because we have chosen any band and the number... of exterior points is infinite." Another student told us: "Although the number of points interior to this band increases... the number of points exterior to this band also increases, indefinitely, that is to say, the number of points outside this band would be infinite." The conclusion of both students was that, for this reason, the studied cloud was not stabilized at the point marked on the grey line on the left.

## 5. Levels of Reasoning

Through the clinical interviews we observed different levels of reasoning in the students' mental construction of the convergence process of a sequence. These were detected from common characteristics of the different ways used to answer the questions considered and from the evolution of the language used by students during the interview. Students utilized the resources of their level and the analysis of this behavior allowed us to find descriptors of each level of reasoning.

### 5.1. Descriptors for Levels of Reasoning.

**Level 0.** Students who are in Level 0 recognize that one point does not have any dimension and that a segment is formed by an infinite number of points. Any segment can be divided into two equal parts and the result of this division is always two segments. The process of successive division of the segment into two equal parts is an endless process; that is, it is potentially infinite.

**Level 1.** Seven descriptors characteristic of students in Level 1 are given below.

*1.1.* Students recognize that clouds of points are potentially infinite processes generated in a certain way. If a point is subsequent to another point, then it must be situated nearer to the grey line on the left.

*1.2.* They observe the images globally and reach the conclusion that the clouds of points shown may exhibit different behaviors.

*1.3.* Bands are utilized as a tool to limit the part of the image that they want to enlarge. Consecutive enlargements are used to study the behavior of clouds of points. The use of enlargements is a tool that allows students to look at arrangements of points in zones where they accumulate, enabling them to perform more and more precise visual approximations.

*1.4.* When looking at the point of stabilization of a cloud, students only consider its global appearance. For example, they observe properties such as the symmetric disposition of the points or the straight path they describe. For the majority of these students, the first points of the cloud are the most important since those points help them see these characteristics clearly.

*1.5.* In their reasoning, students do not use information or properties that do not appear in the image of a cloud. For this reason, they do not notice the utility of the increase/non-increase in the number of points that are situated outside the band. Some students implicitly observe such details. But if they have not recognized what a cloud of points with all its characteristics entails, then they do not use it explicitly and cannot make progress in their level of reasoning.

*1.6.* The definitions of a point of stabilization given by students in Level 1 always refer to the global appearance of the cloud, its form, and the position towards which the points are directed.

*1.7.* (Differentiation from Level 2) Students in Level 1 not achieving Level 2, use bands as a tool mainly to enlarge the image, but do not use the number of points outside the bands. In constructing more reliable conjectures, only visual information, such as consecutive enlargements of the image, is used. Students in Level 2, however, recognize that the number of points outside the bands provides information about the cloud's behavior. This permits them to form more reliable conjectures about the position of the point of stabilization, if it exists.

**Level 2.** The ten descriptors characteristic of students in Level 2 are as follows.

*2.1.* When students begin to reason in Level 2, they recognize that clouds of points have different parts and that each part provides different information.

*2.2.* Students realize that the behavior of a cloud is determined by the latest points that appear in the image. They do this without considering the first points.

*2.3.* The point of stabilization of a cloud is recognized as that point of the grey line on the left towards which the footprints are directed. Students associate its existence with its uniqueness and associate both with the existence of a unique peak in the cloud.

*2.4.* Students recognize that, if the number of points outside a band does not increase, then it is possible that the cloud of points is stabilized. And, in this situation, the point of stabilization must be inside the band.

*2.5.* Students notice that, if the number of points outside a band increases indefinitely, then no point that is inside the band can be the point of stabilization.

*2.6.* Students understand that, when a cloud stabilizes at a point, then the number of points lying outside a band that contains this point cannot increase indefinitely.

*2.7.* Students realize the convenience of drawing narrower and narrower bands to find the point of stabilization. Some of them recognize that only an approximation is obtained. This is one of the descriptors of Level 3 that is implicit in students' reasoning in the Level 2.

*2.8.* Students are not able to think about any band, but rather just a fixed band. In this level, students do not consider the possibility of an infinite process that requires drawing consecutively narrower bands in order to characterize the point of stabilization.

*2.9.* Students are not able to verbalize a definition of a point of stabilization that includes the quantifier "any." Because of that, their provisional definitions will say that the number of points outside of a band containing the point of stabilization must be finite. In their definitions, they will list the properties that they have noticed. And they will think that if these properties hold in some bands, that is enough to characterize the point of stabilization.

*2.10.* (Differentiation from Level 3) Students in Level 2 reason using fixed bands. They do not notice the dynamic nature of the process of drawing bands, and therefore, are unable to carry out reasoning that includes "any band." When they try to characterize the point of stabilization, they say "it is a point such that if we draw one or several bands centered on it, the number of points that are situated outside the bands is finite." In contrast, students in Level 3 are able to carry out reasoning that involves the dynamic nature of drawing bands. In their definitions, they refer in a generic way to narrower and narrower bands, or if they have achieved a higher level of thinking, any band.

**Level 3.** Five indicators of students' at Level 3 are outlined below.

*3.1.* In order to characterize the point of stabilization of a cloud, students of Level 3 perceive that a dynamic process of drawing infinitely many bands is necessary. They impose the condition that the number of points outside these bands should be finite. Since one cannot draw infinitely many bands, they resort to imposing the condition on any band. They know this guarantees that all bands comply with this condition.

*3.2.* Students are able to express a correct informal definition of the point of stabilization of a cloud.

*3.3.* Students are able to express the negation of their definitions in order to assert that the point marked in the shown cloud is not the point of stabilization.

*3.4.* A higher level of reasoning is achieved with our interview by students who express definitions of the following kind: "The point of stabilization is defined as the point that complies with the condition that the number of points from outside any band, centered at the point of stabilization, must be finite." In applications they are able to negate their definitions to tell us that the points marked are not points of stabilization, since at least one band exists in which the number of external points is infinite.

*3.5.* Another characteristic indicating that students have achieved a higher level of reasoning is their acknowledgement of the need for a formal definition of the point of stabilization, one that disregards visual manipulations.

**Level 4.** Students who achieve this level of reasoning not only notice the necessity of a formal definition but are also able to formulate a formal definition that is consistent with their visual definitions, verbalized in the previous level. We think that students are able to formulate this formal definition when they have enough maturity in algebraic and logical manipulations.

**5.2. Detection of the Level of Reasoning through Students' Language.** During our research we carried out 25 clinical interviews, of which 20 were with the present version. Among these 20 students, there were three students whose concept of point was not suitable for continuing the interview. From the answers of the rest, we classified five students in Level 3, seven students in Level 2 and five students in Level 1.

Students' levels of reasoning were detected not only through their methods of solving problems, but also from their progressive refinement of the language they used to express themselves. When they made progress in their levels of thinking and started to reason at a higher level, students felt a need for more precise vocabulary. They wanted more precise expressions as they tried to verbalize, in the most exact way possible, the relations and properties they had just discovered. Many students, after having expressed their own definitions of a point of stabilization, noticed deficiencies and errors in their definitions when we asked them to apply these to specific examples. This caused them to make progress in their levels of reasoning and led them to express new definitions that were generally more precise than previous ones.

Despite the fact that only a few minutes elapsed between the time students expressed their first and second definitions, their progress in level of reasoning gave rise to a notable improvement in the language and expressions used. During the interview it was also noticed that, depending on the level of reasoning that students had attained, different meanings were given to the same words. This was made clear when they were asked to express a definition of the point of stabilization.

For students in Level 1, a definition was a description of the visual characteristics that they were able to recognize in the image.

For students in Level 2, defining meant enunciating all properties they had perceived. In particular, when defining a point of stabilization as points were added to a cloud, they mixed visual properties with the property of the non-increase of the number of exterior points to a band centered on the point of stabilization.

Students in Level 3 tried to express their definitions by looking for properties that exclusively characterized the point of stabilization. They had already noticed that drawing specific bands, despite allowing them to make better approximations, would not characterize the point of stabilization. This led them to look for more general properties. They started to develop abstract properties of a cloud of points, of a point of stabilization, and of relations between the two. Students' search for more precise language led many of them to use logical quantifiers to express their ideas of convergence.

## 6. Statistical Analysis

To enlarge the sample, and in order to confirm the conclusions obtained through clinical interviews, we designed a 35-question multiple-choice test that reflected the images and questions of the individual clinical interviews. Its construction was delicate because we did not want a lot of information to be lost in this change of format (see Appendix A for test questions). The test had the following characteristics:

(1) We selected questions that could not be answered with laconic answers. Long answers were sought so that the reasoning could be understood. This is why many test questions were of the type: What would you think if...? What information is provided to you by...? What is your conjecture? Why do you think this happens?

(2) Each question had five options, four of them chosen from the most representative (right and wrong) answers given by students in the clinical interviews and a fifth open option so that students could answer according to their own opinion if that did not coincide with any of the four options. Moreover, we titled the test *Stabilization of Clouds of Points* so that the students could not relate the test subject with any concept studied in their present academic year and so that they did not think this was an exam.

(3) We chose the questions to cover Levels 1 through 3. We matched the selected questions with questions from each of the three blocks of the clinical interview where the concept had been constructed in a progressive and orderly way. We therefore introduced the elements involved in the convergence process step by step.

The sample consisted of 301 students who were in their first year of university: 61 students from Telecommunication Engineering (University of Sevilla), 72 students from Technical Architecture (University of Sevilla), 96 students from Chemical Engineering (University of Sevilla), 33 students from Industrial Technical Engineering (University School Ford España, Almussafes) and 37 students from Mathematics (University of Sevilla).

Each student completed an answer sheet and, in order to carry out a suitable multivariate analysis, we prepared model answers by choosing the option that would have been expressed by a student of Level 3. We coded each answer by assigning "0," when it did not coincide with the model answer, and "1," when it did.

The analysis procedure chosen was the method of $k$-means. Being a widely used method, it is included in the statistical program SPSS (version 9.0). The $k$-means algorithm starts with an initial partition of the cases into $k$ clusters (determined by $k$ initial centers). It examines each component of the sample and assigns it to one of the clusters depending on the minimum distance. When all objects have been assigned, the positions of the $k$ centroids are re-calculated. The last two steps are repeated until the centroids no longer move. This algorithm produces partitions of the objects into groups in which the variance decreases in each step.

To determine the initial centers where the algorithm starts, we introduced a criterion of initial classification in accordance with both the van Hiele model and the experimental conclusions achieved through the clinical interviews. To this end, we assigned a minimum or maximum number of correct answers in each block of questions, according to the respective level of reasoning.

The criterion chosen was criterion A and we describe it by means of Table 1. The algorithm reached the solution in nine iterations, the program indicated that the distance by which each center had changed in this iteration was 0.00.

TABLE 1. Criterion of initial classification.

| Criterion A | Block 1 | Block 2 | Block 3 |
|---|---|---|---|
| Level 1 | $\geq 2$ | $\leq 5$ | $< 7$ |
| Level 2 | $\geq 3$ | $> 6$ | $< 7$ |
| Level 3 | $\geq 3$ | $> 6$ | $\geq 7$ |

Since the conclusions of the classification obtained through the method of $k$-means depend on the chosen initial centers, and in order to assure the robustness of the study, we used two additional initial criteria. That is, we applied the algorithm with different initial centers. These were labeled Criterion B and Criterion C (see Table 2). The numbers of students and the clusters obtained with these three initial criteria are shown in Table 3.

Criteria A and B provided the same global distribution by assigning all students to the same cluster, while criterion C moved one student to a different cluster. Discriminant analysis carried out subsequently, by specifying the obtained classification using the $k$-means algorithm (criterion A) as the clustering variable, indicated this criterion classified 93.6% of cases correctly and provided us with a similar distribution to that obtained using the $k$-means method.

Our main objective, when carrying out this statistical analysis was to check, with a larger sample, that the percentages of students in the three levels remained similar to those obtained in the clinical interview, and that therefore a methodological proposal based on the test could be implemented in the classroom.

TABLE 2. Additional initial criteria.

| Crit. B | Block 1 | Block 2 | Block 3 | Crit. C | Block 1 | Block 2 | Block 3 |
|---|---|---|---|---|---|---|---|
| Level 1 | – | $\leq 5$ | $< 7$ | Level 1 | $\geq 2$ | $\leq 5$ | $< 6$ |
| Level 2 | – | $> 6$ | $< 7$ | Level 2 | $\geq 3$ | $> 6$ | $< 7$ |
| Level 3 | – | $> 6$ | $\geq 7$ | Level 3 | $\geq 3$ | $> 7$ | $\geq 7$ |

TABLE 3. Results.

| | Criterion A | Criterion B | Criterion C |
|---|---|---|---|
| Cluster 1 | 79 | 79 | 80 |
| Cluster 2 | 128 | 128 | 127 |
| Cluster 3 | 94 | 94 | 94 |

The study indicated that in our sample 26% of students were in Level 1, 43% were in Level 2, and 31% were in Level 3. These percentages are similar to those obtained in the clinical interviews.

The students of the sample came from different degree courses. A comparative analysis between groups indicated that the global level of reasoning reached by each group was also different. By comparing the previous preparation of the students, the subjects they were studying at the time they completed the test, and the results of their tests, we concluded that previous training in reasoning of an infinite kind and experiences in experiments specially designed to think about a specific concept were decisive factors influencing the level of reasoning reached by each student.

## 7. Contributions to Overcoming Cognitive Obstacles

After carrying out our study, we thought it would be interesting to analyze the relation between our conclusions and those obtained in other studies of convergence. We think the designed experience allowed some of the cognitive obstacles mentioned by various authors to be overcome.

First of all, we think that the visualization we used tends to make students capable of managing notions and reasoning of an infinite kind by themselves in a spontaneous and intuitive way, notions that have traditionally been an obstacle to the acquisition of this concept. The ability to enlarge images allows students to manipulate very small quantities in a natural way, without confusing such small quantities with zero, and helps them realize that even smaller quantities exist. In addition, the ability to add as many points as desired enables students to assimilate the potentially infinite nature of certain processes and the existence of infinitely large quantities. This permits students to attain an advanced mental model of convergence and a suitable conception of the infinite.

According to various authors, the word "limit" can lead to confusion between the convergence process and the limit object and this can be an obstacle in the acquisition of the notion of convergence. With our experiment this does not happen because the idea of convergence of a cloud, expressed through the intention of an animal to walk towards a single point, is completely separate from the idea of its point of stabilization.

The convergence definition expressed by students in our study was geometric and visual, but not informal. Not having preconceived ideas with regard to the considered situation enabled students to express their own definitions as a consequence of their own lines of reasoning. Their attempts at a definition led them to an understanding of the meaning of the commonly used symbols in the classic formal definition of the limit of a convergent sequence. This experience helped students use precise language to express the ideas they had constructed along their own lines of reasoning. And, when a high level of reasoning was achieved, it provided them with a formal language in which they included logical quantifiers.

The clinical interviews allowed us to observe the relation between students' mental models, with respect to the notion of sequential convergence, and the level of reasoning achieved. We noticed that the mental models used by students when thinking about convergence differed according to their levels of reasoning. Gradually, these models become part of their concept images and induce them to reason differently. Such models are sometimes big obstacles to attaining an accurate perception of the convergence process of a sequence. However, with suitable teaching strategies, we consider the evolution from one type of mental model to another more suitable model to be both possible and necessary.

**7.1. Mental Models and Levels of Reasoning.** In describing the relation between students' mental models and their levels of reasoning, we will use Robert's terminology (1982, 1983). Students who do not improve beyond Level 1 only reach "primitive" or "dynamic" mental models.

We identified those students who associate convergence with monotonicity by observing their reasoning as they answered the first interview questions that asked them to conjecture about the intention of the imaginary animal whose footprints marked the cloud of points. When the sequences shown were monotone, the students had no doubts; the conjecture was that the animal had chosen a single point on the grey line on the left. However, when convergent non-monotonic sequences (whose image was made up of "waves of decreasing amplitude") were shown, the students' doubts started to appear. Their answers were of the kind: "I think that the animal is walking aimlessly because if the animal wanted to arrive at a fixed point, then it would go towards it by the shortest route."

This type of mental image can be dispelled if the teacher guides students in the correct direction. Computers, as used in our study, can help since they permit the addition of as many points as desired and carry out any enlargements required. Students are thereby confronted with one of the conflicting factors that can impede them from having a suitable concept image of the convergence process.

We think the visualization that we used is very helpful because it allows students to observe the global behavior of sequences and leads them to change their opinions about convergence and non-convergence. A detailed consideration of non-monotonic sequences, whose convergence is visually obvious, accompanied by questions that make students' initial conceptions untenable, may cause a change in their mental images. As a result, students' convergence concepts can be linked with the number of "peaks of the cloud" observed, rather than with monotonicity.

Those students who had dynamic and non-monotonic mental models made reasonable conjectures on the first interview questions. However, they started having problems once clouds of points with several peaks were considered, since they observed the proximity of points of the cloud to different points of the grey line on the left. As a result, they thought there was more than one point of stabilization. If they failed to surmount this kind of mental model, they did not achieve the reasoning of Level 2. In these cases, it is important that students be brought to observe their mistakes through questions of the kind: "Do you think that an animal can walk towards several different points simultaneously?" We have realized that the uniqueness of the limit (present throughout the whole reasoning process) and the necessity of discerning between situations with visually non-obvious differences are factors that can cause the evolution of this mental model towards another more elaborate model.

Students who achieve Level 2 usually have a "prestatic" mental model. In this level, students think that a sequence converges when the terms end up forming a group around the limit (only one peak). In their definitions of a point of stabilization, they usually say, "If a band is drawn with its center at the stabilization point, then the number of points of the cloud that are situated inside it, is infinite while the number of points that are situated outside it, is finite." When such students apply their definitions, they usually draw quite a few bands. Some students insist those bands be very narrow, but the necessity of an infinite process of drawing bands is not observed by them and will not be included in their conceptions of the convergence process. Although they understand that the behavior of the first few terms is not important in the convergence process, these students tend to think that the behavior of the sequence cannot change once they have seen a significant number of points.

It was observed that some students are in a transition phase between the dynamic model and the prestatic model. These students try to describe, on one hand, the idea of "to approach" (a characteristic of the dynamic model), and on the other hand, the idea of "the grouping of the points in a neighborhood around the limit" (a characteristic of the prestatic model). They usually express themselves by saying "If a band is drawn with its center at the stabilization point, the number of cloud points inside this band is infinite." This problem is solved when these students apply their definitions to specific examples, whereupon they usually add elements of more advanced mental models and express a definition as a type of prestatic model, or even as a type of static model, that generally also includes elements of the dynamic model. We think it is important that students' lines of reasoning go through the prestatic mental model stage since this permits them to start to translate the variables $\varepsilon$ and $n_0$ of Cauchy's definition into natural language, although the quantifiers used are not yet correct.

The mental model of many students in Level 2 advances spontaneously towards the static model, or a mixed model, both of which are suitable for leading them to Level 3. If this evolution does not happen in a spontaneous way, we can precipitate it by confronting such students with examples that evoke the contradictions contained in their models. To this end, we ask students to apply their definitions to sequences in which a drastic change in behavior is noticed only after a considerable number of enlargements. Carefully chosen questions and activities can cause a change of a student's mental model towards a more suitable model that considers all elements of the convergence process.

Students in Level 3 usually have static, or mixed, mental models. They have noticed that the terms of a convergent sequence must be grouped around the limit, but successive enlargements of the image may cause them uncertainty about the limit position. Reflection on such uncertainty can improve students' reasoning by causing them to look for a mental model that permits them to resolve this problem. They are led to look for quantifiers that precisely describe their images of the convergence process. These students first observe that the imposition of conditions on a finite number of bands is not sufficient. Since the distance between the points of the cloud and the point of stabilization "progressively decreases," they are led to recall the representation of the dynamic model. The idea that the distance must become "as small as desired" leads them to consider that the band must be as narrow as desired when characterizing the point of stabilization of a cloud.

Students who have mixed mental models tend to consider the image of the progressive approach of the points of the cloud to the point of stabilization, mixing in ideas of the static model. Their definitions of a point of stabilization are of the kind: "It is a point towards which the rest of the points are heading and no matter how narrow the band centered at the stabilization point, there will always be an infinite number of points inside this band and a finite number outside it."

Students who have static mental models do not express the idea of dynamism in an explicit way. Their definitions of a stabilization point are of this kind: "It is the point in which... any band, that is centered at this point, is always going to have a finite number of points outside." When students have a sufficiently advanced mental model, their own lines of reasoning, guided by the questions of the interviewer and selected activities, lead them to use suitable logical quantifiers in order to describe their idea of the convergence process.

During the clinical interviews the convergence concept was constructed in a progressive way, and in the final questions, students considered "clouds of points" as sets formed by an infinite number of points. Their conceptions of the infinite influenced their lines of reasoning about convergence. The images provided by the mathematical assistant (computer) enabled students to connect their geometric intuitions with the notion of the infinite.

**7.2. Concepts of Infinity.** Next, we use the terminology of Sierpinska (1985, 1987) to describe the different conceptions of the infinite that were detected in the lines of reasoning of students at different levels. Students who are "unconscious infinitists" say "infinite" but think "very big." For them, it is enough to see a sufficiently large number of points because they basically think that the process ends. They are sure of their conjectures after observing only a finite number of points. If this conception goes unchallenged, their reasoning will remain visual (Level 1) since they only need to have a finite number of points to reach conclusions that they consider correct and reliable.

Students designated by Sierpinska as "conscious infinitists" think that infinity is something metaphysical whose description is difficult to get at with precise definitions. Such students try to solve problems through lines of reasoning that do not involve the concept of infinity. In our interviews, they looked for geometric properties in the images, such as symmetry that permitted them to use known elements, for example bisectors, in their characterization of a stabilization point. When we asked them for their own definitions of a point of stabilization, they sometimes refused, arguing "that can never be known." They were unable to conceive of any line of reasoning that would describe a process of this kind. Some of these students came to understand that if all points of the cloud, beyond a certain point, lie inside a band, then it is reasonable to think that the stabilization point, if it exists, must be in this band. However, the actual existence of this point, as a result of an infinite process, created serious problems for them. If they did not manage to rid themselves of this conception, they were only able to make conjectures of a visual kind and their lines of reasoning remained at Level 1, although it is possible that some of them used some elements of Level 2 in an implicit way.

"Kinetic infinitists" were able to reach Level 3, if the evolution of their lines of reasoning was appropriate. This was more difficult for the "potentialists" than for the "potential actualists" since they had to change their conceptions and consider clouds of points as sets that have an infinite number of elements. The "potentialists" perceived the convergence process - visualized through the clouds of points - in a correct way until Level 2. They understood that it was necessary to impose conditions on one or various bands. Some of them also observed the necessity of an infinite process of drawing bands. However, if their conception of the infinite did not change, then their definitions included the condition that "bands can be progressively narrower" but they were unable to abstract the convergence process or to express a definition that included logical quantifiers. In contrast, the "potential actualists" were able to express a definition with these characteristics and to achieve a more advanced level of reasoning.

**7.3. Construction of the Sequence Concept.** The RUMEC group analyzed student conceptions of the sequence concept (Mathews, 2001). This work was performed with the usual methodology of this group. These researchers utilized clinical interviews in order to detect different mental constructions of students. They started from students' previous knowledge about the sequence concept; that included the meaning and use of the general term of a sequence, through which students could achieve the cognitive object SEQFUNC. They indicated that the development of this kind of object was related to a student's function schema, and when this schema was poor, students usually had great difficulties. Our method of working was different from that used by the RUMEC group. Although we utilized individual clinical interviews, we focused our experience on the construction of a suitable concept image for convergent sequences, and because of this, students did not need any previous knowledge about numeric sequences.

In the first part, we constructed the sequence concept (cloud of points) through images of the invisible animal's footprints. We carried out the entire interview based on reasoning about such images. The interviewees performed actions on the images, constructed the object "cloud of points," interiorized the convergence process (stabilization of a cloud of points), and some of them encapsulated it (if they achieved Level 3). The students did not need to use or write formulas. They did not utilize the function concept in an explicit way, and the word "sequence" did not appear in the interviews at any time. Our goal was to have interviewees acquire a suitable concept image of the notion of convergent sequence – one that would lead them to verbalize their own definitions. We consider this a suitable way to introduce sequences of real numbers and their convergence, leaving the notation and the formal definition until a later stage.

Harel's Necessity Principle (1998) states: "Students are most likely to learn when they see a need for what we intend to teach them, where by 'need' is meant intellectual need, as opposed to social or economic need." He discusses three forms of intellectual need: need for computation, need for formalization, and need for elegance. In our study, we looked for a way to induce a need for formalization. One of our main objectives was the construction, by students devoid of the logical-algebraic maturity necessary for understanding the concept definition, of a concept image that was essentially coincident with the notion of convergence of sequences. To this end, we used a very simple metaphor, namely a "cloud of points," that provided useful images for dealing with the concept visually.

The first three van Hiele levels of reasoning are prior to formalization. They establish the boundaries between two phases of an acceptance-confidence-crisis, thereby reaching the goal in its optimal level (Level 3). However, even students at Level 3 are conscious that they are conjecturing and not proving. We deliberately did not proceed to Level 4. The last crisis occurring at the end of the computer experience produced a kind of uneasiness in Level 3 students who were unable to decide about convergence with the tools available. This showed the necessity of proceeding further if certainty about the existence of the limit and its exact value is to be sought. We think that, once the words used in the construction of "clouds of points" and "point of stabilization" are identified with the corresponding terms relating to sequences, then the uneasiness felt through the computer experience will remain in the minds of students who will then appreciate the need for formalization in order to attain certainty.

## 8. A Teaching Methodology Proposal

After the interview experience, we instituted a short training in algebraic and logical manipulation with a sample of Level 3 students from University School Ford España. With these students, we noticed a nontraumatic jump from concept image to concept definition. We proceeded as follows.

First, we defined a sequence as an ordering $s(1)$, $s(2), \ldots, s(n), \ldots$ of numbers and emphasized the role of subscripts in positioning the elements of the sequence. The students easily accepted that other sets of indices such as $1, \frac{1}{2}, \frac{1}{3}, \ldots$ can perform the same role.

Once settled in the computer environment, students preferred the parametric representation $(1/n, s(n))$ to the more usual $(n, s(n))$. Everything is visualized in the interval $[0, 1]$ and the only command to use is Zoom. Somehow it was easier to conjecture the possible limit as an accumulation of points rather than as the asymptotic behaviour of the representation $(n, s(n))$ of equally spaced points.

No problem was detected in identifying the sequence formally with what had been done with clouds of points. Translation of convergence from stabilizing clouds to convergent sequences was done without significant trouble: bands were reverted to epsilons, outside points to capital $N$, and quantifiers employed with precision.

The implicit convergence notion buried in our experience (the sequence $1/n$ converges to 0) produced no cognitive obstacles. All students accepted that $1, \frac{1}{2}, \frac{1}{3}, \ldots$ approaches 0 (without even mentioning the fact) and they performed the task of approaching the grey line on the left a finite number of times.

The additional information on our screens about how many points were outside the chosen band was recognized as useful. Although irrelevant in the first stages of the experience, once the manipulation of bands proceeded, it implanted itself in the students' minds. The number of points outside a band changes with manipulation and is crucial to the desired verbal definition of convergence. We think that if we hadn't provided this information automatically, the flow of the interview would have been broken, as several iterations of Zoom In and Out, as well as counting, needed to be performed.

Finally, even in the presence of the metaphor of a "walking animal," no student doubted that they were dealing with mathematics and nothing else. Our study has led us to the conclusion that a teaching methodology used to introduce the notion of convergence should have the following characteristics:

- It should allow the introduction of the concept without requiring students to have any kind of logical or operational skill.
- It should convey the essence of the processes of reasoning of an infinite kind.

- An identifiable hierarchy of levels of reasoning must be present.
- It must prepare the student for the "jump" to the formal definition.

The individual interview that we used to establish the descriptors of the van Hiele levels for the concept of convergence fulfills all these characteristics. It is a suitable methodology by which students can make the mental construction of sequence convergence. Unfortunately, the interview cannot be applied to groups of students. However, the written test is applicable to large groups, and it also has the aforementioned characteristics, although it must be complemented with visualization of the images provided by the interactive computer screen. Appendix A includes the questions of the written test, without images or answer options. Each student must have a computer and follow the indications of the teacher.

We think that construction of the notion of convergence of a sequence should incorporate the following stages:

*Stage 1. Construction of the concept of "a cloud of points."* One should start by explaining the metaphor of the "walking animal" and by introducing the visualization tools. Students should observe clouds of points globally, see the necessity of using the image enlargement tools, and make their own conjectures. In doing so, they become aware that the first points of the image should have no influence on their conjectures.

*Stage 2. Construction of the horizontal bands as tools of reasoning and introduction of the "point of stabilization of a cloud."* At this stage it is important that students associate the existence of this point with its uniqueness. They should associate both ideas with the existence of a single peak in the cloud since this leads them to consider the increase or non-increase in the number of points outside a band as a reliable and useful piece of information.

*Stage 3. Introduction of the band tracing process as a tool of reasoning.* One should show examples in which conjectures are not visually obvious so students realize the necessity of tracing a certain number of progressively narrower bands.

*Stage 4. Abstraction of the concept and search for a definition.* In this part, questions must be formulated without images so that the lines of reasoning begin to be general. The existential and universal quantifiers should be gradually introduced. Students should become conscious of the properties and relations that allow them to express a definition reflecting their idea of convergence. The next step is to ask students to apply their own definitions to different examples.

*Stage 5. Formal definition of the concept of convergent sequence.* Translate the words used in the construction of clouds of points. This should be done through the images and examples used by the student in the interview. Later, the student should study new examples. The translation is made by the identifications: "sequence" with "cloud of points," "numerical value of every term of the sequence" with "height" at which the point representing this term is situated, "convergence" with "stabilization," "limit of the sequence" with "point of stabilization of a cloud," "any $\varepsilon > 0$" with "any band," "$n_0$" with "location of the point after which all points of the cloud stay inside the considered band," and "$n \geq n_0 \Rightarrow |x_n - L| < \varepsilon$" with "all points of the cloud are situated inside the band, except the first $n_0$."

*Stage 6. The elimination of mistakes and the use of the usual terminology.*
After relating the image elements with the formal definition, one must continue asking questions of the kind: "Can the terms of the sequence have the value of the limit?" This question could be translated as "Can the points of the cloud be situated at the same height as its point of stabilization?" And, "If this happens, what relation is there between the value of these terms of the sequence and the value of the limit?" The study of examples having these characteristics should make this situation understood. We think this is the best time to introduce the general term of a sequence. As an example, the sequence proposed by Vinner and Tall, should be shown:

$$s_n = \left\{ \begin{array}{ll} 0 & (n \text{ odd}) \\ 1/2n & (n \text{ even}) \end{array} \right.$$

Students can observe how its cloud of points is generated, and as an exercise, they should describe the behavior of the terms of the sequence. The symbolic expression of the general term of a sequence should then be introduced and related to its visualization.

Our methodology forces students to choose their own answers to the questions on the written test. They study the different examples through the tools provided by the interactive screen. To this end, every student needs to have a computer available on which the mathematical assistant (i.e., the software) has previously been installed. The mathematical assistant can be obtained through e-mail from M. A. Navarro: manavarro@us.es; it is available both in English and Spanish. Furthermore, it would be advisable to have a video projector connected to the teacher's computer so that the running of the interactive screen and the metaphor of the "walking animal" can be explained.

The initial introduction of the concept could be carried out through two sessions of one hour each, during which students answer questions on the test that can be divided into two parts. In the first session, the students must familiarize themselves with use of the screen. The teacher limits the time available for the student to answer and provides the necessary explanations orally or on the screen. We try to replace most of the "information for the student," that appears on the written test, with the explanations of the teacher who answers any questions about the mathematical assistant.

In the second session, the students should already be familiar with the screen and the way in which the points are going to appear in the clouds. However, this should be reviewed since it is very important in order to answer the test questions. During the development of this session, the intervention of the teacher should be kept to a minimum. However, necessary information should be provided and students' doubts resolved.

We think that Stages 5 and 6 of the proposed teaching methodology (i.e., formal definition of the concept of convergent sequence and elimination of mistakes) could be carried out by means of a combination of lectures, the use of the mathematical assistant's images, and small group discussion. To conclude the experience, we consider it appropriate to show students the content of the mathematical assistant and to provide them with the program and some of the sequence expressions used on the test (see Appendix B). Students should then key these expressions into the computer, thereby checking that the same images used on the test are obtained.

After the experience, some students might think that this method is a universal tool to study the convergence of any numeric sequence. It is then necessary to point out that this is not the case. On one hand, they should be made aware that screen precision is limited and this will make it impossible for them to find the limit exactly. On the other, it is possible that, even if the first 500,000 terms of a sequence induce us to think that we are dealing with a convergent sequence, there is no way to guarantee that its tail might not change its behavior completely. Mathematics seeks certainty and the student must discover the necessity of a formal definition, as well as the use of inequalities and logical quantifiers, in order to study the convergence of sequences properly. Our teaching methodology is intended for use prior to the formal treatment of sequences. It is one way of organizing students' understanding of the process of sequence convergence.

## References

Artigue, M. (1995). La enseñanza de los principios del cálculo: Problemas epistemológicos, cognitivos y didácticos. In P. Gómez (Ed.), *Ingeniería didáctica en educación matemática* (pp. 97–135). México, D. F.: Grupo Editorial Iberoamérica.

Asiala, M., Brown, A., DeVries, D. J., Dubinsky, E., Mathews, D., & Thomas, K. (1996). A framework for research and curriculum development in undergraduate mathematics education. In J. Kaput, A. H. Schoenfeld, & E. Dubinsky (Eds.), *Research in collegiate mathematics education. II.* (pp. 1–32). Providence, RI: American Mathematical Society.

Campillo, P. (1998). *La noción de continuidad desde la óptica de los niveles de van Hiele*. Thesis. Universidad Politécnica de Valencia, Valencia, Spain.

Campillo, P., & Pérez Carreras, P. (1998). La noción de continuidad desde la óptica de los niveles de van Hiele. *Divulgaciones Matemáticas, 6*(1), 69–80.

Clements, D. H., & Battista, M. T. (1992). Geometry and spatial reasoning. In D. A. Grouws (Ed.), *Handbook of research on mathematics teaching and learning* (pp. 420–464). New York: MacMillan.

Cornu, B. (1991). Limits. In D. O. Tall (Ed.), *Advanced mathematical thinking* (pp. 153–166). Dordrecht, The Netherlands: Kluwer.

de la Torre, A. F. (2000). *La modelización del espacio y del tiempo: Su estudio via el modelo de van Hiele*. Thesis. Universidad Politécnica de Valencia, Valencia, Spain.

Esteban, P. (2000). *Estudio comparativo del concepto de aproximación local vía el modelo de van Hiele*. Thesis. Universidad Politécnica de Valencia, Valencia, Spain.

Harel, G. (1998). Two dual assertions: The first on learning and the second on teaching (or vice versa), *The American Mathematical Monthly, 105,* 497–507.

Jaime, A., & Gutiérrez, A. (1990). Una propuesta de fundamentación para la enseñanza de la geometría: El modelo de van Hiele. In S. Llinares & M. V. Sánchez (Eds.), *Teoría y práctica en educación matemática* (pp. 295–384). Sevilla: Alfar.

Jaramillo, C. M. (2000). *La noción de serie convergente desde la óptica de los niveles de van Hiele.* Thesis. Universidad Politécnica de Valencia. Valencia.

Llorens, J. L. (1994). *Aplicación del modelo de van Hiele al concepto de aproximación local*. Thesis. Universidad Politécnica de Valencia, Valencia, Spain.

Llorens, J. L., & Pérez Carreras, P. (1997). An extension of van Hiele's model to the study of local approximation. *International Journal of Mathematical Education in Science and Technology, 28,* 713–726.

Mathews, D., McDonald, M., & Strobel, K. (2001). Understanding sequences: A tale of two objects. In E. Dubinsky, A. H. Schoenfeld, & J. Kaput (Eds.), *Research in collegiate mathematics education. IV.* (pp. 77–102). Providence, RI: American Mathematical Society.

Monaghan, J. (1991). Problems with the language of limits. *For the Learning of Mathematics, 11*(3), 20–24,

Navarro, M. A. (2002). *Un estudio de la convergencia encuadrado en el modelo educativo de van Hiele y su correspondiente propuesta metodológica.* Thesis. Universidad de Sevilla, Sevilla, Spain.

Pérez Carreras, P. (2000). *Matemática asistida por ordenador, Cálculo infinitesimal.* Valencia, Spain: Universidad Politécnica de Valencia.

Robert, A. (1982). L'acquisition de la notion de convergence des suites numériques dans l'enseignement supérieur. *Recherches en Didactique des Mathématiques, 3*(3), 307–341,

Robert, A. (1983). L'enseignement de la convergence des suites numériques en DEUG. *Bulletin de l'APMEP,* 431–449.

Sierpinska, A. (1985). Obstacles epistemologiques relatifs a la notion de limite. *Recherches en Didactiques des Mathématiques, 6*(1), 5–67.

Sierpinska, A. (1987). Humanities students and epistemological obstacles related to limits. *Educational Studies in Mathematics, 18,* 371–387.

Tall, D., & Schwarzenberger, R. L. E. (1978). Conflicts in the learning of real numbers and limits. *Mathematics Teaching, 82,* 44–49.

Tall, D., & Vinner, S. (1981). Concept image and concept definition in mathematics with particular reference to limits and continuity. *Educational Studies in Mathematics, 12,* 151–169.

van Hiele, P. M. (1957). *El problema de la comprensión.* Thesis. Universidad Real de Utrecht, Utrecht, The Netherlands.

van Hiele, P. M. (1986). *Structure and insight.* London: Academic Press.

## Appendix A

This questionnaire has been designed to be answered by students while using the mathematical assistant on the computer. The entire questionnaire with multiple-choice answer options (in the original Spanish or English translation) can be obtained from the autors. The questions refer to images of example sequences, whose general terms are given in Appendix B. The teacher must explain how the mathematical assistant works.

**Stabilization of clouds of points.** When we want to split a piece of thread into two equal parts, we usually fold it in half and cut the thread with scissors. In this way we obtain two pieces of the same length. Let's suppose that we take a pair of scissors and a piece of thread and we divide it into two equal parts. Then we take one of these parts and we halve it again. This process is then repeatedly carried out.

**Question 1.** Until when could we continue dividing the pieces of thread we are obtaining in this way?

**Question 2.** When we want to handle mentally the image of a piece of stretched thread, we usually think about a segment and represent it graphically in this way. In Geometry, a ruler and a compass is used in order to divide a segment into two equal parts. We are going to repeat the same process of division with a segment, just as it is shown in the drawing:

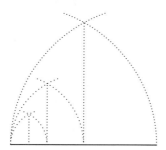

Can we continue repeating this process of segment division as many times as we want?

**Question 3.** If you had a magnifying glass that permitted you to amplify the image as much as you wanted, how many points would you be able to mark using this procedure?

**Question 4.** *(Example 1. The teacher must explain the metaphor of the animal.)* Which is your conjecture with respect to this animal's intention?

**Question 5.** Would you need to see more steps to deduce if the animal has chosen some point on the blue line and if it intends to walks towards it?

**Information to students:**
*The animal's footprints we have observed, make up a drawing where the points are always marked in an orderly way and after each point we can always find a new point, which will be situated closer to the blue line than the previous one. From now on we will call this kind of image a "cloud of points."*

**Question 6.** *(Example 2. The cloud must be generated point by point.)* What differences can you find between this cloud of points and the previous one? Do you believe this animal has chosen a point on the blue line with the intention of walking towards it?

**Question 7.** *(Example 3. The cloud must be generated point by point.)* Do you think this cloud is behaving in a different way to the previous ones? Do you believe the animal has chosen a point with the intention of walking towards it?

**Question 8.** *(Example 4. Only the first six points must be added.)* Do you believe that the animal, which is marking this cloud of points, is walking aimlessly or do you think it has chosen some point on the blue line with the intention of walking towards it?

**Question 9.** *(100 points must be added. The student must mark the point towards which he or she thinks the animal is walking, on the blue line.)* After seeing this image, do you still think the same as on seeing the earlier one? Must the animal have chosen the point marked in red on the blue line?

**Question 10.** *(The previous image must be enlarged.)* After seeing these images, would you change the conjecture you made in the earlier questions?

**Question 11.** *(The enlargements must be removed by pressing "Previous" on the screen. All bands you have drawn must be without enlargement.)* If we removed the first 5 points at the beginning in the image, would it change your conjecture about the intention of the animal or would you still think the same? What if the first ten footprints disappeared from the drawing? Why?

**Question 12.** *(Two paths, one inside the other, and a point situated inside the narrowest path must be drawn by the teacher.)* Do you believe we must find footprints marked within this path? What if the path were narrower?

### Information to students:
*We are studying clouds of points and we are drawing horizontal bands in order to enlarge the images. These horizontal bands will play the role of paths, in whose centre is possibly the point chosen by the animal. Likewise, points of the cloud continue showing the positions of the animal which is always approaching the blue line.*

**Question 13.** *(Example 6. 60 points must be added. The student must think what would be her or his conjecture about the animal marking these footprints. The student must mark the point towards which he or she believes the animal intends to walk and then must draw a band around this point.)* If points situated outside the band disappeared, would your conjecture change? If we add 100 more points to this cloud, do you believe that the number of points outside the green band would increase?

**Question 14.** *(The previous band must be enlarged and 100 points must be added. The student must draw a narrower band around the red point and must enlarge it and add the points.)* Observe that, on adding points to the cloud, the number of points outside the blue band has not changed. Why does this happen?

### Information to students:
*We will say that a cloud of points stabilizes when the animal whose footprints provide the points has chosen a specific point on the blue line, with the intention of walking towards it. The name of this point will be the "point of stabilization" of the cloud.*

**Question 15.** If the behavior of the cloud you have just seen does not change, do you believe this cloud of points will stabilize? If so, must the point of stabilization be inside the blue band?

**Question 16.** *(Example 5. The cloud must be generated point by point, until a conjecture can be made.)* Do you believe that the animal has chosen a specific point on the blue line with the intention of walking towards it? Does it appear to you that the cloud will stabilize?

**Question 17.** *(A band must be drawn around the highest peak and enlarged. The student must add 50 points, add 50 points again and must repeat this operation several times while paying attention to the quantity of points going outside this band.)* If the animal had chosen a point on the blue line inside this band, should the footprints stop appearing outside the band at a certain point in time? If you had only seen the last four images and had observed the increase in the number of points marked outside the green band, which would your conclusion be with respect to the possible point of stabilization of this cloud?

> *Remember: We will say that a cloud of points stabilizes when the animal whose footprints provide the points, has chosen a specific point on the blue line, with the intention of walking towards it. The name of this point will be the "point of stabilization" of the cloud.*

**Question 18.** *(Example 7. The cloud is generated with 100 points. The student must make a conjecture about the stabilization point, must draw a band around this point, and this band must be enlarged; the student must continue adding points.)* Observe that, on adding points to the cloud, the number of points outside the green band does not stop increasing. Why do you think this happens? If the behavior of the cloud of points continues being the same as it was till now, (and the number of points outside the green band does not stop increasing), do you think that this cloud will stabilize at some point situated inside the green band?

> *We are trying to look for a way to distinguish if a cloud of points stabilizes or not.*

**Question 19.** *(Example 8. A first conjecture must be made and the image must be enlarged twice. The student must add points to the cloud and must observe the behavior of the quantity of the points outside this last band.)* On adding points it is observed that the number of points marked outside these bands does not increase. Do you believe that this is sufficient to assure that a cloud stabilizes? Why?

**Question 20.** *(More points must be added and the image must be enlarged as often as necessary in order to answer.)* When we trace a band around a point and we add points to the cloud, what information gives you the fact that the number of points outside this band stops increasing? What would you think if, on adding points to the cloud, that quantity continued increasing and increasing indefinitely?

**Question 21.** *(A band around a peak must be drawn and points must be added. The student must observe that the quantity of points outside this band continues increasing and increasing.)* Do you believe that this cloud stabilizes at a point inside this red band? Why? What differences do you observe between the clouds that stabilize and those that do not?

> *Observe the next cloud of points and think if the red point can be the stabilization point of this cloud. Our objective will be to look*

*for properties of the point of stabilization of a cloud in order to*
*characterize this point in a unique way.*

**Question 22.** *(Example 11. 100 points must be added, a band must be drawn*
*and enlarged, and points must be added until the quantity of points outside the band*
*stops increasing. This process must be repeated at least three times.)* Observe that,
in all bands that you have traced, the quantity of points of the cloud situated
outside them has stopped increasing. If this also happened in the next band you
trace, what could you deduce about the point of stabilization of the cloud?

**Question 23.** If the point of stabilization of the cloud were inside this band,
should there be a point in time from when all the points we added were also situated
inside this band?

**Question 24.** *(A new band must be drawn and enlarged, and points must be*
*added until the quantity of points outside this band stops increasing.)* The number of
points outside this band has permanently stopped increasing. Do we have sufficient
information to assure that the red point is the point of stabilization of this cloud?
Could the point of stabilization be another point close to the one marked in red?
What could we do in order to decide?

**Question 25.**   After seeing the earlier images, it seems clear that we could
continue with this process of tracing narrower and narrower bands centered on the
red point. If, on adding points to the cloud, in each new band the number of points
outside them stops increasing, how many bands of this type do you believe we would
have to draw to be able to assure that a specific point is the point of stabilization
of this cloud?

*Let's suppose that the red point marked on the blue line coincides*
*exactly with the point of stabilization of the last cloud.*

**Question 26.**   In the bands centered on the red point we have drawn, we
have seen that the number of points marked outside them definitely stop increasing. Do you believe the same must happen in any band centered on the point of
stabilization? Must all points of the cloud from one onwards be inside this band?
Why?

*In the next questions we will not have any images on the screen*
*because we are going to imagine that we are studying any cloud*
*of points that we know stabilizes at a specific point. Our objective*
*is to look for properties which characterize this point.*

**Question 27.**   *(One must close the last example and answer without images.)*
On adding points to the cloud, can some band centered at the point of stabilization
exist where the number of points outside this band continues increasing without
stopping? If we think about the complete cloud (with its infinite number of points),
can some band centered at the point of stabilization exist where the number of
points outside the aforementioned band is infinite?

*In the next questions, when we talk about any band centered on*
*the point of stabilization, we do not refer to a specific band, but to*
*whatever band that is going to represent to us all bands centered*
*on the point of stabilization of the cloud.*

**Question 28.** Can there be an infinite quantity of points outside any band centered at the point of stabilization? For any band centered at the point of stabilization, must all the points of the cloud be situated inside it except for a finite quantity?

*We continue our reasoning about any cloud of points that stabilizes at a point.*

**Question 29.** Must there be another point on the blue line, different from the stabilization point, which satisfies the condition that any band we draw around it contains all points of the cloud except a finite quantity?

**Question 30.** You have seen a series of properties of the point of stabilization of a cloud. Taking these into account, what definition would you give of "a point of stabilization of a cloud" so that it is perfectly characterized? When would you say that a cloud of points stabilizes?

**Question 31.** *(Example 1. The student must generate the cloud, must make a conjecture regarding the point of stabilization of this cloud, and must mark the corresponding point.)* Apply the definition you gave in Question 30 to answer this question: Is the red point that you have marked on the blue line the point of stabilization of this cloud?

**Question 32.** *(Example 3. A point, approximately in the centre of the blue line, must be marked.)* Apply the definition you gave in Question 30 to answer this question: Could the red point that you have marked on the blue line be the point of stabilization of this cloud?

**Question 33.** *(Example 5. The point towards which the central peak seems to be directed on the blue line must be marked.)* Apply the definition you gave in Question 30 to answer this question: Is the red point that you have marked on the blue line the point of stabilization of this cloud?

**Question 34.** *(Example 9. You must generate the cloud and observe that there is a red point marked on the blue line.)* Apply the definition you gave in Question 30 to answer this question: Is the red point marked on the blue line the point of stabilization of this cloud?

**Question 35.** Do you think it is necessary to have a definition not of visual type if we want to be able to determine when a cloud of points stabilizes at a point?

## Appendix B

General term of sequences used to generate the figures of this paper and the images of examples for the questions in Appendix A.

*Example 1.* $x_n = \dfrac{n+1}{n}$

*Example 2.* $x_n = \dfrac{n}{25}$

*Example 3.* $x_n = \dfrac{\sin(n/3)}{2}$

*Example 4.* $x_n = \dfrac{\sin n}{n^2}$ (Figures 2, 3, 4, 5, 6).

*Example 5.* $x_n = \sin\left(\dfrac{2n\pi}{3}\right) + (-1)^n \dfrac{\log n}{n^2}$ (Figures 7, 8, 9).

*Example 6.* $x_n = \dfrac{2\sin n \log n}{n^2}$

*Example 7.* $x_n = \left|\sin\left(\dfrac{n\pi}{2}\right)\right| + (-1)^n \dfrac{2\log\left(\dfrac{1}{10} + |\sin n|\right)}{n^{3/2}}$ (Figures 10, 11, 12).

*Example 8.* $x_n = \dfrac{\left|\sin\left(\dfrac{n\pi}{2}\right)\right|}{10^3} + (-1)^n \dfrac{2\log\left(\dfrac{1}{10} + |\sin n|\right)}{n^{3/2}}$ (Figures 13, 14, 15).

*Example 9.* $x_n = \dfrac{n}{2(n+1)} + \dfrac{(-1)^n}{4n}$

*Example 10.* $x_n = \dfrac{\sin n}{10^4} + \dfrac{(-1)^n \log n}{n^2}$

*Example 11.* $x_n = \dfrac{2\sin n \log n}{n+1}$ (Figures 16, 17, 18).

Departamento de Matemática Aplicada I, E.U. Arquitectura Técnica, Universidad de Sevilla, Av. Reina Mercedes 4-A, 41012 Sevilla, España
*E-mail address*: manavarro@us.es

E.T. Superior de Ingenieros Industriales, Universidad Politécnica de Valencia, Camino de Vera s/n 46022 Valencia, España
*E-mail address*: pperezc@mat.upv.es

CBMS Issues in Mathematics Education
Volume **13**, 2006

# Developing and Assessing Specific Competencies in a First Course on Real Analysis

Niels Grønbæk and Carl Winsløw

ABSTRACT. How does one describe, implement, and assess learning goals in university level mathematics? We addressed this problem from the theoretical perspective of *specific competency goals*. In the concrete context of a first course on real analysis at the University of Copenhagen, we conducted a development project and investigated what students were meant to be able to do in the setting of specific content elements. The urgent need for consistency among stipulated course goals, organization of student work, and assessment was in part related to the concrete context of a course on real analysis that posed extraordinary difficulty for students. This need also motivated us to approach local and global aspects of course redesign in a coherent way.

## 1. Introduction

There are several difficult transitions in mathematics education. Much attention has been given to those that coincide with institutional transitions between primary, secondary, and tertiary levels. Nonetheless, within university mathematics programs, certain transitions also cause trouble. Some studies have shed light on the problems that occur in the progression of undergraduate mathematics, from the concrete work of computational calculus and elementary linear algebra to the abstract work of rigorous real analysis and abstract algebra (Selden & Selden, 1995). With their massive need for formal reasoning and conceptualizing, one may speculate that the more advanced courses draw upon different cognitive resources than the elementary courses.

How acute these troubles may be, and in what subjects, undoubtedly varies from institution to institution. Much can be done by careful curriculum planning and special courses that pay specific attention to the obstacles that students encounter in approaching the modern mathematics of the 20th century. Nevertheless, there have been reports that students inevitably meet seemingly insurmountable cognitive leaps and that for the vast majority of students overcoming them requires much patience and help (Burn, Appleby & Maher, 1998). It is our impression that first courses in real analysis are a main site for encountering the need for such leaps.

The main part of this work was done while the second author was working at The Danish University of Education. Both authors are grateful to the university for its support of the project reported here.

Indeed, real analysis has many of the characteristics of modern mathematics that are known to pose difficulties for learners. These include: the frequent need for logical unpacking of informal statements and for construction of statement images (e.g., in the context of $\varepsilon$-$\delta$ formalism (Selden & Selden, 1995)), frequent and non-standard shifts between different representations of objects (Duval, 1995; also see below, Section 2.2), and high demands on quick and accurate reification, such as when dealing with convergence of sequences of sequences (Sfard, 1991).

The existence of thresholds in the process of acquiring professional mathematical insight and competency has been the source of not only didactical but also institutional and organizational difficulties. Although the challenges of learning advanced mathematics are not new, they seem to have gained importance for political and sociological reasons. Universities, in particular Faculties of Science in Denmark and in many other countries, have been facing substantial financial and political pressure due to stagnating student recruitment, operational problems (including high rates of failure), and low production that does not match society's demand for science graduates. In particular, there has been strong pressure on mathematics departments to smooth out steep transitions like the one discussed here in order to facilitate the path to graduation. If this is not to be done simply by lowering standards, then it becomes paramount to deal with important practical and theoretical aspects of the problem.

For students, the problem may be expressed by questions such as: Is this the right course for me? How does it fit in my general study program? Is my background sufficient? Are the goals clear and acceptable to me? Will I be successful? Perhaps a little later in the course the questions become: What kind of participation is required? What am I supposed to be able to do? What are my responsibilities? And, faced with often daunting demands on a student's time, one might ask: What kind of work is it important to do now? How do I deal with the fact that course work and projects are not evenly distributed throughout the semester? Is the assigned course load reasonable when compared with the actual time spent? Does it agree with my expectations?

The same types of questions might be used to express the concerns of the teacher. The felt meaningfulness of the course in relation to a general study plan and the teacher's personal opinion about the importance and relevance of the subject could be central issues. Other instructor questions might pertain to students' willingness and ability to confront challenges posed by the subject matter and the associated difficulties this may cause for the teacher. Likewise, the size and level of tasks could be unclear: What are the teacher's responsibilities for the learning outcome? Depending on external constraints, these might include purely logistic tasks of organizing large classes, establishing learning environments, planning the course, and assigning classes to teaching assistants. Moreover, the teacher has to consider how to balance the tasks related to the course with other duties.

Each university course has its version of these general problems. To reduce frustration, drop-out rates, and the risk of superficial learning it is necessary that these problems undergo discussion and treatment that address specific aspects of a course. Some of this discussion may be specific to the subject matter, some to stipulated constraints of the course in terms of its placement in the general study plan (e.g., prerequisites, other courses based on it), available facilities, number of students and so on. Thus, there is a *need for operational tools to articulate pertinent course goals in a way that minimizes the uncertainty of students and teachers with respect to questions such as those discussed above.* An important condition for these tools to be operational is that they make sense to all parties involved, including students, teachers of the course, teachers of related courses, and study boards. On the other hand, behaviourist patterns of standard performance do not suffice for describing the goals of university education. In addition to descriptions of goals, one needs *tools to monitor learning progression, to measure outcomes, and possibly to improve results.* Everything, of course, is relative to the goals.

It was the objective of our study to devise, implement, and evaluate potential partial responses to these needs within the context of traditional teaching (large lectures and small problem sessions) and with rather strict prescriptions of subject matter. Our key theoretical tool was *specific competency goals* (abbreviated SCGs and introduced in Section 2.1). These formed the basis for the global aspects of this project (Artigue, 1989). In particular, the description of goals presented here is focused on what the students were expected to be able to do and at which level, in pertinent subject matter contexts, rather than being expressed solely in terms of course content. When the use of SCGs was actually carried out in a concrete teaching setting, it had implications for all parts of the agenda for teaching and learning.

Having taken this stand, it seemed necessary to design forms of evaluation (formative as well as summative) allowing students to demonstrate the goal competencies. To give a picture of how the competency agenda might be enacted, without hiding too much of the complexity, we present a rather wide range of selected concerns and observations relating to this agenda and its implementation (Section 3). With this background, we then formulate the basic questions of the study and its methodology (Section 4). The final sections report on results (Section 5) and their analysis (Section 6) with respect to the theoretical framework.

## 2. Theoretical Preliminaries

In this section, we describe the theoretical background for the design of the development project (described in detail in Section 3) and for the discussion of its results (Sections 5 and 6). We consider three levels of inquiry into the problems raised in the introduction:

(1) the level of the *curriculum*, understood as explicit and institutionalized descriptions of the course of study,

(2) the level of *students*, in the sense of *learning* demanded by the curriculum,

(3) the level of *teaching* and *assessment* aimed at fostering and evaluating such learning.

These levels are clearly mutually dependent. We expect that the theoretical considerations presented in this section could also be useful for thinking about these three levels in other contexts of teaching than the one considered from Section 3 onwards.

**2.1. The Curriculum.** Descriptions of an undergraduate course in mathematics might address several very different issues, such as the function of the course within the entire study program, its overall goals, the subject matter topics to be treated, target competencies, the working methods and materials used, and criteria for passing the examinations. Any one of these may be described in more or less detail. But these are not all at the same level. In most descriptions of university courses the subject matter (e.g., linear algebra) comes first, followed by a specification of corresponding topics covered (e.g., matrix representations). In many cases, target competencies – what students should be able to *do* with their new mathematical knowledge – are dealt with in less detail, if at all. Examination instructions may provide more information on this point than the curriculum. For instance, looking at previous final examinations, it might be that there is always an exercise in which one has to determine the $3 \times 3$ matrix of a linear mapping on the space of second-degree polynomials. If this happens to be the only way in which the topic of matrix representations is evaluated, one could speculate that the competencies aimed at were in some sense narrow. Hence, for both internal and external purposes of communication, it would be useful to articulate such aspects of the curriculum.

Customary statements of the goals and means of mathematics teaching may not address the need to consider "the complexity and the specificity of the notions within the context of curricula" (Robert, 1998, p. 142). The close examination of particular domains of knowledge has a long history in mathematics education research (e.g., the German *Stoffdidaktik* tradition as surveyed in Jahnke (1999)). However, many authors have opted to study fundamental or general ideas in broader areas such as calculus (e.g., Fischer, 1976).

In university teaching, the framework of didactical engineering has been particularly influential as a method for integrating analysis of curricula with observations and reflections on innovative teaching practice (Artigue, 1989). The analysis pertains to definite subject matter knowledge to be learned and claims no independence of the mathematical context. In fact, didactical engineering has been concerned with very specific problems related to content, and hence with didactical choices that have been of a more local nature than our focus here: the design of an entire course.

A very different and global viewpoint has been used in the area of curriculum development. At the level of primary and secondary school there have been a number of different and conflicting trends in curriculum development related to mathematics. The scope of these debates has extended much beyond issues directly related to mathematical content. In particular, the roles of mathematics in everyday life and in society have been used to justify new goals and emphases in teaching. This has added both a political and a general or public aspect to the discussion of mathematics curricula.

In particular, the American reform curricula have generated large amounts of research and practical development as well as heated debate. Certain main ideas behind these reforms – constructivism, socio-cultural perspectives, situated learning – have been influential in many other countries as well (e.g., Mansfield, Pateman & Bednarz, 1996). In this reform discourse, attention to specific mathematical content has been reduced and more emphasis has been given to the meaningfulness and the usefulness of mathematics as a human sense-making and problem-solving activity (Verschaffel & De Corte, 1996).

Mathematics education research has been facing the challenge of identifying general learning outcomes for mathematics as a subject. This was particularly visible in the change of focus in two recent international surveys of mathematical performance. While the Third International Mathematics and Science Study

(TIMSS) was organised according to mathematical topics as described in national curricula (Robitaille & Donn, 1992), the Programme for International Student Assessment tested general categories of content (e.g., change and relationships), skills (organised in three competence clusters: reproduction, connections, and reflection), and real-life contexts (Programme for International Student Assessment, 2001). A related theoretical framework, elaborated by a working group under the Danish Ministry of Education, identified a total of eight mathematical competencies to be used in curriculum development at all levels (Niss & Jensen, 2002). The basic idea of this framework was to devise mathematics curricula in terms that were to a large extent independent of a list of topics.

In university teaching there is a need for balance between progress with respect to competency and with respect to content knowledge. However, in the university education of mathematicians, we don't think the general frameworks mentioned above have been very helpful. While it can be shown that extensive exposition of advanced mathematical content does not guarantee a high performance, for example, in problem solving (see DeFranco, 1996), it seems clear to us that general mathematical competencies, such as problem solving skills, could be developed very far in quite limited mathematical contexts (say, number theory). In a sense these two key aspects of the curriculum – content and competency – are orthogonal (see Winsløw, 2005). Progress is needed in *both* directions. An emphasis on goal competencies could serve to counterbalance the pressure to include still more topics in curricula. However, this should not become an unquestioned rationale for reducing our ambitions in terms of content, nor should this reduce our awareness of the complexity of content aspects of mathematical performance.

At the university level, several important facts are intrinsically linked to content knowledge (Robert, 1998). Broadness of knowledge, in terms of mathematical domains, is both of intrinsic value for the professional mathematician and an important factor for performance within a single domain. Mathematical knowledge is highly structured. It is both of intrinsic value and of pedagogical benefit to maintain explicit connections between acquired and new concepts and results. Finally, certain core concepts and methods (e.g., continuity, linearization) are considered successively at higher levels of generalisation and abstraction. This requires careful attention to the level attained and the level required. Thus, even though many general aspects of mathematical activity, like the use of logic, variation of representations, abstraction, and strategies, may be phrased in general competency terms, these aspects appear in different forms depending on the content domain.

Therefore, in our project for designing university level teaching, we certainly needed to specify precise content elements. We neither could nor should have avoided lists of results, topics, and notions. To set up serious and operational learning goals, we needed to say something about how students should be able to handle what was on those lists. More precisely, we needed to consider specific competency, defined as *a person's potential to perform a general type of mathematical action in the context of a specific area of mathematical content*.

We believed that the notion of specific competency was primary, both in the sense of practical use and in a more theoretical sense. In practice, specific competencies can be developed and assessed in teaching contexts. Theoretically, general mathematical competencies might only be meaningfully construed as abstractions of classes of specific competencies, for example, general aspects of problem solving in various concrete mathematical contexts. Stating general aspects of learning outcomes could be useful and important, but they would not replace nor make sense without consideration of specific competencies.

The effect of designing balanced and adequate curricula depends crucially on success in communicating goals to students and teachers, and making our intentions clear, as part of a communal agenda. In particular, there may be limits on how complex and detailed the goals can be. Hence, even if the perfectly explicit curriculum were devised, simplifications and priorities would still have to be made for communication purposes (see Artigue, 1989, p. 297).

For the reasons discussed above, we focus mainly on carefully chosen specific competencies that might also be sensibly evaluated. The core of a course description becomes *specific competency goals* (SCGs): specifications of particular capabilities related to content that students are to acquire during a course. Besides the practical aspect of communication, there is another reason to maintain simplicity in the description of SCGs. Although they should not be too general or vague (like "understanding continuity"), they also should not be so explicit that they reduce competency to behavioural schemata (like "ability to write down the $\varepsilon$-$\delta$ definition of continuity of complex functions"). We present a concrete SCG description of a course in Section 3.2 (also see Appendix A).

**2.2. The Student Learning Required.** In designing, implementing, and revising curricula, it does not suffice to consider ideal targets of knowledge or action potential. One has to construe these as *learning goals*, as outcomes of very complex and demanding processes. Moreover, they have to be considered as components in the institutional context of teaching and assessment.

For the *a priori* analysis of learning goals and the obstacles to be expected, we considered the following three aspects (similar but not equivalent to the dimensions discussed by Artigue, 1989, p. 287–289):

(1) *Epistemological analysis* of the nature and characteristics of the target content knowledge. By the definition of specific competency, this includes not only analyzing content knowledge, but also aspects and conditions of using it in certain contexts.

(2) *Cognitive analysis* of the cognitive functions required for attaining the SCGs and related constraints linked to the target population of students.

(3) *Pragmatic analysis* of how SCGs relate to mathematical practice as it appears to students and teachers in terms of language and concrete activities.

*Epistemological analysis* pertains to the scientific structure of the target content knowledge, including identification of prerequisite knowledge, and to the conditions or situations for its use stipulated by the SCG. Typically, a new piece of mathematical knowledge, such as a notion, a technique, or a result, must be used *in combination* with a whole complex of other knowledge. By indicating the range of such uses, the SCG may add significantly to the epistemological precision of what is to be learned. This part of the analysis emphasizes not only the relations among the elements of mathematical knowledge, but also among associated SCGs. Moreover, an examination of the historical genesis of concepts and theories may shed light on the epistemological analysis of SCGs and provide key examples for illustration and motivation.

*Cognitive analysis* concerns the cognitive demands and constraints for students. In order to conceive of mathematical objects and work with them, requires internalisation of cognitive schemata related to the representation of these objects. This means that a significant cluster of difficulties in learning mathematics may be linked to the cognitive development and mobilization of *systems of representations* (see Duval, 1995, Chapter 2). In particular, a difficulty for learners of mathematics lies in the need to coordinate different representations of the same mathematical object (Duval, 2000). In higher mathematics, where conceptual objects are mostly unrelated to visible or otherwise familiar phenomena, this need for conceiving of objects through various representations may be overwhelming for learners.

*Pragmatic analysis* of SCGs is about the actual course description and its use. Will it make sense to the parties involved? At what stages of the course should the statements of SCGs make sense to students? Can students and teachers relate the SCGs to important mathematical activity in the course? Will students be able to experience, through such activity, achievement of the goals? How can teachers assess the extent to which the SCGs have been attained? A crucial part of communicating SCGs is the use of informal language that is accessible to students and teachers. However, such language use might limit how explicit an SCG can be. In particular, content structure is usually described by mathematicians in terms of objects (like metric spaces) and their properties (like completeness), almost like biologists talk of animals and their behaviour. In a sense this conceals crucial learning difficulties, as students are not faced with objects in the usual sense (immediately apprehensible), rather they are faced with multiple representations of abstract entities that are introduced one after the other over a short span of time. The identification of activities (e.g., exercises) that might develop or assess SCGs, possibly at a basic level, arises as a necessary complement to the goal description.

In order for the above rather abstract categories to come alive for the reader, we present a particular example relevant to our study: the analysis of possible learning goals in the mathematical context of the Cantor set considered as an example of a metric space (Carothers, 2000, pp. 25-29). The example is at a level of specificity which we cannot detail here for all elements of the real analysis course studied.

Historically and structurally, the Cantor set is intrinsically linked to a rather involved set-theoretic construction. It can be loosely described as the subset of the unit interval [0,1] consisting of the intersection of the sequence of subsets beginning with [0,1] itself, and with each of the following subsets being obtained from the preceding set by removing the "middle thirds" of all intervals of which it is the union. To be able to read and use this definition is a necessary, albeit insufficient goal of learning. Viewed in isolation, this goal can be seen as activating specific competencies based on set-theoretic notions and topological properties of the real line. From an epistemological viewpoint, the Cantor set can act as a bridge between prerequisite knowledge (properties of the real line) and target knowledge of metric space properties because it provides several interesting non-trivial examples. As a real analysis course progresses, new knowledge (e.g., about compactness or completeness) may be enhanced when combined with what is known about the Cantor set, forming a rich inventory of concrete instances of new concepts and results.

Sometimes quite different forms of expression are used to define the Cantor set, such as the set of certain base 3 expansions. Technically, this corresponds to representations in different semiotic *registers* (Duval, 1995, p. 21). However, such shifts of representation may not be noticed, explicitly or even consciously, by an expert. A true understanding requires a coordination of systems of semiotic representation that are far from automatic. Of course, notions like register or coordination of representations may rarely be explicitly addressed in teaching. Teachers should be aware of how these shifts of representations occur in teaching and how resulting difficulties might be addressed. The Cantor set could be simply a peculiar collection of points that can be pictured by taking away more and more small segments from the line segment between 0 and 1. It has various properties, such as being infinite, compact, and nowhere dense. The student would have to relate the formal definition to the other ways in which the Cantor set is talked about.

From a pragmatic viewpoint, a main difficulty is gaining a practical sense of how to choose an appropriate representation, depending on situation and purpose. The pragmatic aspect of learning goals is closely related to the choice of *tasks* presented to students, hence to teaching. Students may gain a practical sense of the Cantor set through work with a range of tasks that rely on different representations of the Cantor set If the use of different representations is merely a fact in lectures, students may not reflect on how to make a choice among them or even notice that a choice was made. The students' own work is also important in order to make them realise the difference between an informal, diagrammatic illustration of the Cantor set and a more formal, symbolic one. Both may occur during a lecture, and the student may not notice their different functions (roughly, to support intuition and formal argument, respectively). However, the difference will be perceived very clearly by students when they have to justify their reasoning.

For reasons of communication we may want to express learning goals in terms that are not too technical. In particular, the formulations of SCGs not reflect explicitly much of the epistemological and cognitive analyses Nonetheless, in designing teaching and evaluation, an awareness of these analyses is essential.

**2.3. Teaching and Assessment.** Assuming we have formulated an agenda for a course with explicit targets in the form of specific competencies and that we have further analysed them with respect to epistemological, cognitive, and pragmatic constraints, we must now consider how to implement our goals by appropriate teaching and assessment methods. We wish to emphasize three points. First, a focus on *target competencies* leads naturally to a need to develop certain types of student performance that seem suitable for the enactment of those competencies. Secondly, the complexity of specific competencies does not allow for immediate simple measures of whether they have been attained. Thirdly, we are likely to be faced with external constraints (see Artigue, 1994, p. 32) in a variety of forms: quantity and quality of teaching resources (e.g., teachers and technology), student time and capability, and institutional regulations and traditions. Not all of these are unchangeable, but often one must accept them.

In view of the first two points, it is necessary that the methods of teaching and assessment enable genuine interaction among students and teachers. Also, assuming that we strive to develop a broad range of competencies, it would seem desirable to bring a variety of methods into play (Niss & Jensen, 2002, p. 126). These methods must be consistent with the constraints in the third point. However, it is important to clearly motivate proposed methods by their potential contribution to the target competencies. At universities, where students are adult and teachers are mostly experts, such motivation should be made explicit. For instance, if a certain number of lecture hours is converted into a (probably higher) number of office hours meant for counselling in relation to project work, explanation is appropriate. For instance, enhanced support for project work might be expected to contribute more to the development of target competencies than the extra time for lecturing. The need for clarity in explanations, particularly for changes, may be even greater when it comes to assessment. Indeed, in many contexts, well-described changes in assessment can have considerable effect on teachers' and students' overall activity. Many important improvements in teaching and assessment can be realized without major changes in format (see, e.g., Mason, 2002 on the potentials of lecturing) but clarification and communication about reasons for change and the nature of changes may be even more important for their success.

## 3. Context and a Priori Analysis

Now we turn to the concrete development project, carried out in the Fall 2002 semester. We first describe the context of teaching. Then, based on the theoretical framework laid out in the previous section, we describe *a priori* analysis of learning goals (in Section 3.2) and design of teaching and assessment methods (Section 3.3).

**3.1. Context.** The University of Copenhagen is the largest university in Denmark and the Faculty of Science enrolls about 1000 students each year. The development project was around the real analysis course, *Math. 2AN*. Required of second year undergraduates in mathematics and some areas of physics, it might be taken later by some students. Its prerequisites were first year courses on calculus and linear algebra (at the level of Adams, 1995 or Messer, 1994). Enrollment has averaged about 200 students per year.

Both by reputation and statistically, *Math. 2AN* has been a hard course to pass. In 2000, 215 registered for the final exam, 156 completed the exam, and 134 passed. The exam had two parts: a written examination and an oral examination a few weeks later. In 2000, 184 students took the written part. Results from the written part were made known to students before the oral exam. There was a considerable dropout between the two parts: about 40% The unit catered to students who had at least completed their first year of study so the drop-out rate was high.

As both a formal and a substantial prerequisite for several other undergraduate courses, the conduct of *Math. 2AN* affects other parts of the program. Consequently, there were several external constraints linked to our specifying its content. However, the official description reads *in extenso*:

> Mathematical analysis: metric spaces, continuity, Hilbert space,
> Fourier analysis, partial differential equations.

In other words, decisions on almost all detailed aspects of the course have traditionally been left to the teacher.

The teaching involved two formats: lectures and exercise sessions. The two-hour lectures, for all students, met twice each week for 15 weeks and were used for introducing theory and examples. The three hours per week of exercise sessions were taught mainly by graduate students, with about 30 students per class. For the undergraduate students, the study load of *Math. 2AN* corresponded to about 300 hours of work (including examinations) in addition to 105 hours of scheduled teaching. Teaching resources, student time, and the stipulated content were the main constraints on our development project.

**3.2. Learning Goals.** Besides the short description of topics to be covered, the course previously had no explicit learning goals. It had implicit learning goals. However, a problem with implicit goals is that they seem to be without limits, both for students and teachers. Although we have not tried to document the previous state of the course, we think it is safe to say that teachers often had unrealistic expectations for student performance, and the work of many students was guided by guesses concerning exam requirements. Hence, making a more explicit course description in order to reduce these uncertainties was an aim in itself.

We first give an outline of the four page course description that was communicated to students at the beginning of the course and then discuss our analysis of the specific competencies, according to our framework. The description included:

(1) Overall aims of the course, formulated in general points, briefly related to the undergraduate program in mathematics as a whole.

(2) The content part of the overall aims, summarized in three lists: theoretical frameworks, fundamental concepts, and concrete topics partially known from the first year course. The specific competencies were defined as "skills and knowledge to be built by students in connection with particular topics" and these were exemplified. For example, students would be expected to "check and use the definition of ... [a concept] ... in concrete and abstract situations."

(3) Specific competency goals (SCGs), listed in a one page table related to content matter outline, chapters in the textbook, and exercises in the book exemplifying their enactment (see translation in Appendix A). The SCGs were numbered with symbols like *e2*, where the letter designated a subject matter area (in the case of *e2*, compactness) and the numeral indicated one of the specific competencies related to it. These were conceived of as minimal or basic forms of competency related to a mathematical property or object, formulated in terms that should make sense after one had worked with the notions for awhile.

(4) A presentation of the organization for teaching, consisting of:
   (a) the purpose of teaching: to present new material and to assist students in achieving the SCG;
   (b) an introduction of the thematic projects (see Section 3.3.1), aiming across the SCGs;
   (c) a table of the progression of the semester (see Figure 1);
   (d) a description of the aims of lectures, exercise sessions, and project work in relation to SCGs;
   (e) a list of titles of the six thematic projects (see Section 3.3.1);
   (f) a list of the materials to be used (see Section 3.3.1).

(5) Descriptions of assessment formats and their function with respect to SCGs as well as to the overall aims (see Section 3.3.2).

3.2.1. *Epistemological analysis.* Two very different, and to some extent opposite, approaches, seem to compete in any choice of how to organize a course like *Math. 2AN*. One is the *problem-oriented* approach, starting from problems in concrete analysis, such as seeking solutions to differential equations in the form of trigonometric series (Carothers, 2000, pp. 139-142). The other is the *structural* approach, in which the logical structure of the subject is highlighted, for instance by starting with the most general framework (e.g., metric spaces) and moving on

| Week no. | 36,37,38 | 39,40,41,43,44 | 45,46,47 | 48,49 | 50,51 |
|---|---|---|---|---|---|
| Lectures | INTRO. & BACK-GROUND | METRIC SPACES | FUNCTION SPACES; FOURIER SERIES | HILBERT SPACE | |
| Exercise sessions | start test | *Exercises, occasionally discussion of thematic projects* | | | Old exam |
| guidance | | | 16 hours | | 16 hours |

FIGURE 1. Schematic progression of the semester (translated from Danish).

to more particular cases. The problem-solving approach is more faithful to the historical development of the subject. The notion of metric spaces was introduced by Fréchet in 1906, generations after the work of Bernouilli and Fourier. Such an approach might more substantially motivate the study of abstract notions by relating them to topics that are closer to students' previous studies. However, the structural approach is much more economical. In particular, introducing the axioms of a metric space at an early stage avoids repetitions of definitions (e.g., of continuity) or proofs of theorems. The approach usually taken, and indeed, the one chosen here (based on Carothers, 2000), was inspired more by structural concerns than by the idea of letting notions grow out of problems in familiar contexts. However, as can be seen from Figure 1, the need for grounding abstract notions in familiar instances was addressed through a review of properties of real numbers and functions during the first three weeks. Also, some set-theoretic notions, new or at least unfamiliar to most students, were introduced and applied to the example of the Cantor set. That set later served as a key example for properties of metric spaces. After this introductory material, the course followed the structural approach (see Figure 2). The main exception was that Hilbert space was treated at the end of the course, after the discussion of Fourier series.

As in most real analysis courses like ours, the language of set theory was not given explicit attention. The pragmatic attitude "we are not to worry" (about set theory, see Carothers, p. 34) could be upheld provided students had no difficulties with basic notation and its meaning, often called "naïve set theory." The ability to manage basic set theoretic language was an obvious necessity for most of the course, beginning with the formal definition of a metric. Usually, a mixture of formal and informal phrasing was used. Often, illustrative diagrams were provided to supply insights that were, at best, accurate for very special cases of the illustrated phenomena (e.g., an open covering of a compact set).

From an epistemological viewpoint, many of the SCGs pertained to the *use and combination* of statements about properties. Many important results in metric space theory might be seen as providing alternative ways of stating a property, such as the "the three Cs" (the groups $c$, $d$, and $e$ of the SCGs in Appendix A). Only a few of the SCGs ($d3$, $f1$, $f3$, $g2$, and $h4$) were based directly on central theorems. In contrast with concrete mathematics, which appears to bear on rather specific objects, abstract theories often leave students with a feeling that mathematics is only the art of saying the same thing in different words (Russell, 1969). For instance, a theorem may provide four different ways to state the continuity of an abstract function. For the beginner this may seem pointless. The purpose of these different

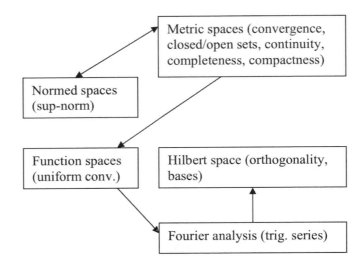

FIGURE 2. Structural approach to organisation of the course with respect to rough levels of abstraction.

wordings is not only of a rhetorical nature. For a given purpose, one formulation of continuity may be much more convenient than another. The frequent insistence in the SCGs on "use in situations," as in an exercise, was meant to indicate this. Only the confrontation with particular aims and results of rephrasing makes the general possibility a potential for action (rather than something observed in a few standard uses, typically in proofs). None of the SCGs (besides perhaps *g2*) required students to be able to reproduce entire proofs from the text. On the contrary, emphasis was on the ability to explain and use relations, concepts, and methods. As mentioned, the SCGs were conceived in accordance with the mainly structural approach taken in teaching.

A main novelty for students was that *methods* were to be understood in a much more theoretical way than in first year courses. In particular, there were no methods of an algorithmic nature, as in calculus or linear algebra. Very few problems could be solved by simple transformations, for example, of a symbolic expression. Even the simplest methods taught in *Math. 2AN* involved at least some reasoning. For example, students would use an equivalence theorem of type $\forall \omega \in \Omega \colon A(\omega) \iff B(\omega)$ for a specific object $\rho \in \Omega$ to infer $A(\rho)$ from $B(\rho)$ in basic tasks relating to SCGs *b3*, *c1*, *d1*, and *e1*. Like the basics of set theory, the elements of logic needed to make appropriate inferences were not explicitly discussed unless students displayed persistent difficulties. On the other hand, unpacking the logic of informal statements was by no means assumed to be automatic (Selden & Selden, 1995).

Tasks relating to a single SCG (such as *b3*) could be quite difficult because of the complexity of relevant concept definitions. This is why many of the SCGs were written to insist on *simplicity* in the situations and arguments through which they should be realized. Each SCG drew upon a variety of sources (previous knowledge, other competencies). Dually, each one might be viewed as a point from which other SCGs unfolded, for instance as ordered in Appendix A. The presentation in Appendix A is not meant to imply that the SCGs are to be achieved in a strictly linear fashion, but merely to indicate that many of them, in order to be realized in a given situation, may rest on previous ones (e.g., $c1$ could rely heavily on $b3$).

3.2.2. *Cognitive analysis.* The general cognitive requirements implied by the SCGs can be summarized as necessitating students to develop the ability to

- use several systems of semiotic representation (registers), including some with which students might be less familiar: set theoretic, algebraic and logic symbolism, informal diagrams, and coordination among these;
- unpack the logic of rather complicated formal and informal statements, most stated in natural language;
- read and construct simple sequences of reasoning based on formal definitions of properties and on theorems establishing equivalence of properties.

In the analytic language of (Duval, 1995) this required students, respectively, to

- integrate new semiotic registers into their cognitive architecture;
- understand and use conversions between new and old registers;
- understand and use forms of discursive expansion and reasoning that may be distinct from the discursive actions of natural language.

As noted above in Section 2.2, for working with the Cantor set the essential cognitive challenge resides in *coordinating different representations* of a subset of $[0,1]$. One difficulty could be that some of these representations are in a partly unfamiliar register, such as the set-theoretic representation $\bigcap_n \bigcup_k I_{n,k}$. Students' difficulties often consist simply in relating familiar types of representations of the same object. While a naïve geometric representation is helpful for understanding the construction of the Cantor set, coordination with symbolic representations, including algebraic ones, may be essential for showing even the simplest properties of this object.

The conceptualization of this and similar concrete examples would fall under SCG $a3$, and might be a prerequisite for tasks relating to most SCGs in groups $b$ through $e$ in the study of topological properties of concrete point sets. Indeed, the use of involved set theoretic expressions may also occur in completely abstract settings for most notions related to SCG groups $b$ through $e$, where an informal diagram is typically the only alternative to a symbolic representation. Likewise, in the SCGs pertaining to functions and function spaces ($e2$ through $f3$), a variety of representations available to the students in the context of functions must be coordinated. Similarly, coordination is required for $h1$ through $h4$. The Hilbert spaces considered may contain functions or (geometric) vectors that mix previously known concepts in new ways, such as with orthogonal functions in a basis.

A main challenge of the course was the requirement that students understand and construct formal reasoning, ranging from demonstrating properties of a concrete example to simple proofs. It is implicit in several of the SCGs that representations have to be mobilized in reading and constructing reasoning about properties of the objects represented. This means that representations of mathematical objects would need to be invoked in discursive expansion aimed at bridging the gap between assumptions and conclusions, and that a choice of representation would have to be made to ensure the coherence of the discourse (Duval, 1995, p. 129). Making sense of such discourse and, to a much greater degree, constructing it requires not only the successive mobilization of representations and the recognition that a given inference is formally correct. It also requires one to conceive of a proof as a whole, as a semantic bridge between statements, made of statements.

3.2.3. *Pragmatic analysis.* The formulation of SCGs in Appendix A was guided by the intention that they make sense to the students who were to achieve the goals. Goals that are explicit with respect to content, such as SCGs, may not be fully understandable to learners *before* studying the content because the goal statements are likely to refer to unknown mathematical concepts. Nonetheless, goals were formulated with the purpose that students could recognise their presence, understand their significance, and see them in relation to success with concrete activities in the course such as following a lecture, reading a proof, or doing an exercise. After completing a chapter in the textbook, the aim was that goals be accessible to students as a checklist for outcomes of the work they had done. Moreover, SCGs must be reflected in course assessments and comprehensible to students as well as teachers (guided perhaps by the further description mentioned in the beginning of Section 3.2). The SCGs used in our development project talked about dealing with mathematical objects and their properties in the same colloquial way as textbooks and exercises do. To exemplify what an SGC meant, we referred to exercises from the textbook as examples of their enactment. This was done in order to make clear what was meant, for instance, by being able to"check and use the definitions of metrics and norms" (*b1*).

One could criticize our formulation of SCGs from several points of view. First, there is the balance between general and specific: it could be proposed that our SCGs are too specific (dealing with very special tasks) or too general (hence, impossible to evaluate). Also, one might criticize the fact that they represent requirements that are very *different* in nature (e.g., applying a specific theorem (*d3*) as well as requiring general command of a set of notions *d2*)). Some might disapprove of the *repetitions* (e.g., that concepts should be checked and used in different formulations and contexts, as in *c1* and *d2*). In principle one would like the specific competencies to be quite explicit about what action potential to develop and how to evaluate it in relation to the mathematical topics. On the other hand, too long and detailed a list would be both non-operational and at risk of being mistaken for a list of simple routines to be acquired. In our opinion, the SCGs listed in Appendix A can be clearly identified in concrete tasks. They

- express the basic ways in which a beginner should be able to handle the topics at hand,
- add significantly to a mere list of topics by specifying what a student should be able to do,
- and seem, to us, reasonable given the preparation and level of our students.

To sum up, the appropriateness of the list of SCGs depended on balancing a number of different needs. For the first attempt, we had to rely on our own judgement as teachers, supported by the *a priori* analysis just presented. The SCG description is a piece of communication, so its appropriateness must be evaluated on the basis of how it is acted upon by the students. We return to the SCGs in the light of experience (*a posteriori*) in Sections 5.1 through 5.3 and 6.1.

**3.3. Teaching and Assessment.** In *Math. 2AN*, in Fall 2002, we implemented several major and minor changes in teaching and assessment that were related to our wish to further and evaluate the goals discussed above. Of course, these changes did not follow automatically from the new course description. Just like the description itself, these changes had to be evaluated empirically.

3.3.1. *New formats in teaching.* The main changes implemented were:

(1) Adoption of a textbook (Carothers, 2000). Previously, lecture notes produced by teachers of the course had been used. A textbook generally contains context in addition to essential theorems, proofs, and examples. Lecture notes tended to offer no side roads for students to explore or dismiss. The final elaboration of SCGs shown in Appendix A was done after the textbook and the parts to be used in the course had been chosen.

(2) Exercise sessions were organized to develop SCGs one by one as the course progressed by focusing on students' solutions to carefully chosen exercises (see Appendix A for examples). Moreover, as many of the exercises in the textbook were rather abstractly formulated, almost all exercises were given a concrete counterpart that the students could use, either as an alternative to or as a hint for the abstract task. For instance, instead of showing that any polynomial $p$ may be approximated by a polynomial with rational coefficients in the sup-norm on [0,1], students were asked to show it for $p(x) = ax^n$, with $a$ real.

(3) As part of the coursework, six thematic projects were proposed to students. Each was formulated with an introduction explaining the specific competencies to be used and developed in the project and an ordered list of open tasks to be completed. These tasks ranged from "Prove that ..." to "What can you say about ...". In contrast with the exercises solved before and during class sessions, the project tasks were meant to integrate the SCGs. However, to enable students to work on some projects early in the course, most were designed so that they might be done with only a certain part of the SCGs. Students were given a table showing the earliest date at which each project task could be approached and were encouraged to do the projects in groups of at most four students.

(4) The lecturer (the first author of this paper) urged and helped students to form study groups for their work on the thematic projects. Some hours of lecture were converted into office hours during which students could get help and feedback on their project work.

(5) The lecturer did not attempt to cover all the central proofs. Rather, lectures served to introduce and discuss general ideas and results in the textbook, to point out difficult and important steps in arguments, and to make introductions and connections to the thematic projects. In short, the lectures were appetizers for individual student work rather than "feeders" (saving students from having to read the text).

All of these changes were planned to happen with the usual teaching formats and resources. The exchange of 8 hours of lecture for 32 hours of project counselling was done internally. Only the adoption of a new text required acceptance by the study board. The main innovation was the six thematic projects, with the following titles:

- The Cantor set
- The Riemann integral
- Existence and uniqueness of solutions to differential equations
- The Hilbert space $L_2(I)$
- The heat equation
- Convergence of Fourier series

The description of each project was two to three pages long. The students were expected to produce a written report, but this was not marked (mainly due to constraints on resources, see Section 2.3). The project reports served as the basis for an oral examination, more on this in Section 3.3.2.

The projects were intended to serve several didactic purposes, primarily to:

- further creative and independent student work in order to build competency, not just knowledge,
- highlight the coherence of SCGs and the need for combining them,
- enhance general communication and reasoning skills.

These functions can be difficult to achieve with lectures or exercise sessions alone, especially when new material has to be introduced at a quick pace (which was the case here). In particular, the exercise sessions dealt mainly with simple closed tasks related to subject matter that was introduced in lectures the week before. They reflected a conception of the SCGs as linearly ordered or even independent.

3.3.2. *New formats in assessment.* The exam for *Math. 2AN* traditionally had two parts: a written, three hour, test (books, notes, etc. allowed) and an oral examination (a 30 minute examination on a randomly selected topic following 30 minutes preparation with the use of any written aid). The final grade was given as a kind of average of the grades from the two parts, but no official grade was given for each part.

Before our project the usual procedures had been, roughly, as follows: the *written* exam consisted of three or four longer exercises of varying difficulty (according to many students these were "mostly difficult"). At the *oral* examination, it had been the practice to randomly select a question from a list of topics provided to students in advance and, after preparation, the student would present that topic; typically this was a theorem and its proof. For some students this provided a good exercise in the understanding and communicating involved in mathematical arguments. However, weaker students tended to deliver memorized "proof recitations" with little sign of understanding. After the presentation of the selected topic, it was customary to ask a few supplementary questions on another topic. An external examiner was present at the oral exam.

The changes in assessment announced at the beginning of the course were:

(1) The *written* exam would consist solely of simpler exercises, with each question pertaining to one or a few SCGs (ideally, as in the exercise sessions).
(2) At the *oral* exam, one of the six thematic projects would be selected randomly and presented after the usual preparation time of 30 minutes. The student would bring three copies of each project to the exam, giving two of the copies of his report for the selected project to the examiners.

The first change was an adjustment of the written exam to reflect the minimal elements of the course goals, so these corresponded to the equally adjusted practice in the exercise sessions. The second change was quite important. Providing external motivation for students to work hard and seriously on the thematic projects was one objective. But just as important was the shift of focus from presenting the books' arguments to presenting one's own arguments. In particular, because many parts of the projects could be addressed at different levels and with different means, the students were meant to demonstrate their competencies to work with topics of real analysis, rather than to understand and deliver a classical proof.

Two potential problems with the second change were considered in some detail before the course: How could extensive, counterproductive report-sharing among students be avoided, and how could students be given a just amount of help and feedback for their work? For the first problem, we realized that a similar problem already existed with presentations from the book when weak students tried to memorize the work of a brighter fellow. We opted for emphasizing the importance of presenting and defending one's own (group's) work. Students were asked to sign a sheet before the exam that they would do so. For the second point, the main constraint was that we did not have the means to grade the students' reports during the semester, and so the students could not obtain information on the satisfactory state of their work. Also, we did not want teaching assistants to provide complete answers to parts of the projects, for example, during exercise sessions. Students were able to ask questions of the type: "How do I get started on this? Is this argument okay?" but they still maintained the overall responsibility for their work.

## 4. Questions and Methodology

In the introduction, we mentioned some general questions raised by students and teachers that motivated this study. To be treated systematically, they needed to be sharpened and focused. The theoretical framework of SCGs and changes in the course provided a basis for more precise research questions and for the methodologies to investigate them. In this study, not all questions were systematically investigated; so these constitute a wider research agenda.

**4.1. Questions to Address.** The first questions related to the proposed description of the course:

**Q1:** Do students find SCGs formulated as in Appendix A meaningful and helpful for their work?

**Q2:** How does the explicit form of SCGs work as a tool for teachers?

The next questions concerned the effects of the changes on teaching methods:

**Q3:** Can SCGs be used to improve the planning and running of exercise sessions?

**Q4:** How does student work on thematic projects contribute to the achievement of specific and general competencies?

**Q5:** How could the emphasis in lectures of ideas and connections, at the expense of content covering, be realized? How could it affect students' work?

Finally, some questions on assessment:

**Q6:** Can SCGs be measured on written tests and exams?

**Q7:** What are the consequences of using thematic projects as the basis for oral evaluation of the students' grasp of theory?

**Q8:** Can the explicit competency descriptions help students work in ways that help them meet the formal requirements of the assessment? Or at least make the requirements more transparent?

Most of the questions were of a qualitative nature that would benefit from several types of data and final answers should not expected. All of these questions are of general interest, apart from the specific context of *Math. 2AN*. However, as this study was part of a didactical engineering project, observations that might lead to improvements of the didactical design were considered, too.

**4.2. Qualitative Methods.** We collected qualitative evidence in several ways during the course. The planned data collection included

- audio-recorded interviews with a focus group of four students who where interviewed individually three times: in the second and tenth weeks of the course and after completion of the exam. These interviews addressed, among other things, research Questions Q1, Q4, Q5, Q7, and Q8;
- field notes from regular observation by the second author of lectures and exercise sessions, usually followed by a discussion with the teacher on various aspects of the event. Here, the focus was mostly on Questions Q2, Q3, and Q5.

Important information that could bear on any of the eight research questions was also collected in less systematic ways. For example, from students' responses to the official course evaluation (this included written comments on the teaching) and from more informal or even spontaneous statements of teachers and students during the course (e.g., through emails to the lecturer). In some cases we also invited such comments, for example, by writing to all students and examiners after completion of the exam. This mainly provided "voices" whose significance and intended meanings may be unclear. Finally, the complete collection of the students' written projects also represents a body of data from which evidence could be extracted.

**4.3. Quantitative Methods.** The collection of quantitative data was mainly done through two multiple choice tests. The first test at the beginning of the course served chiefly to pinpoint students' prerequisite knowledge (further details are given in Section 5.6). The second, a post-test related directly to Question Q6, was given at the end of the course before the two-part examination. Post-test items were specifically phrased as a diagnostic test of the SCGs (see Appendix B). The written exam (Appendix C) could, with reservations arising from the differences in context and design, be considered a third test. Although it was not designed as a test of the SCGs, the the written exam exercises did conform to the intention of testing one or a few SCGs. Additionally, the official student evaluation of the course provided some rough quantitative indications mostly relating to research Questions Q3, Q4, and Q5. Interpretation of the evaluations was assisted by extensive comments provided by many students. Finally, certain aspects of the examination results, including comparisons with previous years, provided quantitative data that, at least internally, were of some interest.

## 5. Results

The presentation and discussion of results is organized according to research questions Q1 through Q8. This means that informally and systematically collected data appear together, and that we strive to emphasize a coherent presentation of the main outcomes of the first cycle of the project, rather than an exhaustive presentation of the data.

**5.1. Students' Perception and Use of SCGs.** The list of SCGs was accessible to the students at the website of *Math. 2AN* along with the rest of the course description (Appendix A). The description was frequently referred to in lectures. We did not systematically ask all students how much, or in what ways, they made use of the description or the list of SCGs. However, this question was raised during the last two interviews with the focus group.

At the second interview, at the end of October, three of the four students were aware of the existence of the list, but only one of them had looked at it. This student reported using it before the fall break (six weeks into the course) for the purpose of "evaluating how far [she] had come." This was a pleasant experience for her as she felt comfortable with the goals pertaining to the parts of the course already completed ($a1$ through $c2$ in Appendix A). According to her, these had been "intensively trained." She also said that she felt she could "do more" than what was indicated in these SCGs. She added that to realize that one amply meets the goals can be "nice when you are frustrated" – a reference to the excessive work load that all four students complained about at this time. As for the other students, one had decided to drop the course because he felt he was "too much behind schedule" (in fact, he had not worked on the course for several weeks). The two remaining students were provided with the list of SCGs $a1$ through $d3$ during the interview and expressed various degrees of "uncertainty" about the extent to which they felt they had reached the SCGs. The student who had not previously noticed the list was rather dismissive about the pertinence of such a description. He said: "if you follow the course properly you don't need it; otherwise you don't understand it."

Judging from various remarks in informal interactions (such as Figure 3), this attitude might well have been rather common among the students. It was our general impression, though, that the discourse about competencies (e.g., in lectures and in the formulation of the theme projects) did contribute to a gradual shift of focus among many students, from content consumption to competency development.

It seemed that students initially had very traditional perceptions of standards and requirements. This was expressed, for instance, in their anticipation that "solutions" to thematic projects would be assessed the same way as textbook proofs, or that they were somehow required be "at the same level." At the last focus group interview after the exam, one of the students said that the list of SCGs had been useful as "clear goals" relating to the written exam. Yet, in the second interview (about midway through the course), he had found it difficult to relate to them, in particular, to assess his current status with respect to them. The nature and development of these aspects of student attitude would be worth a more systematic investigation.

**5.2. Teachers' Perception and Use of SCGs.** The lecturer, the first of the authors of this paper, was directly involved in the entire project and made extensive and explicit use of the new description of the course, both in organizing and giving the lectures and in planning exercise sessions. In lectures, to be further discussed in Section 5.5, the main impact was probably the general shift of focus from covering content to exhibiting ideas and forms of action to be acquired by each student. For planning exercise sessions, the use of the SCGs was significantly developed during the course (see Section 5.3).

Teaching assistants were not involved in the design of and analysis behind the competency-based descriptions for the course. They were presented with the description one week before the beginning of teaching. Thus, it is perhaps not surprising that they found it difficult to use and relate to the SCGs. Perhaps even more than for students, the teaching assistants' habitual conceptions about course goals meant that new descriptions were only implemented gradually and partially. We quickly realized that merely providing and explaining the descriptions to the teaching assistants did not, in itself, lead to immediate change in their practice.

Hi Niels

   Now that the exam is well behind us, I'd like to tell you how the
total experience of Math2AN has been.  We have learnt a lot, but it
was hard.

   I have read Carothers very carefully, even if it appeared that
one was not required to know all the proofs from the book.  My impression is that it is difficult to understand the abstract arguments in the thematic projects if one doesn't study the textbook in
depth.  Therefore the thematic projects felt like an extra
work load, and my feeling is that they were the straw that broke the
camel's back for many of the other students.

   There were things about the course which could be improved:
First, there were several errors in the formulation of the
thematic projects.  Clearly it can't be avoided, especially the
first time.  But even minor points of doubt can become big troubles
when you try to understand a task.

   Secondly, the coordination of session exercises was bad in the
beginning, later it improved, as it was clarified what exercises
were important.

   I don't quite know what I think about the competency goals.  They
were a bit unclear, and I feel that to a large extent you can see
for yourself what is important in the course.

   But the proces of working with the thematic projects was very
profitable.  I have learnt something about defining a problem, to
solve it, and to express things precisely.  Altogether I think the
format of the course, with thematic projects, was succesful, and
that the hard work was rewarded in the end.
Sincerely,

-----

FIGURE 3. An email from a student, sent spontaneously to the
lecturer after completion of the course (translated from Danish).

Indeed there was a considerable uncertainty among the teaching assistants with
regard to what the new descriptions meant for the running of exercise class sessions.
An instance of this occurred several weeks into the course, after observation of an
exercise session where almost an hour had been spent on elaborate, partially wrong
calculations in an exercise whose only pertinence to the course was related to SCG
*b1*. A discussion with the teaching assistant following the session indicated that
the teaching assistant had no idea of how the exercise was related to the SCGs,
and indeed was quite unfamiliar with the framework of Appendix A. On a different
occasion, another teaching assistant declared openly that he found the degree of
explicitness of the SCGs to be "overkill."

In their reports back to the lecturer on how the exercise sessions went, the
teaching assistants consistently used two indicators of success: the extent to which
the exercises had been covered and to what extent these had been presented by
students. Thus, although there was, generally speaking, an awareness of the importance of student activity and performance in exercise sessions, the real difficulty
was in providing tools to assess and further this activity from a *qualitative* viewpoint. The given descriptions clearly did not suffice to mitigate certain constraints
which were often observable in problems arising in class sessions, such as the great
variation in students' preparation, interest, and performance, and the constraints
on the teaching assistants in terms of time as well as their own pedagogical and

mathematical experience and training. Some of the teaching assistants had recently passed the course themselves and some of the exercises were evidently at the limit of their own competencies. Some of them found themselves in the role of teacher for the first time, with at most a one-day workshop on methodology as pedagogical training. Despite a growing awareness of certain particular student difficulties, like the task of clarifying what a given exercise "asks you to do" (i.e., "unpacking" in the sense of Section 3.2.2), students' ability to present correct solutions to the exercises remained a dominant concern for most of the teaching assistants.

**5.3. Use of SCGs in Planning and Running Exercise Sessions.** From the beginning of the course students expressed confusion about the relation between the exercise sessions and the rest of the course. They had noticed the stipulation that the exercises would serve to prepare them for the written exam. The concrete counterparts of exercises were often ignored, despite the fact that in numerous instances the majority of students failed to solve or even understand the abstract form of the exercise. This led us to give precise indications of the competency goals associated with the exercises posed each week. For instance, in the instructions for class sessions pertaining to Chapter 7 of the book (see Appendix A): *Please focus on the use of different methods to determine completeness [of a space], and on how the notions developed in the chapter work within abstract arguments.* This rephrased SCGs *d1* and *d2*.

The main concerns of students as well as of teaching assistants often pertained very directly to the nature of the exercises that would show up in the written exam. We repeated in a guide for the exam what was said in the description; namely, that the written exam would assess the SCGs, both in concrete and abstract contexts. Two sets of examples of "exam like" exercises were also provided, with the last one (published two weeks before the end of the term) containing simpler examples of how the SCGs would be tested. We strove for simplicity and to avoid deliberately the use of complicated examples. This was the result of development in our own thinking about how the SCGs could be adequately tested. It was also, in part, a reaction to the massive difficulties the students reported having with the first test (see Section 5.6 below), and with many of the exercises posed for class sessions. Thus, the discussions about exam requirements with and among students and teaching assistants were vital to the development of tools for implementing the SCGs as motivation and meaning for the activity in class sessions. As the exam approached, these discussions became increasingly related to the issue of testing.

We think that SCGs may be used as part of an effort to improve the planning and running of exercise sessions, but this would require a much clearer and direct use of SCGs as the intended focus of each exercise. In particular, it would need to be made clear that an SCG can be reached at different levels of abstraction and complexity. This would challenge the view of many students and teaching assistants that an abstract form of an exercise is necessarily more worthwhile to solve. Moreover, the SCGs should guide the lecturer in his choice and formulation of exercises pertinent to the set goals. In fact, the teacher might be tempted to remove exercises that caused difficulties for students. Without explicit SCGs this might well result in an undesired reduction in the development of some specific competencies. Also, for establishing the didactical contract (see Brousseau, 1997, Chapter 5), there seems to be considerable potential in explicit and transparent messages that emphasize the status of SCGs as criteria for what is measured on the written exam. Not making full use of this potential, especially in the beginning of the course, was at the root of many deliberations that might have been avoided, or at least given a sharper focus, with clearer and more direct use of SCGs.

**5.4. Thematic Projects and Competency Development.** Among the modifications introduced, the thematic projects produced the most immediate and spectacular effects. It was clear to all students that at the oral exam they would have to present one of these instead of the usual presentation of proofs from the textbook. Two strong reactions were noticed very early in the course. First of all, students were quite enthusiastic and positive about the idea. For instance, they willingly formed study groups with the aim of collaborating on the projects. Secondly, a host of questions were raised: What criteria would be used in this new form of assessment? How much help could the students expect to get from the lecturer and from the teaching assistants?

The last question was also immediately raised at our first meeting with the teaching assistants, who felt they needed more detailed instruction than what we initially offered: "You are not to present the students with complete solutions, but you may give hints for students to get started on a task, and if the students feel uncertain about the pertinence of a part of their work, you may provide feedback (but not complete alternative solutions)." A certain scepticism and anger was implicit in several follow-up questions: Could the teaching assistants be accused of examination fraud if they helped students more than they were supposed to? What if students "shopped around" between assistants in order to get the answers "for free"? How would we deal with extensive or even organized copying of projects among the students? In response, two weeks into the course we gave new written instructions to the assistants, stipulating that their *main* task was of a logistic nature: to help students work together and to facilitate discussion; and that, if they felt it was difficult to enter discussions about the project without providing solutions, they could simply stick to these logistic tasks and leave the substantial advising to the lecturer (who provided project counselling). This seemed clearer to the teaching assistants, although refraining from giving solutions seemed awkward to them. After all, this was perceived to be the main task of the teaching assistant!

One of the strongest and clearest messages of several of our data sources – interviews, students' evaluations, informal interaction – was that many students felt that the work load required for the thematic projects was a problem. In the student evaluation, 83% of the students ranked the difficulty of the thematic projects as 4 or 5, on a scale from 1 to 5. Out of 71 students who completed the questionnaire, 55 provided qualitative comments about the thematic projects, which in itself was unusual. Generally, students' comments indicated rather strongly felt reactions. Despite the variation naturally present in these 55 comments, two messages were very clear:

(1) The idea/principle of thematic projects was good (47 comments!) both for learning (explicit, often with other words such as "understanding" in 16 comments) and as a basis for the oral exam (discussed in 17 comments).
(2) The projects were very (or too) time consuming (25 comments).

A rather typical comment was:

> I think it is an interesting idea. It is a better way to learn the material than to just rewiew for two weeks in January. But

> of course it gives you more work. In fact the only thing that
> irritates me about the thematic projects [is] that they are not
> evaluated by a more competent person during the semester.

Indeed, many of the students, including three of the four focus group students, felt that the workload was excessive. The thematic projects were seen by many as something "extra" (e.g., see Figure 3) introduced in addition to the usual tasks of students to read the book and do (or at least "observe") exercise solutions. The idea that the projects could be seen as preparation for the oral exam, in lieu of preparing a textbook presentation, did catch on to some extent. In spite of this and despite the widespread student opinion that one could learn more from doing the projects, it became obvious to us about midway through the semester that the requirements would be really excessive if the students were expected to do all six projects *in extenso*. It was therefore decided to modify them in two ways by reducing the number of mandatory tasks in each project and by allowing each student to remove one of the six projects from those he could draw at the oral exam. We discuss the use of the projects in the actual exam in Section 5.7.

Based on our data, it was difficult to estimate the overall contribution of the thematic projects, as a part of the students' work, to competency development. Many of the students' own assertions, of which Figure 3 may be seen as typical, seemed to indicate that it was considerable. What can be affirmed with certainty is that the format made students work harder and that, judging from the five written projects delivered by each student, the vast majority of those who made it to the oral exam had in fact worked autonomously with non-standard and pertinent tasks, including constructing and presenting pieces of mathematical reasoning in a precise and correct way. Although only a minority of students directly consulted the lecturer for advice on the projects, the questions asked were generally of high quality, and more or less clearly the last resort of a group of students who had tried hard to solve the problem. These questions and answers were published anonymously on the course website during the course.

A general feature of the project work was the rather high demands on certain general competencies, in particular those pertaining to communication, reasoning, and problem solving. The positive effect of this, beyond the specific agenda of the course, was noted in several student comments (e.g., Figure 3), as well as by one of the external examiners (see Section 5.7).

**5.5. Change of Focus in Lectures: Action and Result.** It is important to realize that lectures represent a tradition that students were well acquainted with before *Math. 2AN*. The most common use of them, implicitly and explicitly referred to in many of our interchanges with students, was to provide a coherent and rather formal presentation of new mathematical theory, occasionally illustrated by simple examples. The possibilities of interaction between the lecturer and the students were more limited than in small-class teaching. A problem was to find a suitable pace and level of explanation, given that students' interests and abilities often seemed to fall into two almost disjoint clusters. Trying to hit an average level might have meant, at least in the experience of the lecturer, hitting right between the two clusters and hence satisfying very few.

It is against this background that the efforts of focusing lectures on ideas and methods that would serve to enable student activity (mostly outside the lecture hall) should be seen. First of all, lectures were always related to some piece of the text. It was made clear from the beginning that they were meant to enable the students' own reading of the text, rather than to provide an alternative by spoon feeding or simplifying the text. Some of the techniques employed by the lecturer in order to create variation from a linear presentation of theory:

(1) Almost every lecture began with a survey of key ideas and notions. Sometimes this was in the form of concept diagrams on an overhead that was returned to later, to clarify the connection between details and overview.

(2) In order to exemplify the work meant to be done in reading the textbook, the lecturer sometimes showed an overhead with a short excerpt from the book (e.g., a technical lemma and its proof) and discussed it in detail. The goal was to make explicit the necessity of unpacking mathematical text to the point of transparency relative to one's own understanding.

(3) The lecturer made meta-comments on the nature of techniques used in examples and proofs (e.g., "standard," "known from . . . ," "clever," and "original") to indicate what students were expected to do on their own (e.g., in exercises or theme projects, with explicit references to these).

(4) To establish interaction with students, the lecturer posed more or less challenging questions and allowed sufficient time for students to reflect and react. Such interaction involved a minority of the students, but one positive effect was that as the semester progressed more students felt able to spontaneously pose questions and offer remarks.

(5) Connections among theory, the tasks worked on by students in exercise sessions, and the thematic projects were emphasized whenever possible.

Attention to each of the above was subject to time constraints and the need to maintain coherence and recognizability of genre. The final student evaluation of the lectures was quite positive. However, the qualitative comments provided by students were mixed and seemed to reflect differences in students' needs and beliefs. Nonetheless, it did appear from the students' comments, and from the more detailed assertions of the focus group, that most students agreed that lectures should not just repeat or reword the main theoretical development of the book. A large majority credited the lectures with providing them with good "overview" of the material.

A specific and central issue we noticed during the semester was the need for explicitly addressing the *coordination of semiotic representations*. In particular, the case of the Cantor set provided an early example of massive student confusion due to the need for frequent and flexible change of representation in reasoning about it. It emerged that an unease with rather simple forms of representational change was often at the root of students' difficulties in understanding explanations provided in lectures and informal interaction. Consequently, during some lectures, attention to representational change was an explicit focus of presentation. Unlike the text, in lectures several channels of communication were available, including writing, drawing, speaking, pointing and so on. The lecturer was a professional in the use of mathematical signification. When he attempted during lecures to be conscious and explicit about the relations he was trying to emphasize, he could add a significant "meta-level" of instruction, making students aware of the need to make connections among representations in the different registers that are almost always tacitly assumed in texts. A main difficulty for the lecturer in assuming this task (of fostering attention to representational flexibility and for assessing outcomes of efforts addressed towards it) was the inability of students to express their need for, and benefit from, such meta-comments. For some students the meta-comments were merely helpful explanations. For others, the lecturer's deviation from a linear presentation was described as an irritating, "mess of confusion," especially when students were trying to take notes.

**5.6. Measuring SCGs on Tests and Exams.** In the introduction, we mentioned the quest for tools to evaluate student's progress and achievements with respect to the course goals. Given the size and heterogenity of the student population of *Math. 2AN* we felt that this required quantitative methods in the form of tests as a complement to analysis of individual student performances. As mentioned in Section 4.3, three tests were given. However, only the second or "post-test" (Appendix B) was a test aimed at the SCGs. The first test, given at the beginning of the semester, was meant to pinpoint students' prerequisite competencies while the last was the written exam (Appendix C) taken about one month after the post-test.

5.6.1. *Pre-test.* At the first exercise session in the semester, one hour was used to let the participating students (110 in total) do a multiple choice test with 15 questions on topics from the prerequisite courses. All test items involved notions and techniques that were more or less assumed in *Math. 2AN*. The topics were simple ODEs (1 item), basics of real numbers and sequences (5 items), basic notions of set theory (2 items), integration (2 items), continuity (3 items), sums and series (2 items). According to the teaching assistants, almost all students completed the test well before the hour passed.

We shall only provide some main conclusions from this pre-test. Although the average score (about 8 correct answers out of 15) was perhaps a little disappointing, the most informative result was the pattern in the types of items which caused more, or less, difficulty. Overall, students fared better with items pertaining to rote knowledge (e.g., formal definitions and relations among abstract notions) and less well with questions that required computation or reasoning. For instance, items addressing the abstract statement of continuity of a real function were completed correctly by between 60% and 75% of the students. In particular, 73% of students chose the right version of the $\varepsilon$-$\delta$ definition among four similar expressions made up of logical symbols. By contrast, only 7% succeeded in choosing from among four options the correct rewritings of an integral obtained by linear substitution. One half of the students chose the correct value of the sum $\sum_{n=1}^{\infty} \frac{2^n}{3^{n+1}}$ or of the integral $\int_0^{2\pi} \cos(mx) \sin(nx)\, dx$. More curiously, 59% of the students chose the statement "$x$ is infinitely small" as the best description of a real number $x$ satisfying the inequality $|x| < \frac{1}{n}$ for every natural number $n$ (this is a classical "trick" question, see Artigue, 1999, p. 1379).

The results from the pre-test made us seriously reconsider our ideas of what could be taken for granted. They provided an indicator that certain prerequisite competencies had to be explicitly addressed, even if these could not be treated extensively within the course. The sheer difficulty of some of the questions was a genuine surprise to us, the teaching assistants, and other colleagues with whom the data were discussed.

5.6.2. *Post-test.* The post-test was voluntary and taken by 78 students in the last exercise session of the semester (see Appendix B for the test). Before discussing the results, a few words on reliability are called for. First, as a fraction of the students enrolled in the course were present at this last session, one could object that somehow they were not representative of the entire population, for instance in

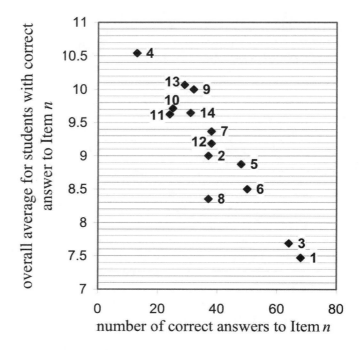

FIGURE 4. A rough test on the reliability of the post-test. Numbers at plotted points refer to test item numbers (Appendix B).

the sense of being more serious than "average." In what follows, when we talk about "the students," we only refer to this group. Secondly, one may wonder if students really did their best, or whether some tried to guess despite the instructions not to do so given preceding the test. Indeed, for many items, each of the seven possible answers were provided by at least some students. This may simply reflect the difficulty of the test. One would like to be able to assert that the test really gave a picture of the individual student's attainment of the tested SCGs. For this, we used a method described by Rasch (1960, pp. 63-64). In Figure 4, each item, identified by number, is plotted with coordinates $(x, y)$ where $x$ = number of correct answers to Item $n$, and $y$ = overall average for students with correct answer to Item $n$.

One might expect a decreasing graph, indicating that students who can do the difficult items do well on the test overall and that students who do poorly on the test have their right answers among the easy questions. That is, the degree of difficulty is a good indicator of performance. In this sense the test seemed to be reasonably reliable. Some questions, in particular Item 8, were atypical. This could be an indication that they should be revised or at least considered further before reuse.

An overview of the results is provided in Figure 5 where the following information is shown for each of the 14 post-test items:

**Row 1:** SCG from Appendix A at which the item aimed.
**Row 2:** Percentage of completely correct answers offered by students.

| Item | 1 | 2 | 3 | 4 | 5 | 6 | 7 | 8 | 9 | 10 | 11 | 12 | 13 | 14 |
|---|---|---|---|---|---|---|---|---|---|---|---|---|---|---|
| SCG | a1 | a2,e3 | a3 | b1 | b2 | b3 | c1 | c2 | d1 | e1 | e1* | e3 | f1 | g1 |
| % correct | 87 | 47 | 82 | 17 | 62 | 64 | 49 | 47 | 41 | 32 | 31 | 49 | 37 | 40 |
| % one correct, no false | 87 | 69 | 82 | 17 | 62 | 64 | 64 | 75 | 68 | 32 | 31 | 66 | 62 | 53 |
| % 1 correct | 90 | 77 | 95 | 68 | 87 | 81 | 75 | 83 | 70 | 67 | 65 | 75 | 73 | 56 |
| % don't know | 3 | 5 | 0 | 4 | 1 | 10 | 21 | 14 | 29 | 23 | 18 | 21 | 24 | 40 |

*In fact, Item 11 also addressed the relation between compactness and continuity, not explicitly mentioned in the SCGs.

FIGURE 5. Summary of the results from the post-test.

**Row 3:** Percentage of student answers including at least one correct option *and* no false ones.

**Row 4:** Percentage of student answers that included at least one correct option (though solutions might include additional incorrect options).

**Row 5:** Percentage of students answering "don't know" to the item.

As a rough overall summary, we observed that

(1) There was a clear decreasing tendency in performance on items according to their number (from Item 1 to Item 14). The ordering of the items corresponded roughly to the progression of the course. Also, note that we had no indication that the allotted test time was insufficient.

(2) For all questions, a majority of students produced an answer that included at least one correct option (see Figure 5, Row 4). That is, none of the questions had a majority of students declining to answer or giving a completely wrong answer.

(3) For all questions except Items 4, 10, and 11, a majority of students provided an answer which included at least one correct option and no false one (see Figure 5, Row 3).

(4) Only Items 1, 3, 5, and 6 were completed correctly by most students.

(5) Fewer than half of the students gave correct answers to at least half of the questions (see Figure 5, Row 2).

The first observation above was consistent with students' indications in interviews that they felt more comfortable with material covered early in the course; or in more negative terms, that the work load often made them feel behind schedule. It also fit the theoretical idea (see Section 3.2.1) that SCGs built on and developed each other, especially previous ones, according to structural order and the timeline of the course.

We expect observations (4) and (5) are the most eye-catching. The image of "half success, half failure" can also be expressed through the average for all students of 6.8 correct answers on the 14 items. In the cases where two of the options provided were true, a "partially correct answer" could be interpreted from student solutions in which *one* false and *one* correct statement were discarded and *one* correct statement was accepted (Items 2, 7, 8, 9, 12, 13, and 14; see Figure 5, Row 3). These items had the majority of "wrong" answers. One reason could be that the students tended to answer either that a single option was correct or "don't know." The kind of answer referred to above in observation (2) (see Figure 5, Row

| Exercise | 1(1) | 1(2) | 1(3) | 2(1) | 2(2) | 3(1) | 3(2) | 3(3) | 4(1) | 4(2) |
|---|---|---|---|---|---|---|---|---|---|---|
| SCG | b3 | b3 | d1,e1 | b1 | f2,b2 | f2,f1 | f2 | f2/g2 | h1 | h3 |
| Average points (% of max.) | 70 | 56 | 75 | 81 | 71 | 69 | 61 | 45 | 88 | 49 |

FIGURE 6. Summary of the results from the written exam.

4) could also be said to be "partially correct," but avoiding the completely wrong answers (as well as "don't know") would be a fairly weak measure of success.

Looking into the details behind Figure 5, while taking into account the content of the questions and the options provided, was of considerable interest. For instance, the failure of a majority of students to provide a correct solution to Item 2 may surprise. Working the problem requires one to notice that $f(x) = x^2$ is continuous but *not* uniformly continuous on $[0, \infty)$ and to use the $\varepsilon$-$\delta$ definition in a straightforward manner. Recall that 73% of the students doing the pre-test proved capable of recognizing the correct symbolic form of the $\varepsilon$-$\delta$ definition of continuity. For Item 2 on the post-test, 26% of the students recognized the abstract form of the definition in option (2) but *not* the form in which $\varepsilon = 1/1000$, in option (1). Hence, while 73% did accept answer option (2) as correct – this was exactly the percent who could recognize the abstract definition on the pre-test – about a third of these students seemed not to have acquired the "concrete" part of SCG *a2* to the extent of using the $\varepsilon$-$\delta$ definition for a concrete function *and* for a particular value of $\varepsilon$.

Another area of interest was the bottom scoring question, Item 4. One should not interpret the result as "only 17% of the students could decide on whether an object was a metric space." In fact, Item 4 was the only one where a *wrong* set of statements, namely, $\{1, 3\}$, was most frequently chosen by students (45%). We conjecture that many students only checked the axioms for the two last options, while overlooking that the first one (i.e., $d(x, y) = x - y$) was not quite the usual distance between reals. Although to the professional this is a triviality, it may be a genuine student trap that was mainly avoided by the best-performing students (their average score was 10.5, as opposed to the general average score of 6.8).

5.6.3. *Written exam.* As promised in the course description, the written exam consisted of fairly simple exercises, each typically aiming to assess one SCG. A total of 114 students took the written exam. Their results, for each of the exercises on the written exam, are given in Figure 6 in the form of average percentages (e.g., average points out of the total assignable points for each exercise). These results include credit for partially correct answers. A main difference between the written exam and the multiple choice pre- and post-tests considered above was that the availability of students' detailed work for these exercises enabled a finer assessment (even if expressed quantitatively by the percentages).

One cannot compare Figures 5 and 6 directly as measurements of particular SCGs: the context (test/exam) and the exercises (multiple choice/explicit reasoning) were different in nature. In assessing the written exam, more general types of competency could be observed, particularly those related to written mathematical communication and the use of symbolic representations.

As with the post-test, there was some tendency that items pertaining to topics covered late in the course (e.g., Exercises 3(3) and 4(2)) seemed more difficult.

However, the relatively weak performance by students on Exercise 1(2) suggested that a better interpretation would be that SCGs that had been worked on extensively were better attained. Indeed, the concept of *interior* was not used anywhere except in the introduction to metric spaces, whereas *convergence*, as dealt with in Exercises 2(2) and 3(1), was central throughout the course.

With appropriate caution, the numbers in Figure 6 could be used by teachers of subsequent courses, taking into account *how* the SCGs were evaluated. For instance, the SCG *h3* on handling orthogonality in a simple concrete situation was, on average, only "half attained" (Exercise 4(2)). This indicated that subsequent treatments of Hilbert spaces would need to review such basic concepts.

5.6.4. *Summing up.* We now try to sum up what can be said regarding research question Q6: Can SCGs be measured on written tests and exams? The answer is not definitively "yes" or "no." Our attempts point to three problems. First of all, it is difficult to construct questions which are both noise-free and non-trivial in the sense of addressing only the SCG aimed at and still placing it in a meaningful context. Secondly, because most of the SCGs can be tested at various levels of abstraction and many of them focus on more than one notion, it is obvious that no single question could pretend to check all aspects of an SCG. Indeed, this poses the problem of the extent to which questions can be said to be representative of the SCGs. Thirdly, especially for multiple choice tests, the interpretation of wrong answers must be considered seriously. To deal decisively with these problems would probably require a more sophisticated design, including more items assessing each SCG. On the other hand, neither the complexity of design nor the test size escapes the constraints of time and resources available in ordinary teaching or even in a research project. In fact, merely processing the data of the fairly simple tests presented here was a considerable task.

An important issue relating to research question Q6 is what *use* could be made of measuring the achievement of SCGs. Even if we had a working diagnostic test of individual SCGs, one might still miss an important feature of SCGs. Namely, that they are structurally interdependent and that apt performance in most situations is a synergetic result of several specific competences. Hence, while it seems possible to get a reasonable picture of the status of separate SCGs, and indeed to some extent this was achieved in this study, such a picture is not in itself a sufficient basis for an overall evaluation of students' achievement. The diagnostic testing of individual SCGs could, at best, be used for partial evaluation (when combined with other forms, such as the oral exam given here) and for orienting the work of students and teachers. For example as indicators like, "we need to work more on this SCG" or "this SCG could now be worked on at a higher cognitive level." Nevertheless, using SCGs as a substantial element in assessment seems important for developing and maintaining the didactic contract, assuming SCGs are a central part of it.

**5.7. Thematic Projects as Basis for the Oral Exam.** We have already touched upon the effects on the course of having thematic projects as a significant part of the basis for the oral exam (Section 5.4). We now turn to how this influenced the oral exam itself, understood as an evaluation of students' grasp of theory in a more global sense than individual SCGs.

One of the significant differences from earlier years was that a fairly high proportion of the students who had registered for the exam did not actually complete the written part (59 out of 174), and another 20 students dropped out between the written and the oral exam. Of the remaining 94 students, 12 failed and 82 (about 47% of the students registered for *Math. 2AN*) passed with a high average mark. The fact that only about half of the students who had registered actually passed the exam was not unusual for this course (although it was a bit lower than for the previous year). However, the number of students who "failed themselves" by dropping out was higher than usual. In an email after the exam, students were informed of the pass-rate and exam-taking facts and asked for feedback. We were especially eager to hear from students who had registered but not completed the exam. Two factors were cited in the rather few responses received. Some, as was usual, decided against taking the oral exam after being informed of a poor result on the written exam. But a more important factor seemed to have been the failure to do the thematic projects properly, due to time constraints. One of the students added that he felt the requirement to do a certain number of exercises prevented students from "studying" (i.e., working independently as university students, rather than like school children). Others were positive about the form of the exam, as were most students before the exam.

Two of the three focus group students who passed the exam were interviewed after the exam. They had both achieved about average marks. One had finished the projects before the end of the course and felt it was nice to be able to concentrate on preparing his presentation after the written exam. He felt he "really mastered this material." The other student had only managed to do two of the projects during the semester and finished the mandatory parts of the last three projects (out of the five required) in the week before the oral exam. Nonetheless, she still felt confident that "what she did was right," especially as she "knew all the references" for the results used. She had, however, heard about students who were examined on projects they had done partially wrong. She described this form of exam as "tuned to the elite" because it required "real understanding" and "self-assessment" rather than just repetition. Indeed, this could be a possible explanation for the fact that so many students decided for themselves that they would not pass the exam.

For the examiners, the format of the oral exam was new. A main difficulty seemed to have been to avoid doing just as they had in previous years. When the task of the student was to present a well known proof, there was a tendency among examiners to count down from a perfect presentation by looking for elements in the student's performance that deviated from it. In the new format, on the contrary, it was explicit in the official description that the examiners were to try to look for "what the students were able to do, at their own level." Indeed, the students were meant to make choices about approach, including level of ambition. Especially in the case of relatively weak students, the enactment of this principle was not always obvious. On the one hand, one could not let serious mistakes pass without trying to help a student correct them; on the other hand, one could not take over the presentation of the student by dwelling at length on such points. Probably, a better preparation of examiners for this new form of exam is desirable, perhaps in the form of actual training, one of the aims being a raised awareness of how to assess the explicit competency goals given at the beginning of each project.

Two of the external examiners offered detailed descriptions of their impressions. One, who acted in this function for the first time, said that it was difficult to assess the benefit to students of working on the thematic projects, but that some students had demonstrated a "good ownership" of the material covered by them. The other, who had extensive experience as an external examiner, was quite enthusiastic about the format, saying, "Perhaps, with some modifications and extensions [to prerequisite courses], here is a possible path for renewing both the instruction and the exam." Several of his comments actually echoed those of a majority of the students. According to his impressions, the project work helped students to study, and the students who did so were able to "run the show" (i.e., the oral exam). In contrast to the view of the less experienced examiner, his impression was that students felt more confident with this form. Should this be true, it might be related to the above discussion about those students who dropped out before the oral exam.

Finally, we make a few comments on another source of insight into students' work: the written projects delivered by the students at the time of the oral examination. First, the worry about whether students would extensively share, or even trade, projects among each other was unnecessary. The projects delivered by different groups were really different, not only in wording but also in approach (naturally, more so for relatively open questions). Moreover, students did take advantage of possibilities for choosing different levels of explanation or investigation (e.g., from giving a concrete example to formulating and proving a general theorem). Responses to questions that, in principle, could be answered by copying excerpts from the book, such as "Explain at least two ways to describe the Cantor set," varied greatly. Some of these were among the best examples of autonomous mathematical expression in the projects. Most of the projects did contain mistakes or shortcomings, including some at a quite elementary level. This confirmed the students' asserted needs for feedback during the semester.

**5.8. Competency Descriptions and Student Achievement.** We discuss here the final and most basic research question of this study:

> Q8: Can the explicit competency descriptions make students work in ways that would help them meet the formal requirements of the assessment? Or at least make the requirements more transparent?

We were surprised by how little interest the students took in the descriptions of the course goals. It became obvious to us that their availability was a necessary but not sufficient condition for their efficacy. Our first formulation of them (and Appendix A is just the core part) was mostly a teachers' version. The SCGs simply cannot be fully understood by students prior to taking the course. But, we believe that experience with exercise sessions, especially in the later parts of the course, and with the projects, suggested that rather similar formulations might be useful if emphasized to students in relevant contexts. Useful questions to pose to students would include: What are the goals of the exercises this week? Of this lecture? Of this project? In other words, clear and transparent statements of local goals could be a tool for providing (or even negotiating) meaning in the interaction between teachers and students.

The SCG "master plan" might be of some help to some students in order to perceive the coherence and meaning of the course. However, first and foremost, it is a tool for the reflection and development of the teacher as didactical engineer. As some written and oral student remarks suggested, for many students the competency discourse appeared to be an added complexity. On the other hand, it appears from our experience that successfully conveying to students how the SCGs are actually behind what is tested, particularly on the written exam, may make the collection of SCGs useful to some students in the final phases of the course.

## 6. A Posteriori Analysis and Perspectives

We now summarize our main conclusions from this first "run" of the engineering cycle, in the form of some preliminary conclusions and their immediate consequences for further development of the design described in Section 3. These develop the reflections initiated in the previous section, taking into account the *a priori* analysis of Section 2. At this stage, it seems premature to talk of (non-) validation of hypotheses (see Artigue, 1989, p. 297), but some of the points suggest sharpening and extending the *a priori* analysis for use in further developments.

**6.1. SCG Descriptions Are Necessary but Not Sufficient.** This must be understood with respect to three types of agents: students, teachers in general, and teachers directly involved in the engineering. For the *students*, local goals are at least initially more likely to be meaningful and should be emphasized in the context of teaching where they are relevant (instructions for exercise work, project formulation, etc.). Such an approach is consistent with the ideas underlying the SCG: that *competency goals must be related to situations and content* in order to be meaningful and (even partly) achievable by those who are meant to work towards them. In other words, they must be amenable to comparison with performance, not in the naïve sense of describing target behaviour, but in the sense that performance can be conceived of as more or less satisfactory evidence of progress towards the goals. As students are very sensitive to assessment, one should provide specific and explicit examples of how solving a concrete task will be interpreted as such evidence. These examples could begin even rather early in the course.

For teachers in general, there is a need to make the use of SCGs explicit with respect to concrete teaching tasks. This is similar to what we just described for students. Also, the role of the teacher in more complex tasks such as facilitating rather than managing or directing project work or handling assessment of the goals, needs to be more thoroughly described. The experiences and developments from the first run need to be capitalized on in explicit instructions to teaching assistants, while not neglecting the need for and benefits of a continuous dialogue with them.

Finally, for those of us involved in the engineering itself, there is a need to re-consider both the formulation of the SCGs and their function in our thinking about the objectives of the course. A main benefit for the lecturer was the enhanced and much more extensive interaction with students and teaching assistants regarding course goals. The above considerations pertain mostly to the need for enhancing clarity, but the basic advantage of focus on *students' work* as objective should not be lost by undue emphasis on procedures and measurements. And, in stating SCGs, there is a need for systematic yet simple tools to describe levels and scope for goal attainment. For instance, this could contribute to enhancing both teachers' and students' recognition of the fact that concrete or contextualised enactment (e.g., of mastering a definition) is, as the test data suggest, often *harder* for students than enactment in a formal, abstract task. Also, the general importance of coordinating different representations must be emphasized in the competency descriptions and made explicit by examples in lectures and exercises.

**6.2. Thematic Projects Should be More Manageable.** A main engineering product that could result from this work is the format of thematic projects to develop students' integrated, independent and collaborative work, on SCGs. The oral evaluation is, in the context of *Math. 2AN*, an important motivation for the projects; not only as an incentive to work on them, but also to legitimize their presence in the course as an alternative to the presentation of textbook arguments. However, as appears very clearly from the data presented in Section 5, the projects we actually proposed were too demanding, even after it was made clearer that not all questions had to be solved or solved at the highest level of generality or completeness. Also, the projects must be more similar in structure, beginning with simple and closed tasks and ending with one or two open problems. The link between the projects and the textbook material (including worked exercises) should be more apparent to avoid what some students perceived as a "necessity" to search for supplementary material. It is important not to discourage students with overly high expectations, particularly in terms of autonomous work. This must be balanced with previous experiences of the students. In this case, that experience had been with solving standard exercises and reading/presenting simple proofs.

**6.3. Provide Timely Posited Checkpoints for Projects.** Another technical deficiency of the way in which we sought to implement the thematic projects was the absence, due to limitations in assigned teacher time, of comprehensive review and comments during the semester on students' project work. We stated in Section 3.3.2 that we did not have the means to grade the students' reports during the semester. This was true under the assumption that students would, as usual, be given the opportunity to have sets of exercises (of the type used in class sessions) graded every week. This practice is meant to help students' written work, especially the written part of the exam. However, the option was not used by all students; it was used more frequently by the best performing ones. A main critique of the thematic project format, both by students and teaching assistants, was the absence of feedback on actual student work during the semester. Hence, in future cycles of the development project we will substitute, in some weeks, a possibility of having a certain project reviewed and commented on by the teaching assistant. This should be done every second week, from about six weeks into the course, and correspond to a suggested completion schedule. This option would be available only in the given week and for the given project, both to smooth out the teaching assistants' workload and to motivate students to work regularly throughout the term on the projects, rather than mostly at the end of the course. It will be important to provide the teaching assistants with precise guidelines and sufficient help (in particular, this would still not include the provision of solutions).

**6.4. Tools for Measuring Competency Goals.** The post-test (Appendix B) contributed to our thinking about the problem of measuring SCGs, as well as how to devise the written exam. However, these two issues should not be kept apart. In order to help students understand the meaning of SCGs, it could be helpful to provide them with sets of exam-like exercises that exemplify different ways of enacting each SCG. This could also strengthen teachers' focus on the merits of maintaining a clear link between goals and assessment. For the purpose of diagnostics (i.e., for obtaining a quantitative, albeit partial, picture of student performance at a given point), it may still be desirable to formulate such items in a multiple choice format.

**6.5. Curriculum and the Notions of Competency.** Several of the above conclusions concern local choices about teaching and assessment with the aim of improving the implementation of a course description based on competency notions. It is now time to return to this description and the underlying theory: Can the SCGs be improved or amended by making use of our observations?

In Section 2.1 and elsewhere, we mainly considered students as *cognizing subjects* who, in principle, strive to achieve the set goals. Aspects which were not extensively considered in the theoretical underpinnings, but which proved unavoidable in order to interpret and manage the reactions of students and teachers, include affective and meta-cognitive factors; such as students' and teachers' rational and articulated (pre)conceptions about what it means "to study mathematics." Artigue (1989, III.2) points out that these factors represent an important but often ignored area posing delicate methodological problems. There is an extensive literature on mathematical beliefs. Although much of it is concerned with primary and secondary level students and teachers, a significant part of it pertains to university students and should be taken into consideration (e.g., works by Schoenfeld, 1985, and Tall, 1991). Thus, we could consider a third complementary dimension of learning goals, orthogonal to content and competency, namely that of students' *socialization into mathematical ways of thinking and doing* (see Winsløw, 2004). We believe this dimension cannot in practice be separated from, or reduced to, students' development in mathematical competency and content knowledge. Indeed, this dimension was assessed in this study, in particular in our analysis of students' responses to questionnaires and of the interviews with focus groups of students.

Any curriculum contains goals which are chosen by others than those who are supposed to achieve them. Ironically, the many accepted meanings for the word "competency" illustrates the inherent tensions in this situation. The sense employed in this paper, as well as in Niss and Jensen (2002), is that of competencies one has with respect to acting mathematically. In other contexts, competency is related to more personal qualities, like flexibility or analytical skills. Another sense is that of formal competency or *qualification* (e.g., to exercise a certain profession, such as teaching). While this report is concerned almost exclusively with the first sense, some of the situations encountered in teaching may be interpreted as evidence that students are often more concerned with the latter two senses: to experience personal growth by becoming personally competent rather than achieving external competency goals, and to acquire formal recognition of competency.

A basic motivation in our construction and analysis of specific competency goals was to mediate the implicit inconsistencies between a professional mathematician's understanding of mathematical competency and the practices in teaching and assessment as experienced by students. Indeed, the need for explicitness and accessibility of goal descriptions turns out to be a major point in this effort. We hypothesize that pursuing concerns about students' socialization into mathematical practice may reveal an equally important tension between the new competency goals and those implicitly pursued by students.

It is unproductive and most likely illusory to postulate an outright conflict between teacher goals and student goals. We affirm that course goals in undergraduate mathematics must be based on the professional's insight. But the *a priori* analysis of course goals might become more fully reflective of these insights if it also addressed the social dimension of learning mathematics more explicitly, not so much in the global sense as more or less learned descriptions of mathematical maturity, but in a local, specific sense: What kind of experiences of substantial increase in mathematical outlook and power should the course provide (e.g., in the context of metric space)? What examples and connections to earlier parts of the curriculum could be a vehicle for fostering such experiences? Such questions are often implicit in the use of project work, but they could also be relevant for the design of didactical situations in exercise sessions as well as for the revised purposes of lectures.

At least some of the SCGs could be enriched by examples of activities that are not only chosen to illustrate what actions the SCG corresponds to, but also to allow students to experience and reflect on the purpose and benefits of those actions. The criteria for succes of such activities would not only be related to personal experience, they would also be related to aims of socialization. Such activities could strive for *shared* experience, involving both students and teachers. For instance, themes for "scientific debate" (Legrand, 2001) in a particular mathematical context such as different ways to approach a problem or to do a proof might be a valuable complement to the formal statements of the SCGs and corresponding exercises (as in Appendix A). They could illustrate their functions in "mathematical socialization."

## Acknowledgments

We are grateful to the students, teaching assistants, and colleagues who have offered time and wisdom to this project. In particular, we wish to thank our colleague Inge Henningsen of the Institute for Mathematical Sciences at University of Copenhagen for explaining to us a method of Rasch (1960). We also thank Annie Selden, editor of RCME, for a great number of corrections and suggestions for improving the English of the text.

## References

Adams, R. (1995). *Calculus: A complete course.* Don Mills, Ontario: Addison-Wesley.

Artigue, M. (1989). Ingénierie didactique. *Recherches en Didactique des Mathématiques, 9*(3), 281–308.

Artigue, M. (1994). Didactical engineering as a framework for the conception of teaching products. In R. Biehler, R. Scholz, R. Strässer & B. Winkelmann (Eds.), *Didactics of mathematics as a scientific discipline* (pp. 27–39). Dordrecht, The Netherlands: Kluwer.

Artigue, M. (1999). What can we learn from educational research at the university level? Crucial questions for contemporary research in education. *Notices of the American Mathematical Society, 46*, 1377–1385.

Brousseau, G. (1997). *Theory of didactical situations in mathematics* (N. Balacheff, M. Cooper, R. Sutherland, & V. Warfield, Eds. and Trans.). Dordrecht, The Netherlands: Kluwer.

Burn, B., Appleby, J., & Maher, P. (Eds.) (1998). *Teaching undergraduate mathematics.* London: Imperial College Press.

Carothers, N. L. (2000). *Real analysis.* Cambridge, England: Cambridge University Press.

DeFranco, T. C. (1996). A perspective on mathematical problem solving expertise. In J. Kaput, A. H. Schoenfeld, & E. Dubinsky (Eds.), *Research in collegiate mathematics education. II* (pp. 195–213). Providence, RI: American Mathematical Society.

Duval, R. (1995). *Sémiosis et pensée humaine. Registres sémiotiques et apprentissages intellectuels.* Bern, Switzerland: Peter Lang.

Duval, R. (2000). Basic issues for research in mathematics education. In T. Nakahara & M. Koyama (Eds.), *Proceedings of the 24th Conference of the International Group for the Psychology of Mathematics Education: Vol. 1* (pp. 55–69). Hiroshima, Japan: Hiroshima University.

Duval, R. (2002). *Quel ou quels point(s) de vue pour analyser la connaissance scientifique dans une perspective d'enseignement et d'apprentissage?* Unpublished manuscript.

Fischer, R. (1976). Fundamentale Ideen bei den reellen Funktionen. *Zentralblatt für Didaktik der Mathematik, 8*(4), 185–192.

Jahnke, T. (1999). Zur Kritik und Bedeutung der Stoffdidaktik. *Mathematica Didactica. Zeitschrift für Didaktik der Mathematik, 21*(2), 61–74.

Jensen, J., Niss, M., & Wedege, T. (Eds.). (1998). *Justification and enrolment problems in education involving mathematics or physics.* Roskilde, Denmark: Roskilde University Press.

Legrand, M. (2001). Scientific debate in mathematics courses. In D. Holton (Ed.), *Teaching and learning of mathematics at university level: An ICMI study* (pp. 127–135). Dordrecht, The Netherlands: Kluwer.

Mansfield, H., Pateman, N. A., & Bednarz, N. (Eds.). (1996). *Mathematics for tomorrow's young children. International perspectives on curriculum.* Dordrecht, The Netherlands: Kluwer.

Mason, J. (2002). *Mathematics teaching practice: A guide for university and college lecturers.* Chichester, England: Horwood.

Messer, R. (1994). *Linear algebra: Gateway to mathematics.* Boston: Addison-Wesley.

Niss, M., & Jensen, T. H. (2002). Kompetencer og matematiklæring. Idéer og inspiration til udvikling af matematikundervisning i Danmark (in Danish). *Uddannelsesstyrelsens temahæfteserie: Vol. 18.* Copenhagen, Denmark: Undervisningsministeriet.

Programme for International Student Assessment (2001). *Knowledge and skills for life: First results from the OECD Programme for International Student Assessment.* Paris: Organisation for Economic Co-operation and Development.

Rasch, G. (1960). Probabilistic models for some intelligence and attainment tests. *Studies in Mathematical Psychology I.* Copenhagen, Denmark: Danmarks Pædagogiske Institut.

Robert, A. (1998). Outils d'analyse des contenus mathématiques à enseigner au lycée et à l'université. *Recherches en Didactique des Mathématiques, 18*(2), 139–190.

Robitaille, D., & Donn, J. (1992). The third international mathematics and science study (TIMSS): A brief introduction. *Educational Studies in Mathematics, 23*, 203–210.

Russell, B. (1969). *Autobiography of Bertrand Russel*. London: Allen and Unwin.

Schoenfeld, A. S. (1985). *Mathematical problem solving*. New York: Academic Press.

Selden, J. & Selden, A. (1995). Unpacking the logic of mathematical statements. *Educational Studies in Mathematics, 29*(2), 123–151.

Sfard, A. (1991). On the dual nature of mathematical conceptions: Reflections on processes and objects as different sides of the same coin. *Educational Studies in Mathematics, 22*(1), 1–36.

Tall, D. O. (1991). (Ed.). *Advanced mathematical thinking*. Dordrecht, The Netherlands: Kluwer.

Verschaffel, L., & De Corte, E. (1996). Number and arithmetic. In A. Bishop, K. Clements, C. Keitel, J. Kilpatrick, & C. Laborde (Eds.), *International handbook of mathematical education* (pp. 99–137). Dordrecht, The Netherlands: Kluwer.

Winsløw, C. (2005). Two dimensions in the conception of mathematics in tertiary education. In C. Bergsten & B. Grevholm (Eds.), *Conceptions of mathematics: Proceedings of NORMA 01, the 3rd Nordic Conference on Mathematics Education* (pp. 250-258). Linköping, Sweden: SMDF.

Winsløw, C. (2004). Hvad skal vi med matematikdidaktikken?. In K. Schnack (Ed.), *Didaktik på kryds og tværs* (pp. 325–344). Copenhagen, Denmark: Danish University of Education Press.

## Appendix A: Specific Competency Goals as Presented to Students
(translated from Danish)

The following table appeared in the materials distributed at the beginning of the course.

| Content | Text* | Competency goals | Exer-cises* |
|---|---|---|---|
| **a. Background *(about 3 weeks)*** Review of Math. 1, and related stuff: properties of real numbers and functions (e.g. sup, inf, limsup, liminf, Bolzano-Weierstrass), equivalence and cardinality of sets, Cantor-Bernstein's theorem, the Cantor set | C1 C2 | **a1)** apply inf, liminf etc. in simpler arguments concerning real numbers and functions **a2)** understand and use various definitions ($\varepsilon/\delta$, sequence based) of continuity and convergence **a3)** explain the concept of 'countability' and the most important examples in $\mathbb{R}$, and be able to handle simple unseen examples | 1.35 1.45 2.16 2.22 |
| **b. Metric spaces and normed spaces *(about 2 weeks)*** Metrics, norms, convergence, open and closed sets, closure, induced metrics | C3 C4 | **b1)** check and use the definitions of metrics and norms **b2)** use definitions of Cauchy-seq. and convergent seq. **b3)** apply different formulations of openness and closedness in simple arguments | 3.5 3.18 3.37 4.5 4.46 |
| **c. Continuity *(about 1 week)*** Continuous mappings, homeomorphisms, introduction of the space $C(M)$ | C5 | **c1)** apply different formulations of continuity, both in concrete and abstract arguments **c2)** explain the meaning of homeomorphisms and check concrete examples | 5.8 5.30 5.36 5.48 |
| **d. Completeness *(about 1 week)*** Total boundedness, Bolzano-Weierstrass, complete metric space, absolute convergence, Banach's fixed point theorem, completion | C7 | **d1)** use different methods to show completeness of concrete metric spaces **d2)** use the notions in simple abstract settings **d3)** Apply the fixed point theorem to concrete mappings and functional equations | 7.12 7.18 7.36 |
| **e. Compactness *(about 1 week)*** Compactness as related to the notions in d), sequential compactness, the covering theorem, uniform continuity | C8 | **e1)** apply different formulations of compactness in a variety of situations **e2)** explain the meaning of uniform continuity **e3)** check and use uniform continuity in simple concrete and abstract situations | 8.17 8.27 8.48 8.68 |
| **f. Function spaces *(about 1½ week)*** Uniform convergence, the linear spaces $C(X)$ and $B(X)$, the supremum norm, Weierstrass' approximation (w/o proof) and relation to trigonometric polynomials | C10 C11 | **f1)** explain the relation of uniform convergence to continuity **f2)** check and use uniform convergence in concrete situations (including series) **f3)** know and use some fundamental properties of $B(X)$ and $C(X)$, including those that depend on $X$ | 10.9 10.14 10.23 10.26 10.34 11.8 |
| **g. Fourier series *(about 1½ week)*** Convergence of FS, particularly Fejer's theorem. Complex form of FS. Proof of Weierstrass' approximation theorem. | C15 | **g1)** determine the FS of a given function, and its properties of convergence **g2)** explain when, how and why a FS of a function $f$ converges to $f$ **g3)** convert between real and complex forms of FS | 15.7 15.9 |
| **h. Hilbert space *(about 2 weeks)*** Review of Euclidean spaces, definition and examples of H-spaces (particularly $\mathbb{C}^n$, $\ell^2$, completion of $C(I)$), convexity, projections and bases in Hilbert space, relation to FS | N | **h1)** know and check concrete examples of H-spaces **h2)** explain and use the concept of basis of H-spaces **h3)** carry out simple arguments using convergence and orthogonality in Hilbert spaces **h4)** explain the abstract Hilbert space isomorphism $x \mapsto \hat{x}$ and (for $L^2(\mathbb{T})$) its relation to FS. | ** |

\* Refers to chapters and exercises in (Carothers, 2000), but "N" refers to distributed notes.
\*\* Notes to be published later in the semester. [The notes on Hilbert space were not finished at the beginning of the course, so this field was empty in the handout produced at that time.]

## Appendix B: Post-Test (translated from Danish)

First we gave a brief explanation of how to indicate one's choice of a subset from among the three options provided for each item. Students were asked to choose answer **G** for items they felt uncertain about. For each test item, students chose among the following answer choices and wrote the letter in a box (e.g., choice A meant options (1) and (2) were both correct answers).

**A:** $\{1, 2\}$ **B:** $\{1, 3\}$ **C:** $\{2, 3\}$ **D:** $\{1\}$ **E:** $\{2\}$ **F:** $\{3\}$ **G:** don't know

**Item 1.** Consider the sequence of real numbers given by $x_n = (-1)^n + 1$. Which of the following assertions is/are correct?

    (1) $\limsup x_n < \liminf x_n$

    (2) $\limsup x_n > \liminf x_n$

    (3) $\limsup x_n = \liminf x_n$

**Item 2.** Let $f \colon [0, \infty[ \to [0, \infty[$ be given by $f(x) = x^2$. Which of the following assertions is/are correct?

    (1) There is $\delta > 0$ such that $|x - 2| < \delta \implies |f(x) - 4| < \frac{1}{1000}$.

    (2) For every $\varepsilon > 0$ there is $\delta > 0$ such that $|x - 2| < \delta \implies |f(x) - 4| < \varepsilon$.

    (3) There is $\delta > 0$ such that for all $a \in [0, \infty[$ we have: $|x - a| \implies |f(x) - f(a)| < \frac{1}{1000}$.

**Item 3.** Which of the following sets is/are countable?

    (1) The interval $[0, 1]$.

    (2) The set of prime numbers, $\{p \in \mathbb{N} \mid p \text{ is a prime number}\}$.

    (3) The unit circle, $\{(x, y) \in \mathbb{R}^2 \mid x^2 + y^2 = 1\}$.

**Item 4.** Which of the following maps $d \colon \mathbb{R}^2 \to \mathbb{R}$ is/are a metric?

    (1) $d(x, y) = x - y$.

    (2) $d(x, y) = |x^2 - y^2|$.

    (3) $d(x, y) = |x^3 - y^3|$.

**Item 5.** Let $(x_n)$ be a bounded sequence of reals. Which of the following assertions can be infered from this?

    (1) $(x_n)$ is a Cauchy sequence.

    (2) $(x_n)$ has a convergent subsequence.

    (3) $(x_n)$ is convergent.

**Item 6.** Which of the following subsets of $\mathbb{R}^2$, equipped with the usual metric, is/are open?

    (1) $\{(x, y) \in \mathbb{R}^2 \mid y \geq x\}$.

    (2) $\{(x, y) \in \mathbb{R}^2 \mid x^2 + y^2 \neq 1\}$.

    (3) $\{(x, y) \in \mathbb{R}^2 \mid y > 0 \land x \geq 0\}$.

**Item 7.** Let $(M, d)$ be a metric space and let $f \colon M \to \mathbb{R}$ be continuous, where $\mathbb{R}$ is equiped with the usual metric. Also, let $(x_n)$ be a convergent sequence in $M$. Which of the following assertions is/are correct?

    (1) If $-\frac{1}{n} \leq f(x_n) < 0$ for all $n \in \mathbb{N}$, then $f(\lim x_n) < 0$.

    (2) If $f(x) > 0$, then there exists $\delta > 0$ such that $d(x, y) < \delta \implies f(y) > 0$.

    (3) $\{x \in M \mid |f(x)| < 1\}$ is open in M.

**Item 8.** Which of the following assertions is/are correct? (The sets indicated are to be considered as subspaces of $\mathbb{R}$ with the usual metric.)

    (1) $]0, 1[$ and $[0, 1]$ are homeomorphic.

    (2) $[0, 1[$ and $]0, 1]$ are homeomorphic.

    (3) $\{\frac{1}{n} \mid n \in \mathbb{N}\}$ and $\mathbb{N}$ are homeomorphic.

**Item 9.** Which of the following metric spaces is/are complete?

    (1) $\{(x, y) \mid y \geq 1\}$ as a subspace of $\mathbb{R}^2$ with the usual metric.

    (2) $]0, 1]$ with the usual metric.

    (3) $\{(x_n) \in \ell_2 \mid x_1 = 1\}$ as subspace of $(\ell_2, \|\cdot\|_2)$.

**Item 10.** Which of the following metric spaces is/are compact?

    (1) $\{0\} \cup \{\frac{1}{n} \mid n \in \mathbb{N}\}$ as subspace of $\mathbb{R}$ with the usual metric.

    (2) $\{0\} \cup \{\frac{1}{n} \mid n \in \mathbb{N}\}$ as subspace of $\mathbb{R}$ with the discrete metric.

    (3) $\{\frac{1}{n} \mid n \in \mathbb{N}\}$ as subspace of $\mathbb{R}$ with the usual metric.

**Item 11.** Let $A \neq \emptyset$ be a closed subset of $\{(x, y) \in \mathbb{R}^2 \mid x^2 + y^2 \leq 1\}$, the closed unit disc in $\mathbb{R}^2$ with the usual metric. Which of the following assertions about $A$ is/are correct?

    (1) There is $(x_0, y_0) \in A$ such that $y_0 \leq y$ for every $(x, y) \in A$.

    (2) There is a continuous function $f \colon A \to \mathbb{R}$ such that $f(A) = [0, 1[$.

(3) There is a continuous function $f \colon A \to \mathbb{R}$ such that $f(A) = [0, \infty[$.

**Item 12.** Which of the following functions is/are uniformly continuous?
  (1) $f \colon [0, 1] \to [0, 1]$ given by $f(x) = x^2$.
  (2) $f \colon \mathbb{R} \to \mathbb{R}$ given by $f(x) = x$.
  (3) $f \colon \mathbb{R} \to \mathbb{R}$ given by $f(x) = x^2$.

**Item 13.** Let $f_n(x) = (1-x)^n$ and $f(x) = \begin{cases} 0, & x \in ]0, 1] \\ 1, & x = 0 \end{cases}$ Which of the following assertions

is/are correct?
  (1) $f_n \to f$ pointwise on $[0, 1]$.
  (2) $f_n \to f$ uniformly on $[0, 1]$.
  (3) $f_n \to f$ uniformly on $[\frac{1}{2}, 1]$.

**Item 14.** Let $f(x) = \cos(x) + x^2$, $x \in [-\pi, \pi]$. Which of the following assertions is/are correct?
  (1) The Fourier series of $f$ converges uniformly to $f$ on the interval $[-\pi, \pi]$.
  (2) There is a trigonometric polynomial $T$ such that $\|f - T\|_\infty < 6 \cdot 10^{-23}$ in $C([-\pi, \pi])$.
  (3) The Fourier series of the derivative $f'$ is uniformly convergent.

## Appendix C: Written Exam (translated from Danish)

*It may be used without proof that the series $\sum_{n=1}^{\infty} \frac{1}{(2n-1)^2}$ is convergent with sum $\frac{\pi^2}{8}$.*

**Exercise 1** (30%). Let $A = \{(x, y) \in \mathbb{R}^2 \mid y = x^2\}$.
  (1) Show that $A$ is a closed subset of $\mathbb{R}^2$ with the usual metric.
  (2) Show that the interior $A^\circ$ is empty.
  (3) Is $A$ compact? Is $A$ complete? The answers must be justified.

**Exercise 2** (20%). Let $(M, d)$ be a metric space and let $(a_n)$ be a convergent sequence in $M$ with $\lim_{n \to \infty} a_n = a$. Define the function $f \colon M \to \mathbb{R}$ and for every $n \in \mathbb{N}$ the functions $f_n \colon M \to \mathbb{R}$ by
$$f(x) = d(a, x) \text{ and } f_n(x) = d(a_n, x).$$
  (1) Show that $|f_n(x) - f(x)| \le d(a_n, a)$ for all $x \in M, n \in \mathbb{N}$.
  (2) Show that $f_n \to f$ uniformly.

**Exercise 3** (30%). Consider the Fourier series $\sum_{n=1}^{\infty} \frac{1}{(2n-1)^2} \sin(nx)$, $x \in [-\pi, \pi]$.
  (1) Justify the fact that the series is uniformly convergent and that the sum function, $f$, is continuous.
  (2) Prove that $\int_{-\pi}^{\pi} f(t) \cos(t) \, dt = 0$.
  (3) Let $F(x) = \int_0^\pi f(t) \, dt$, $x \in [-\pi, \pi]$. Show that
$$\frac{1}{\pi} \int_{-\pi}^{\pi} F(t) \cos(nt) \, dt = -\frac{1}{n(2n-1)^2}$$
  for $n \in \mathbb{N}$.

**Exercise 4** (20%). Let $x_n = \frac{1}{2n-1}$ for all $n \in \mathbb{N}$.
  (1) Show that $(x_n) \in \ell_2$ and determine $\|(x_n)\|_2$.
  (2) Exhibit a vector different from the null vector which is orthogonal to $(x_n)$.

University of Copenhagen, Institut for Mathematical Sciences, Universitetsparken 5, DK-2100 Copenhagen Ø, Denmark
  *E-mail address*: `gronbaek@math.ku.dk`

University of Copenhagen, Centre for Science Education, Universitetsparken 5, DK-2100 Copenhagen Ø, Denmark
  *E-mail address*: `winslow@cnd.ku.dk`

CBMS Issues in Mathematics Education
Volume **13**, 2006

# Introductory Complex Analysis at Two British Columbia Universities: The First Week - Complex Numbers

Peter Danenhower

ABSTRACT. In this paper I consider problems that students have with multi-representations of complex numbers. I found that students have many misconceptions and difficulties with the basic representations: $z = x + iy$, $z = (x, y)$, $z = re^{i\theta}$, and the symbolic representation, in which $z$ is used directly. In addition, I studied how well students were able to judge when to shift from one representation to another. Where possible, I identified different levels of understanding of the various representations. Students were competent with the $z = x + iy$ and $z = re^{i\theta}$ representations, and most made reasonable shifting decisions between these two forms, but there was little evidence that students understood the symbolic representation. In addition to the question of shifting representations, I considered several specific problems that students had with the various representations, such as understanding that $x$ and $y$ are real, cancelling a common factor in a complex fraction, and finding the argument in the polar form. From the data I identified four characteristics of understanding a given representation that are consistent with results reported in the literature on representations of fractions. The paper concludes with suggestions for further research and suggestions for teaching.

## 1. Introduction

In British Columbia, complex analysis is typically a third year university course, with subsequent fourth year and graduate courses sometimes available. The third year version is usually oriented towards science and engineering students and has a strong focus on applications.

**1.1. Research Rationale.** I expected the subject to be an excellent area of research in post-calculus mathematics education for the following reasons:

(1) Third year complex analysis includes a variety of topics, some of which depend on a thorough development of simpler concepts, while others do not. I conjectured that this would enable me to study basic material for its own sake, as well as to study the effect that such misunderstandings have on later work. In addition, I expected to be able to study some advanced

The work reported here was done in partial fulfillment of the requirements for the doctorate degree under the direction of Harvey Gerber and Rina Zazkis at Simon Fraser University.

topics independently of the extent to which students understood basic material.

(2) I expected to encounter several types of problems that students might not have confronted in lower level courses. For example, at Simon Fraser University students do not generally have to construct proofs until their third year. The standard second year analysis course is not a prerequisite for complex analysis at either of the universities studied. In addition, success at complex analysis requires a greater synthesis of concepts from algebra, trigonometry, calculus, geometry, and topology, and more symbolism than other courses at third year level. Finally, the shift from thinking "real" to thinking "complex" is not trivial.

*Feasibility.* From a practical perspective, enrollment is high enough and course offerings frequent enough in third year mathematics courses at Simon Fraser University and the University of British Columbia to attract acceptable samples of students for study.

*Research Objectives.* The objective of the research reported in the present paper was to determine what problems students in an introductory complex analysis class were having in understanding complex numbers. I was particularly interested in those aspects of complex numbers and their representations that are important prerequisites for further development of the course.

**1.2. Background.** I was unable to find any references to work on complex analysis in the mathematics education research literature. I did find a few references on the subject of complex numbers. For example, Tirosh and Almog (1989) studied 78 high school students in Israel and found that their students had a very difficult time understanding that complex numbers are numbers. Many students insisted that a number had to represent a quantity. In addition, students had great difficulty understanding that the usual ordering relation on the real numbers does not hold for the complex numbers. In a post-test administered by Tirosh and Almog, 95% of the students agreed that $i < 4 + i$. I did not pursue research reports on high school students further, since I assumed that third year university students would have a different set of difficulties when learning complex analysis.

Indeed, I found no evidence during interviews or tutoring sessions that the students studied did not recognize complex numbers as numbers. They did have problems understanding that the ordering relation, $<$, does not extend to the complex numbers, but most of their difficulties were with what it means to extend $<$. The students studied had had no instruction on ordering at the time of the interview. So, although there was some superficial similarity between the responses of high school students and the students in the present study, careful scrutiny showed distinctly different problems of understanding.

Given that there was virtually no prior research on complex analysis to serve as a foundation for the research reported here, I relied extensively on research in other areas of mathematics education to develop a research framework. These are described in detail in Danenhower (2000).

**1.3. Epistemology of the Complex Numbers.** In this section I give some general comments on what is meant by understanding each representation in terms of the reification\encapsulation model discussed in detail in Section 3. I discuss the algebraic extension form, $z = x + iy$, the vector form, $z = (x, y)$, the polar form,

$z = re^{i\theta}$, and the symbolic form in which complex numbers are simply represented by $z$.

The objective of this section is to give readers some idea of what kinds of difficulties I expected students to have. Since there is no literature in mathematical education (at third year university level) on the topics discussed in this section, the claims made are necessarily speculative. They are based on my experience as a student and tutor of complex analysis, discussions with colleagues, and the discussion and emphasis in several texts on complex analysis such as Alhfors (1979), Churchill and Brown (1990), and Saff and Snider (1993).

**The Algebraic Extension Representation: $z = x+iy$.** The algebraic form $z = x+iy$ emphasizes the complex numbers as an algebraic field extension of the real numbers. The advantages of this representation are that it is relatively easy to learn and to apply since it does not require much preparation other than algebra. The main disadvantage of the algebraic form is that its simplicity makes calculations involving large exponents very tedious. In addition, the multivaluedness of the complex numbers is not easily represented in the algebraic representation so that finding roots is not easy.

The main conceptual difficulties with the algebraic form are understanding what $iy$ means and understanding that the complex operations of $+$ and $\times$ are defined in terms of the $+$ and $\times$ operations for real numbers in a way that retains all the usual rules. The surprising ease with which one can extend the usual rules for real numbers masks the fact that one is working in a new system of numbers, and some errors can be traced to thinking in terms of real numbers as opposed to thinking in terms of complex numbers .

In viewing the data in terms of the process-object distinction, I expected to identify a process understanding of the algebraic extension representation as a preoccupation with the details of computations, whereas an object understanding would be revealed as a fluency with computations and an awareness of when to shift to another representation.

**The Cartesian Vector Representation: $z = (x, y)$.** The Cartesian vector representation emphasizes the complex numbers as the vector space $\Re^2$ over $\Re$ with the vector product defined as $(x, y)(u, v) = (xu - yv, xv + yu)$. The main advantage of this representation is that any reference to roots of negative numbers is avoided. The complex numbers can be understood as a formal system in this representation. Thus, any conceptual problems that students have accepting roots of negative numbers can be avoided, at least in the beginning. There is no need to understand $i$, since $i$ is not used in this form.

The formal approach is also the main disadvantage of the Cartesian vector form; students may find it hard to understand and think about. In addition, the Cartesian vector form suffers from the same problems as any rectilinear form; complicated products are much easier to calculate using the polar representation and multivaluedness is hard to express.

In addition to computational skill, a primary indicator of understanding of the Cartesian vector form is the skill level with the formalism. Does the student understand that $i$ is not a scalar in the Cartesian vector representation?

**The Polar Representation: $z = re^{i\theta}$.** The polar representation of complex numbers, $z = e^{i\theta}$, is a vector representation in polar coordinates. One can use

Euler's equation, $e^{i\theta} = (\cos\theta + i\sin\theta)$, as a convenient computational aid; there is no need to develop a general theory of complex exponents to understand $e^{i\theta}$ in this special case. The rules for calculating products and integral powers are derived from the corresponding rules for sine and cosine.

As with the other representations discussed so far, I expected that the primary indication of a process understanding would be a preoccupation with computational details. However, this indicator was expected to be confounded, in some cases, by difficulties with trigonometry and use of exponents. I expected an object understanding to be indicated by fluent use of the polar representation, good decisions about when to shift representations, and possibly incorporation of geometric arguments.

**Symbolic Representation of Complex Numbers.** In the symbolic representation, one treats complex numbers as $z$. The connection between the complex numbers and the real numbers plays a much less prominent role than in the other representations.

Since the level of experience with complex analysis needed to make the meaning of the symbolic representation completely clear is far beyond a beginning complex analysis course, I expected to see little more than a process understanding of the symbolic form. However, since a student who is attempting to use the symbolic representation has most likely already mastered one or more of the other representations, I did not expect to see students struggle with symbolic computations. Instead, I expected different skill levels to be revealed by the number of basic operations (modulus, conjugation, applications of geometry) that a student could use.

## 2. Methodology of Data Collection and Analysis

**2.1. Data Collection.** I collected data primarily from attending three introductory courses in complex analysis, two at Simon Fraser University and one at the University of British Columbia. In addition, I conducted informal study sessions with students and held 54 clinical interviews with 21 students.

**Data Collection: Pilot Project - Class 1.** I began with a pilot project in the Summer Semester of 1996 at Simon Fraser University. Participants were recruited from the regular Math 322 class, Introduction to Complex Analysis. The textbook for the course was *Introduction to Complex Analysis* by Priestley (1990). The instructor followed the topics covered in the textbook closely, but frequently used his own presentation.

Six students volunteered for the pilot study. I conducted two separate sets of interviews of about one hour duration each. The first set was conducted during week eight of classes, and the second was conducted after the final one hour test. A total of seven interviews were conducted. In addition, I gave extensive help to all six students.

All interviews were qualitative in nature. In designing the clinical interviews, I followed the recommendations of Ginsburg (1981), Lincoln and Guba (1985), Howe and Eisenhart (1990), Asiala et al (1996), Hunting (1997), and Zazkis and Hazzan (1999). Interviews consisted of students working through problems on a prepared worksheet, while explaining their thoughts orally. Interviews were tape recorded. I did not hesitate to ask students what they were thinking about, and I freely helped them if they were stuck. The actual questions were usually designed to explore difficulties that I had noticed during tutoring sessions with students, but some questions were intended to explore student understanding of important concepts in the course. For more details, see Danenhower (2000). I obtained enough good data to include the pilot project in my overall results.

**Data Collection: Main Study - Class 2.** The first part of the main study took place in Fall 1996 at Simon Fraser University. Again, the class was Math 322, Introduction to Complex Analysis, but the textbook was changed to *Complex Variables and Applications* by Churchill and Brown (1990). The material covered in the course consisted of the first sixty sections of the text.

I recruited six participants from Class 2 and conducted six sets of interviews of about one hour duration each. Interviews began in the third week, and subsequent interviews were held approximately every two weeks after that, up to the last week of classes. In addition, I gave extensive help to three of the students.

**Data Collection: Main Study - Class 3.** The second part of the main study took place in Spring 1998 at the University of British Columbia (UBC). The class was Math 300, Introduction to Complex Analysis, and the textbook was *Fundamentals of Complex Analysis for Mathematics, Science, and Engineering* by Saff and Snider (1993). Math 300 is normally a year long course at UBC, but it is also possible to take it as a one semester course. I studied the one semester version. The material normally covered in this version of Math 300 consisted of the first six chapters of Saff and Snider. I recruited nine students from Class 3. My plan was to obtain more detailed data on the theme of multirepresentations.

I conducted two sets of interviews of about one hour duration each. The first interview was at the end of the third week of classes, and the second interview was at the end of the fifth week of classes. In addition to my usual interview technique, I challenged students more aggressively than in the previous two classes because I was interested in testing the robustness of their beliefs.

**2.2. Data Analysis.** The research was conducted within a broad theoretical framework as described in Romberg (1992). In particular, a research framework should include a teaching model, a theory of learning, and a theory of mathematical knowledge. However, in the present paper, I confine the discussion to specific aspects of this framework, namely, the method of letting the data determine themes, and two specific processes of learning mathematics - reification and APOS. For a description of the entire framework, see Danenhower (2000).

I allowed the data to determine the theoretical organization of the analysis. Hazzan (1999), following the ideas of Glaser and Strauss (1967), used this method of analysis to analyze data collected while studying abstract algebra students. As Hazzan has pointed out, allowing the data to shape the theoretical organization is especially suitable for research done in a new field. Hazzan found that a broad theme, namely Reduction of the Level of Abstraction, could explain most of her data. Note that Glaser and Strauss' idea is that not only do the hypotheses come from the data, they are devised during the course of the research as the data are collected. So the hypotheses and research themes generally evolve in this method.

I chose the theme of multiple representations for this paper primarily because this was the single theme with the most supporting data. This restriction means that much of the data collected was not used. Several additional themes emerged from the data, such as "thinking real, doing complex" and "multivaluedness." These themes were especially interesting, but have been omitted for reasons of clarity of analysis and to keep this paper at a reasonable length.

Most of the data included in the analysis for this paper can be understood in terms of reification and APOS theory. For a discussion of reification, see Sfard (1994), Sfard and Linchevski (1994), and Kieran (1992; Kieran discusses the same ideas but uses different terminology). Reification is essentially the learning process of moving from a process understanding of a concept to an object understanding of the concept. APOS stands for Action, Process, Object, Schema and is an expansion of the idea of interiorization due to Piaget. The Research in Undergraduate Mathematics Education Community (RUMEC) has been especially active in APOS-based investigations. See, for example, Asiala, et al. (1996) or Breidenbach et al. (1992). In this paper I am concerned with the first three stages of APOS theory: *actions* formed by manipulating previously constructed mental or physical objects, *processes* formed by interiorizing actions, and *object* understanding from the encapsulation of processes.

Although APOS is a four stage theory and reification has only two stages, there are certainly similarities between the two. This has been noted by Sfard and Linchevski (1994). Also, see Meel (2002) for an in-depth review and analysis of reification and APOS theory within the general context of theories of understanding. I have used Sfard-Linchevski-Kieran reification to analyze what is meant by understanding in Section 3 and have sometimes also used APOS when there was enough data to make the identification of the process of interiorization possible.

In summary, I collected data on three beginning complex analysis classes held at Simon Fraser University and the University of British Columbia. Collected data were from attending classes, individual study sessions with students, and clinical interviews. I identified the theme of multiple representations of complex numbers as having the most supporting data. I analyzed the data in terms of students struggling with the learning process of reification or interiorization and encapsulation.

## 3. Results and Analysis

In this section I present the results of analysis of the data pertaining to the theme of multiple representations. I found that the task of learning the different representations, and when to use them, was largely taken for granted during instruction of the three classes studied. However, I also found that students mostly struggled, relying on only one or two representations.

### 3.1. Shifting Representations.

**Interview Questions and Results.** To study students' ability and willingness to change representations, I asked students in Classes 1 and 2 a series of questions designed to see how they would employ the algorithm for converting the form $\frac{a+ib}{c+id}$ to the standard forms $x + iy$ or $re^{i\theta}$. The questions were as shown on the next page.

Five and six students were questioned, respectively, in Classes 1 and 2. In addition to shifting, I wanted to study what difficulties students had working with whatever representation they chose. I expected that students would probably shift to the polar form, $z = re^{i\theta}$ at Question 6, and that they would definitely attempt to do Question 8 using the polar form. In addition, Questions 4 and 5 can be done symbolically or using geometric methods, and Question 3 can be done by cancellation. The results are tabulated in Table 1.

*Classes 1 and 2, Interview 1, Questions 1 through 8*

Simplify the following, i.e., express them as $a + ib$, or as $re^{i\theta}$ (whichever you prefer).

1. $\dfrac{2+i}{2}$         2. $\dfrac{2}{1+i}$         3. $\dfrac{2+2i}{1+i}$

4. $\dfrac{a+ib}{a-ib}$         5. $\dfrac{|a+ib|}{a+ib}$

6. $\dfrac{-2+2i}{(1+i)^3}$         7. $-\dfrac{8}{(\sqrt{3}+i)^6}$

8. $\dfrac{2\cdot(1+i)^4}{(-2\cdot\sin 15^o + 2\cdot i\cos 15^o)^5}$         (Class 1)

8. $\dfrac{2\cdot(1+i)^4}{(-2\cdot\sin\left(\frac{\pi}{12}\right) + 2\cdot i\cos\left(\frac{\pi}{12}\right))^5}$         (Class 2)

In Table 1, the columns refer to question numbers, and the row entries give the numbers of students who attempted to use the $z = x + iy$ or $z = (x, y)$ form by employing the method of realizing the denominator in order to simplify the given expression. It is worth noting here that the term "realizing" the denominator is not in common use in complex analysis, but it should be. The term that is usually used is "rationalizing" the denominator. Multiplying a complex number by its complex conjugate always produces a real number, but does not necessarily produce a rational number, so the accurate term is "realizing" the denominator. For example $(e + i\pi)(e - i\pi) = e^2 + \pi^2$, which is not rational.

TABLE 1. Numbers of Students who Attempted to Simplify Using $z = x + iy$

| Question: | 1 | 2 | 3 | 4 | 5 | 6 | 7 | 8 |
|---|---|---|---|---|---|---|---|---|
| Class 1 ($n$=5) | 5 | 5 | 3 | 3 | 3 | 1 | 1 | 0 |
| Class 2 ($n$=6) | 6 | 6 | 6 | 6 | 6 | 4 | 3 | 1 |

Table 1 shows a clear trend to shift to other methods on the last few questions. One student in Class 2 used the vector form, $z = (x, y)$, for Questions 3 through 8; this was counted as the $z = x + iy$ form for the present purpose. In Class 1 students began to shift to other methods as early as Question 3. Class 2 students did not shift to other methods until Question 6. It is interesting that Question 6 also appeared to be a turning point for Class 1 students. Two Class 1 students shifted methods at Question 6. Thus, all but one Class 1 student had shifted methods by Question 6, whereas it was not until Question 8 that all but one Class 2 students had shifted.

**Shifting Representations: Analysis.** The data can be interpreted to indicate that students in these two classes were not able to access symbolic or geometric methods, in effect, the shifts were almost exclusively between algebraic or Cartesian vector methods and polar vector methods. For example, every student who attempted to simplify Question 5, $\frac{|a+ib|}{a+ib}$, using the $z = x + iy$ or $z = (x, y)$ representation, worked as follows:

$$\frac{|a + ib|}{a + ib} = \frac{\sqrt{a^2 + b^2}(a - ib)}{(a + ib)(a - ib)} = \frac{a\sqrt{a^2 + b^2}}{a^2 + b^2} - \frac{ib\sqrt{a^2 + b^2}}{a^2 + b^2}.$$

Some students made mistakes, and some left $\sqrt{a^2 + b^2}$ as $|a + ib|$, but none cancelled the common factor of $\sqrt{a^2 + b^2}$. This fact probably comes from being trained in high school that the final expression, above, was in "standard form." Two students, one in each class, noted that $\frac{|a+ib|}{a+ib} = \frac{|z|}{z}$, but did not pursue this idea further. This was a good observation, since one can save work:

$$\frac{|a + ib|}{a + ib} = \frac{|z|}{z} = \frac{\bar{z}}{|z|} = \frac{a - ib}{\sqrt{a^2 + b^2}}.$$

A geometric solution is also possible, but no one attempted it. Thus, in these questions, the shift in representations made by all but one student was from algebraic or Cartesian vector form to polar vector form.

The strong differences between the two classes should not be taken too seriously because Class 2 students were asked these question in the third week of classes, whereas Class 1 students were asked between the eighth and twelfth weeks of classes. Analysis of student worksheets revealed some evidence that students became more comfortable with the polar form as the course progressed.

I believe that the common factor $1 + i$ in the numerator and denominator of Question 3, $\frac{2+2i}{1+i}$, suggested to two students that they shift. One appeared to notice that the numerator and denominator of this number were parallel, as vectors, in effect have the same argument. In fact, I would classify the observation of "parallel numbers" as a geometric method of solution, so this is the one instance of a geometric approach that I observed for these questions. For the same two students, the arguments of the numerator and denominator were the key in Question 4, $\frac{a+ib}{a-ib}$, and Question 5, $\frac{|a+ib|}{a+ib}$, as well. Both students quickly realized that the moduli of the numerator and denominator would cancel (in polar form), and concentrated their attention on how to find the final argument (having established that the modulus of the given expression was 1). Students who did not shift until Question 6, $\frac{-2+2i}{(1+i)^3}$, appeared to be motivated by a desire to save work. None of the students who shifted at Question 6 said that the arguments of the numerator and denominator were related or that this fact could be exploited to save work. Instead, they appeared to base their shift on noticing that the polar form was useful for a division problem involving an expression with an exponent. There was a sequence of increasingly sophisticated thinking on these problems that might be described as follows:

*Stage 1.* Multiply the numerator and denominator by the complex conjugate of the denominator. In other words, the "realizing" algorithm is routinely applied.

*Stage 2.* The Cartesian form is too hard; the polar form should be easier for a division problem.

*Stage 3.* The arguments of the numerator and denominator appear to be related (either equal or negatives), so cancellation or polar form should be easier.

*Stage 4.* (For Question 6) The argument of $-1 + i$ is 3 times the argument of $1+i$, but the $1+i$ is cubed, so the arguments will cancel and only the moduli need to be considered. But these are equal, so the answer is 1. Here Stage 3 thinking is combined with a good grasp of the geometry of the plane, so that the exact relationships between the arguments of the numerator and denominator can be visualized and used to assess whether switching to polar form is likely to be helpful.

In Question 7, $-\frac{8}{(\sqrt{3}+i)^6}$, all but one of the students who used the polar form quickly recognized that the argument of the denominator was $\pi$. Nonetheless, I do not think this should be taken as evidence of Stage 4 thinking, because all of the students appeared to stumble onto this result. They appeared to be motivated to use the polar representation because the Cartesian method looked very tedious. Specifically, the exponent 6 in the denominator strongly suggested that Cartesian methods would be difficult, whereas exponents were easily handled in the polar representation. Thus, the data give no indication that students were doing anything more than applying the rules of the polar representation in a routine way, which I have classified as Stage 2 thinking. In the case of Question 7, it was not entirely clear how Stage 4 thinking would manifest itself, in effect, how Stage 4 thinking would be clearly different from Stage 2 thinking. The problem was that in Question 7 the Cartesian method was much harder than the polar form, so the process of deciding to shift to polar form was very brief.

Lesh, Post, and Behr (1987) have identified three criteria of understanding the idea "1/3" that are relevant to the present discussion. These are referred to below as the LPB scheme.

(1) Recognition by the student of an idea embedded in a variety of qualitatively different representational systems.
(2) The ability to flexibly manipulate the idea within given representational systems.
(3) The ability to accurately translate from one representation system to another.

Other authors have noted that the ability to shift representations is an important aspect of understanding representations in mathematics. For example, Dreyfus (1991) identified four stages in the learning process of a general representation in mathematics including integrating representations and flexible switching between them. However, I am not aware of any attempt to analyze the LPB or Dreyfus schemes in terms of reification or APOS. Certainly the question of what it means to understand the criteria identified in these schemes warrants such an analysis.

My data indicated that the students studied could accurately use the algebraic and polar vector forms, although the polar form was hard for some students. And most could accurately represent a complex fraction in either of these forms. Thus, Criteria 1 and 2 in the LPB scheme were met in a limited way, ignoring the geometric aspects of each of the vector representations and ignoring the symbolic representation completely. The main difficulty students had shifting from Cartesian to polar form was not knowing basic trigonometry, such as which arguments were possible when the cosine was $\frac{\sqrt{3}}{2}$ or realizing that $\arctan\left(\frac{b}{a}\right)$ was not single valued and did not uniquely specify the argument of $a + ib$. Thus, Criterion 3 was not as clearly met as Criteria 1 and 2 by most of the students studied.

The students were able to work with the Cartesian form, the polar form, and could even translate reasonably well from one form to the other, but 5 of 11 students interviewed did a question as hard as Question 6, $\frac{-2+2i}{(1+i)^3}$, using Cartesian methods. This indicated that the ability to decide when to shift representations was an important and separate aspect of their conceptions of complex fractions.

TABLE 2. Types of Understanding of the Simplification Problem
in Terms of the Sfard-Linchevski-Kieren and APOS Models.

### Reification

**Process understanding:** Concerned with the details of simplification. Uses the realizing algorithm routinely. Applies the polar form routinely. Multiplies $(a+ib)(a-ib)$ in the denominator (i.e., does not recognize or is unable to apply the fact that the whole point of multiplying by the complex conjugate of the denominator is that the product will be the modulus squared, which is a real number.

**Object understanding:** Uses prior knowledge, such as $(a + ib)(a - ib) = a^2 + b^2$. Looks for a better way to do the problem, such as cancellation. Uses geometry to simplify the algebra where possible. Is able to use the polar form without the formalism, e.g., can simplify an expression such as $(\sqrt{3} + i)^6$ without converting to the form $r^6 e^{6i\theta}$ explicitly. Generally, indicates the ability to see simplification of a complex fraction as achievable in several ways and shows evidence of consideration of more than one method.

### APOS

**Action understanding:** Concerned with the details of the realizing algorithm for each problem. Applies algorithms routinely and apparently without thought of other possibilities (absence of evidence of a thought process leading to a decision to perform the calculations).

**Process understanding:** Shifts to the polar representation at some point, but is still primarily concerned with the details of the simplification. Possibly uses some "short cuts," e.g., uses $(a + ib)(a - ib) = a^2 + b^2$ without calculation. Generally fluent technique, i.e., is able to do the simplification correctly with few obstacles. Shows that he/she is doing the calculations by decision, rather than because "it is what you do in this kind of problem".

**Object understanding:** Uses prior knowledge freely. Indicates consideration of more than one method of solution. Uses geometry to aid calculations. Freely modifies realizing algorithm or polar method to save work.

For each of the first three stages of thinking about shifting representations, I was concerned with identifying evidence of process–object understanding using the Sfard-Linchevski-Kieren reification model. However, where possible, I included a characterization of the simplification problem into action–process–object understanding using the APOS model. General characteristics of process and object understanding (reification) and action, process, and object understanding (interiorization and encapsulation in APOS) are given in Table 2. In the analysis below, I do not distinguished between the two models when discussing an object understanding. I also have not specified which model I mean when referring to a process understanding, even though this concept is not quite the same in the two models. I have referred to an action understanding in the APOS model when needed.

I expected to see all levels of understanding, in either model, at each of the four stages. This is because at each of the four stages there are several concepts to understand, and different levels of understanding are possible depending on whether the reification or the APOS model is used to describe each understanding.

*Stage 1.* Thinking about *shifting representations* in Stage 1 can be characterized as "multiply the numerator and denominator by the complex conjugate of the denominator," referred to as the realizing algorithm, or just the algorithm, in the following discussion. This stage is actually a null stage; students who use this algorithm for putting a complex fraction in standard form are at best demonstrating that they have the ability to manipulate a single representation, as described in Criterion 2 of the LPB scheme, and at worst are showing no ability to judge when to shift. The data contained a range of thinking by students who applied the standard algorithm. Although Table 2 shows rather distinct shift points, student thinking was more complex.

For example, none of the interviewed students did Question 1, $\frac{2+i}{2}$, using the algorithm: they all divided 2 into 2 and 2 into 1 to get $1 + \frac{1}{2}i$. Thus, all of recognized that they did not have to apply the algorithm unless the denominator was complex. One student had doubts about using the algorithm on Questions 4 and 6, but could not think of anything else to do. Another student treated each question as a multiplication problem, doing Question 3 essentially, as follows:

$$\frac{2+2i}{1+i} = (2,2)(1,1)^{-1} = (2,2)\left(\frac{1}{2}, \frac{-1}{2}\right) = (1+1,0) = 2.$$

In this calculation, the realizing algorithm is hidden in the formula for the inverse, so I did not count this as a shift in representations. Nevertheless, this student was thinking about this problem a little differently from one who routinely applied the algorithm. This is because using the formula for the inverse is slightly easier than applying the algorithm, since using the formula for the inverse avoids having to multiply the denominator by its complex conjugate. Almost all students who applied the realizing algorithm in a routine way also knew that the final denominator would be the sum of the squares of the real and imaginary parts of the original denominator, indicating that students were attempting to be as efficient as possible.

Even though most students applied the realizing algorithm to Questions 2 through 5, some wondered whether there was a better way, and almost all incorporated some prior knowledge in the form of other formulas to make the method simpler. Applying prior knowledge to make a procedure more efficient shows that these students were well past having an action understanding (APOS). These students either had, or were well on their way to obtaining, an object understanding of the realizing algorithm for simplification of complex fractions. By an object understanding, I mean the ability to think about the algorithm as a whole, recognize that the algorithm is inefficient or not the best way to do a particular problem, improve the algorithm, or compare the algorithm with other strategies such as cancellation. I believe that an object understanding of the simplification methods for each representation is necessary for efficient shifts to occur because students need to have an overview of the problem under consideration, irrespective of how it was posed, and assess the merits of each representation before choosing a simplification strategy. Thus, Stage 1 thinking ranges from routine application of the simplification algorithm (action understanding in APOS) to full object understanding of the simplification algorithm in Cartesian vector form.

Within the context of the classes studied, it is hard to imagine a student acquiring an object understanding of the Cartesian representation without learning something about the polar representation, so that a student would have moved to Stage 2 thinking. Still, I have included the possibility of an object understanding of the realizing algorithm in Stage 1. Most of the students I studied appeared to either have a full object understanding of the realizing algorithm, or be close to obtaining such an understanding. Certainly, none of the students studied was thinking at the action level of understanding (APOS).

*Stage 2.* I have described Stage 2 thinking of shifting representations as "the Cartesian form is too hard, the polar form should be easier for a division problem." This type of thinking appeared to be the justification of all students who shifted at Question 6, $\frac{-2+2i}{(1+i)^3}$. As already noted, students made a number of minor mistakes when using the polar representation, such as not defining a branch of $\tan^{-1} \theta$, and struggled with basic trigonometry. Thus, their thinking in the polar form seemed to be less advanced than in algebraic or Cartesian representations. They mostly had a *process* understanding of simplifying a complex fraction using the polar representation, since almost every student was preoccupied with the details of the calculation. From the data, I identified a process understanding as: concerned with the details of the calculation, does not look ahead to see if the calculation is likely to work, does not consider possible "short cuts" in the calculation, and does not bring prior knowledge to bear on the present calculation. For students to exhibit an *object* understanding of the polar form, they would have to show recognition of how closely the polar form is related to geometric methods, in effect some sign of what I have identified as Stage 3 and Stage 4 thinking. Some students did use pictures to find the angles in Questions 6 and 7, but I believe this is still Stage 2 thinking in a process mode, because they did not use their pictures to analyze the problem or otherwise obtain the answer. The purpose of their pictures was solely to find the angle using trigonometry. The best evidence of a process understanding is that those students who drew a "trigonometric" picture did not realize that their pictures could be used as part of a geometric solution. For these students, trigonometric pictures were simply devices to find the arguments needed to do the rest of the computation.

Almost all the interviewed students had a process understanding of the polar form; none had a clear object understanding. I did not attempt to further identify action and process (APOS) understandings in Stage 2 because the data are somewhat obscured by the difficulties that students had with trigonometry. For example, a student might have had a good conceptual understanding of the polar form, but have appeared to be applying the rules routinely because he or she was unsure of the trigonometry. Thus, from the data, it would have been unclear whether such a student was exhibiting action or process understanding in the APOS model.

*Stage 3.* Previously, Stage 3 thinking was identified as "the arguments of the numerator and denominator appear to be related (either equal or negatives), so cancellation or polar form should be easier." One student did Question 3, $\frac{2+2i}{1+i}$, using cancellation, and two others began their solutions to Question 5, $\frac{|a+ib|}{a+ib}$, using a symbolic approach. This indicates some ability, even good ability to shift representations wisely, as well as a knowledge of some of the symbolic methods, if one counts cancellation as a symbolic method. However, these students were unable to use symbolic methods, except for cancellation, to finish the questions. Thus, these students appeared to have an awareness of symbolic methods and some knowledge of when to shift, even though they could not actually use symbolic methods. More details on the method of cancellation are reported in Section 3.2.

The only conclusion the limited data on Stage 3 thinking support is that the LPB or Dreyfus criteria are not ordered. A student might have any level of understanding of any of the criteria. For example, the student who showed the most advanced thinking as far as shifting representations was concerned was also the least able to do calculations in a particular representation.

**Shifting Representations: Summary.** The interviewed students were proficient with Cartesian and polar forms, were somewhat less proficient in translating from one to the other, and nearly half did not have good judgment about when to shift to another form. Most students had an object understanding of algebraic and Cartesian vector methods, but only a process understanding of polar methods. Further, the data support the LPB criteria for understanding representations.

**3.2. Cartesian Calculations with $x$ and $y$ Real.** In this and the subsequent two sections, results about student understanding of the Cartesian, polar, and symbolic representations are presented. In Class 3, I investigated students' understanding of several aspects of the $z = x + iy$ and $z = (x, y)$ representations. I was interested in how firmly students understood that these representations treat the complex numbers as a vector space V over a field K, where $V = \Re^2$, and $K = \Re$. Complex multiplication is then just a vector product that defines a field structure.

I report on two questions about Cartesian calculations. The first question was intended to study how well students understood that $x$ and $y$ were real in the $z = x + iy$ and $z = (x, y)$ representations. The second question was intended to study student understanding of simplifying a complex fraction by cancellation of a common factor from the numerator and denominator of the fraction.

**Calculations with $x$ and $y$ Real: Interview Questions and Results.** The following question, Question 2, was asked in the third week of classes among students in Class 3. There were nine respondents. The results are tallied in Table 3.

*Interview 1, Class 3, Question 2*

2. If we use the $z = (x, y)$ representation for complex numbers, which of the following are correct statements? (Circle the correct ones)

| | |
|---|---|
| a. $3 + 6i = (3, 6)$ | e. $3 + 6i = (2 + i, -i + 5)$ |
| b. $3 + 6i = 3i(-i, -2)$ | f. $3 + 6i = (1 + 2i)(1, 2)$ |
| c. $3 + 6i = 3i(-i, -2i)$ | g. $3 + 6i = 3(1, 2)$ |
| d. $3 + 6i = (3, 6i)$ | h. $3 + 6i = (2 + i, 1 + 5i)$ |

In Table 3, the entries in the row denoted "Correct" show the number of students who said the indicated part was correct. There were essentially three reasons given by students for rejecting an expression as incorrect:

(1) The right-hand side of the equation did not simplify to the left-hand side.
(2) The expression was ruled invalid because $x$ or $y$, or both, were imaginary or complex.
(3) The expression was rejected by inspection because the coordinates obviously were not equal.

Also, in Table 3 the entries in the rows denoted "Incorrect - Simplified," "Incorrect - Complex," and "Incorrect - Unmatched," record the number of students who chose these answers, respectively. The "Incorrect - Other" row is the number of students who rejected the expression for some other reason, such as a miscalculation, or they did not know. Note that the only parts of Question 2 that are correct are *a.* and *g.*

TABLE 3. Results of Interview 1, Class 3, Question 2, $n=9$

| Part | a. | b. | c. | d. | e. | f. | g. | h. |
|---|---|---|---|---|---|---|---|---|
| Correct | 9 | 0 | 7 | 0 | 3 | 0 | 9 | 0 |
| Incorrect-Simplified | 0 | 9 | 0 | 3 | 0 | 7 | 0 | 4 |
| Incorrect-Complex | 0 | 0 | 1 | 5 | 3 | 1 | 0 | 3 |
| Incorrect-Unmatched | 0 | 0 | 0 | 0 | 3 | 1 | 0 | 2 |
| Incorrect-Other | 0 | 0 | 1 | 1 | 0 | 0 | 0 | 0 |

The interviewed students demonstrated a wide variety of thinking on Question 2. There was considerable variation in the exact sets of answers selected as correct, as well as in their justifications of these answers. I have included two interview excerpts to illustrate how students thought about some parts of this question.

*Excerpt from Interview 1, Class 3, Student K.*

K.   This one here, ... I think there's something wrong with this [h.] statement where it's got 1 plus $5i$, because, for it to be in the imaginary, complex plane, both these numbers have to be real.

Int.   Hmhm.

K.   Because, otherwise it's just, it has no meaning, because this would have, this only has real numbers here, and this only has real numbers here [Student K is indicating the $x$ and $y$ components of $z = (x, y)$], and you have complex numbers in both.

Int.   OK. So h. is wrong because these numbers [$2 + i$ and $1 + 5i$] aren't real?

K.   Yes.

Int.   OK. So what about some of the other ones, they aren't real in this one either [c.].

K.   But when you multiply them together they end up being real.

Int.   Oh, OK, as long as there's a factor of $i$ out there that makes them real?

K.   Yes, that's fine.

Int.   OK. ...I'm not sure if at this point you're rejecting all of these, just because they are not real, but, when you did this one [part f.], you rewrote that as 1 plus $2i$, and then multiplied, and I'm just curious why you didn't do that with some of the other ones, like you could write this [part h.] as 2 plus $i$, plus 1 plus $5i$ all times $i$.

K.   ...Um, I guess because I also think there is only an $(x, y)$ representation on the complex plane, and this one here, um, because the... , the... get real, in a real system, but this one here is not in the vector form, the $(x, y)$ representation.

Int.   You're saying that you can just throw that [h.] out the window because they [the $x$ and $y$ components] are not real.

K.   Yes.

Int.   You just don't even consider it any further, right?

K.   Yes.

Int.   h., we don't consider any further because they aren't real.

K.   Yes.

*Int.*  But in *f.*, $x$ and $y$ were real, you can put it in the form 1 plus $2i$? Is that correct?

*K.*  Um, yeh, um, I guess, ...I guess you could put it into this form here, ...algebraically, but this [part *h.*] wouldn't have any meaning geometrically. Is what I'm trying to say. Whereas this one here [part *f.*] does have meaning geometrically. You can put this in the form $(1, 2)$ and $(1, 2)$ and both these are real, and you can multiply them together, multiply the modulus and add the angles.

*Int.*  OK, so, I just want to get... on this, so you could put them in here just as an algebraic rule, but what would that mean? Is that what you're saying?

*K.*  Well, I guess you could get some sort of feeling for what that meant geometrically, when you did that, then you could interpret that back into what this function means geometrically.

*Int.*  Hmhm.

*K.*  I'll try that... You still get a wrong answer. Is that right?

*Int.*  OK.

*K.*  And now if you have, so you end up with, negative 3, one, two, three, and positive 2. Yeh, I guess, I guess you just have some,

*Int.*  OK, so now you're saying this fails because you don't get the right number?

*K.*  Yes.

As can be seen, Student $K$ vacillated between believing that $x$ and $y$ had to be real and allowing them not to be real under certain circumstances. The next interview excerpt gives further evidence that students did not understand whether the scalars in the vector representation were real or complex. This following excerpt with Student $R$ refers to *f.* $3 + 6i = (1 + 2i)(1, 2)$.

*Excerpt from Interview 1, Class 3, Student R.*

*R.*  ...Ah, same goes in here, ...um , if you take a complex number and multiply, and you multiply by, I suppose it's like multiplying, like multiplying a vector by a scalar, except the scalar is a complex number, so then for f. when you multiply your complex scalar into your vector, which is made-up of real numbers, you end up with complex parts in the $-$ , $x$ and $y$ $-$

*Int.*  OK.

*R.*  $-$ notation.

*Int.*  Are you concluding in f., you're rejecting it because you're multiplying it by a complex number?

*R.*  By a complex number.

*Int.*  Not a real number.

*R.*  Yeh. A real number by a complex number is going to be a complex number.

*Int.*  If I rigged these [the values in the question] so that the product was real, would it be OK?

*R.*  As long as it was $(3, 6)$, yeh.

*Int.*  OK,

*R.*  ...um, yeh, then that would work.

Student $R$ did not object to multiplying by a complex scalar, but rather argued that the right-hand side of $f$. could not be equal to the left-hand side because the right side of the equation, after multiplication, would have complex numbers in the two slots of $z = (x, y)$. As a result, it could not be equal to $(3, 6)$. This provided further evidence that students did not fully understand that $x$ and $y$ had to be real in the Cartesian vector form.

**Calculations with $x$ and $y$ Real: Analysis.** It is possible to look at Question 2 in several ways, as a question about manipulative skills or as a question about the structure of a vector space. I considered this to be a question primarily of form. Did students understand the $z = (x, y)$ representation, in particular, did they understand that the scalars in this representation have to be real? That is, the numbers in the ordered pair are real and any scalars outside the parentheses are real. Without an understanding of the vector space structure, students might still understand the structure of the form. I also looked for signs that the interviewed students might have an object understanding of this representation.

I found almost no mathematics education literature on form in the sense I asked. Most research seems to be concerned with student understanding of the significance of the form in question (e.g., limits), or how to improve student understanding of the significance of the form (Confrey & Smith, 1994; Dubinsky, et al., 1994; Thompson, 1994; White & Michelmore, 1996; Williams, 1991). Thus, I devised my own classification of how students think about form. Using the data on Question 2, I identified four general stages of thinking about form.

    *Stage 1. A mechanical understanding.* Students who thought along this line were concerned with whether or not the right-hand side of each expression could be translated, using whatever knowledge they could bring to bear on the problem, and simplified to the left-hand side. These students were unconcerned about questions of form or the nature of the $z = (x, y)$ representation. Thus, this type of understanding corresponds to an action understanding in the APOS model.

    *Stage 2. Simplification to $(3, 6)$.* Students who tried to simplify the right-hand side to $(3, 6)$ showed slightly more insight than a mechanical understanding. These students appeared to bring outside knowledge to bear on the $z = (x, y)$ representation, as opposed to doing the problem however they could. Whereas, Stage 1 thinking ignores the form of the $z = (x, y)$ representation, Stage 2 thinking recognizes that a new form is being introduced, as shown by an attempt to put answers in the $z = (x, y)$ form. In the APOS model, this stage corresponds roughly to the process of *interiorization*, that is, a student in this stage is acquiring a process understanding of the formal structure of the $z = (x, y)$ representation.

    *Stage 3. Process understanding.* In this stage, students recognized that most expressions on the right-hand side were not correct (except for $a$. and $g$.) and attempted to correct them. These students were clearly struggling with what appeared to be incorrect usage of the $z = (x, y)$ representation. But rather than dismiss such expressions, they tried to make sense of them. This showed a more advanced understanding than Stages 1 or 2, but also showed that these students did not understand that mathematical form cannot simply be altered at will. There are precise rules that need to be followed. These student did not have an object understanding of the formalism, but they had acquired a working knowledge, so Stage 3 corresponds to a process understanding in the APOS model.

*Stage 4. Encapsulation.* Students in this stage realized the form was not correct, so the corresponding expressions (except for *a.* and *g.*) were not correct. The one student who exhibited Stage 4 thinking throughout Question 2, rejected parts *e.*, *f.*, and *h.* without any calculation or further consideration because of the *i*'s appearing in these expressions. Other students showed Stage 4 thinking in their answers to one or two of the parts. The one student who rejected several parts because of improper form likely had an object understanding (in the APOS model) of the formalism of the $z = (x, y)$ representation, whereas the students who rejected just one or two parts because of improper form were likely in the process of *encapsulation.*

These stages can be used to analyze the results of Question 2. Numbers of correct and incorrect answers for all parts are given in Table 3, but to see examples of the four stages of thinking, one needs to look at interview transcripts and worksheets. All nine students first determined that *b.* $3 + 6i = 3i(-i, -2)$, was incorrect by simplifying the right-hand side to get various answers not equal to $3 + 6i$. None of the students objected to the multiple occurrences of $i$ on the right-hand side of the equation when they first attempted it (one student reconsidered it later). Thus, all but one student thought about this as a problem of simplification, with apparently almost no understanding of the structure of the $z = (x, y)$ form. However, seven of the nine students interviewed did simplify to the $z = (x, y)$ form (the coordinate pair $(3, -6i)$ was typical), thereby exhibiting Stage 2 thinking.

The complexity of Question 2 affected the level of thinking. Seven or eight students used Stage 2 or 3 thinking for parts *a.*, *b.*, *c.*, and *d.*, while three students used Stage 1 thinking for parts *e.* and *f.* Presumably this is best explained as reductionism: faced with complexities in unfamiliar notation, some students returned to the notation they knew best.

Another variation of Stage 2 thinking can be seen in the second interview excerpt above. Student $R$ did not object to multiplying by a "complex scalar" when the result was $(3, 6)$. However, in the vector representation of complex numbers there are no "complex scalars," the complex numbers are vectors.

A good example of Stage 3 thinking appeared in the first interview excerpt above. Student $K$ switched back and forth and was concerned with having a geometric interpretation of the misuse of the $z = (x, y)$ form. Nevertheless, $K$ did not quite reject the misuse of the $z = (x, y)$ form altogether. Rather $K$ tried to find some reasonable interpretation of the representation with $x$ and $y$ complex and also tried to interpret multiplication by a complex scalar.

Another example of what might be Stage 3 thinking is the case of part *e.* $3 + 6i = (2 + i, -i + 5)$. Three students decided that *e.* was incorrect because $2 + i$ could not equal 3, and left it at that. These students did not reject $2 + i$ outright because of the $i$, as Stage 4 thinking would require, but appeared to have changed their thinking from previous parts, where they had attempted to simplify the right-hand side. The data do not indicate exactly why students chose a comparison approach on part *e.*, when they had simplified the previous parts. It might be because they simply did not look deeply enough to find a way to simplify the right-hand side to $3 + 6i$. In fact, I had expected some students to say the right-hand side was equal to $2 + i + i(-i) + 5i = 3 + 6i$, and four students did do this.

In addition to the one student who did all parts of Question 2 using Stage 4 thinking, some evidence of Stage 4 thinking can be seen in answers to part $d$. $3 + 6i = (3, 6i)$. All of the students who rejected $d$. by inspection did so specifically because of the $i$ in $6i$ on the right-hand side, not because $6i$ is not equal to 6. This is evidence of Stage 4 thinking. However, perhaps these students just had not thought of some way to simplify $(3, 6i)$. Indeed, only one student rejected $e$. because of the multiple occurrences of $i$ on the right-hand side. The rest either said $e$. was correct or rejected it because $2 + i$ is not equal to 3. Thus, it is not clear whether the answers to part $d$. show Stage 4 thinking.

It is interesting to look a little further at parts $d$. and $e$. Seven out of the nine students changed their thinking between parts $d$. and $e$. Six students who rejected part $d$. because of the $i$ in $6i$ on the right-hand side said $e$. was incorrect because $2 + i$ was not equal to 3 and there was no way to make it equal, say by multiplying by a scalar. One student said part $d$. was wrong by calculation, but said part $e$. was wrong because of the $i$ in $2 + i$. I think that the six students who went from Stage 4 thinking to Stage 2 or 1 between parts $d$. and $e$. probably did so because of the new complications introduced in part $e$. In part $e$., $x$ and $y$ are complex numbers, rather than real or pure imaginary. Moreover, there is no scalar factor. Apparently, there is a clear conceptual leap between pure imaginary and complex components even at third year university level. I think Question 2 could be modified in several ways to gain considerably more understanding into how students are thinking about such ideas. For example, the order of the parts could be changed, or say, a question like part $e$. with a scalar factor could be added.

My analysis ignored whether the interviewed students' answers to Question 2 might have been confounded by any difficulties they might have had with the equal sign. It is known that students sometimes treat "=" as though it is the word "is", or treat "=" as though it was a vertical line between debit and credit ledgers in bookkeeping. I ignored this possibility because I saw no evidence in the data that interviewed students did not correctly understand the equal sign. That is, it appeared that all nine of the interviewed students understood that the task was to see whether the two sides of the equation were exactly the same complex number. There were, of course, various levels of success of integrating this understanding (of the equal sign) with other aspects of Question 2, such as how to handle the multiple occurrences of $i$. Nevertheless, it would be interesting to investigate the matter of the equal sign further, or possibly revise Question 2 so as to exclude the equal sign.

**Calculations with $x$ and $y$ Real: Conclusion.** I analyzed the data collected on Cartesian calculations, with $x$ and $y$ real, as a mathematical form. I identified four stages of thinking about this particular form that may be generalizable to other mathematical forms. In any case, this particular question has opened a door to questions of mathematical form that warrant further research.

**3.3. Cartesian Calculations with Cancellation.** The second type of questions I asked about Cartesian calculations, that shed light on students' understanding of the $z = x + iy$ representation, involved the possibility of cancelling a common factor from the numerator and denominator of a complex fraction.

**Calculations with Cancellation: Interview Questions and Results.** The questions were as shown on the next page. In these questions I was interested in whether or not students would realize that they could factor out a real number from the numerator and then cancel a common complex factor from the numerator and denominator. For example, I wanted to see if students would simplify $\frac{2+2i}{1+i}$ as follows: $\frac{2+2i}{1+i} = \frac{2(1+i)}{1+i} = 2$. Put another way, I was interested in determining if students realized that $\frac{a+ib}{a+ib} = 1$.

*Classes 1 and 2, Interview 1, Question 3*

Express the following as $a + bi$. $\qquad \dfrac{2+2i}{1+i}$

*Class 3, Interview 1, Question 3*

a. Put the following into $a + bi$ form.

   i) $\dfrac{2+4i}{1+2i}$ $\qquad\qquad$ ii) $\dfrac{-6+3i}{1+2i}$

b. Three students did the following problem three different ways. Which of them are correct? (Circle the correct ones)

   i) $\dfrac{6-9i}{2-3i} = \dfrac{6-9i}{2-3i} \cdot \dfrac{2+3i}{2+3i} = \dfrac{12+27}{4+9} = \dfrac{39}{13} = 3$

   ii) $\dfrac{6-9i}{2-3i} = \dfrac{6-9i}{2-3i} \cdot \dfrac{2+3i}{2+3i} = \dfrac{12-11i+9i+27}{4+9} = \dfrac{39-2i}{13}$

   iii) $\dfrac{6-9i}{2-3i} = \dfrac{3(2-3i)}{2-3i} = 3$

c. Put the following into $a + bi$ form.

   i) $\dfrac{-4+2i}{2-i}$ $\qquad\qquad$ ii) $\dfrac{6+2i}{-3-i}$

In addition, the interview questions seemed to bring forward student confusion about basic facts. To see this, I give two interview excerpts from Class 3, taken from the discussion of 3.b.iii) $\frac{6-9i}{2-3i} = \frac{3(2-3i)}{2-3i} = 3$. In the first excerpt, Student $L$ was uncertain about whether or not it was possible to cancel a complex factor from the numerator and denominator of a complex fraction. In the second, Student $C$ was not sure what the scalars were in the $z = x + iy$ form of complex numbers.

*Excerpt from Interview 1, Class 3, Student L, Question 3.b.iii.*

*L.* $\quad$ Alright, so part 3?... OK, same quotient of complex numbers, ...so I look at the numerator, and they have factored the numerator, they've taken a 3 out of it

*Int.* $\quad$ Hmhm.

*L.* $\quad$ And they've done that correctly. So I see that, this 2 minus $3i$ cancels in the numerator and the denominator, to get the right answer of 3. So part 3 is correct as well.

*Int.* $\quad$ OK, summarizing, in [part] 3?

*L.* $\quad$ Summarizing in [part] 3, it was not necessary to multiply by the complex conjugate of the numerator, because we could factor the numerator, into some, and it cancelled the denominator.

*Int.* $\quad$ And that's, that's allowed? To cancel a complex number from top and bottom?

*L.* $\quad$ ...Um, ...yes. A complex number divided by itself is 1.

*Int.*  OK, are you sure?

*L.*  No, I'd have to work out a general form $a + ib$, and then...

*Int.*  OK.

*L.*  Intuitively, I'd say yes,

*Int.*  OK, so you,

*L.*  I would need to, I would need to check that, in a general form.

*Int.*  OK, you can check it on the side there if you want.

(Student $L$ works to the side)

*L.*  ...So, I would say that part 3 is correct. You can cancel the complex numbers.

During this section of the interview, L attempted to supply a proof of the cancellation rule. Student $L$'s "proof" had errors, though $L$ thought it correct. Thus, it is apparent from this excerpt that $L$ was not entirely sure of the basic rules of Cartesian calculations and cancellation, in particular. Having verified the rule (at least $L$ thought so), Student $L$'s confidence improved, as shown by the last line. In the next interview excerpt, there is more evidence that students were not sure what the scalars are in the $z = x + iy$ form of complex numbers.

*Excerpt from Interview 1, Class 3, Student C, Question 3.b.iii.*

*C.*  And this is not correct. Oh, they didn't, they um, they factored, and ...this is also correct.

*Int.*  OK. So in this one, what did they do again?

*C.*  They factored the numerator, they just took the scalar value of 3, and then multiplied, they're left with $2 - 3i$, if you multiply through you get $6 - 9i$, and then you can just cancel these $[2 - 3i]$ out, and you get the answer.

*Int.*  OK. Can I just ask you, you referred to 3 as a scalar, can you elaborate on what you mean by that?

*C.*  Oh, ah, this is just a real number, You can just multiply, 6 and 9 have a common factor of 3.

*Int.*  Hmhm.

*C.*  And then you pull that out of the complex, the rectangular expression for, ... a lot like vectors, so this is like multiplying a vector by 3.

*Int.*  OK.

*C.*  And that's legal as far as I know. [laughs]

*Int.*  Can you tell me what the allowed scalars are? Are they just real numbers or complex numbers?

*C.*  No, [This "no" refers to: "Are they just real numbers?"] you should be able to factor, you should be able to factor even complex numbers. ... both the imaginary and the real component.

*Int.*  OK. So, if I found some way to factor this or another problem as two complex numbers, and then cancel that would be OK?

*C.*  Yes...

From this excerpt, one can see that $C$ included complex numbers amongst the available scalars in the $z = x + iy$ form.

**Cancellation: Analysis.** The results for Question 3, Classes 1 and 2 and Question 3, Class 3 are summarized in Table 4. The entries in Table 4 give the

TABLE 4. Numbers of Students Who Used Cancellation to Simplify Various Questions

| Question | Yes/Total |
|---|---|
| Class 1, Question 3 | 2/5 |
| Class 2, Question 3 | 0/6 |
| Class 3, Question 3.a. i | 1/9 |
| Class 3, Question 3.a.ii | 0/9 |
| Class 3, Question 3.c.i | 3/9 |
| Class 3, Question 3.c.ii | 5/9 |
| Correct use of cancellation | 6/6 |

numbers of students who simplified by cancellation (rather than realizing the denominator) followed, after the slash, by the total who attempted the question. Students from Class 3 were all able to recognize the correct uses of realizing the denominator and cancellation. The answers given for Question 3.b were the same for all nine students: i) and iii) were correct, and ii) was incorrect.

According to Student $L$'s written work on the Questions worksheet, and the section of the interview omitted above, $L$ had "proved" the cancellation rule for complex quotients by reasoning as follows:

$$\frac{a+ib}{a+ib} = \frac{a+ib}{a+ib} \cdot \frac{a-ib}{a-ib} = \frac{a^2+b^2}{a^2+b^2} = 1.$$

This is not a valid proof because one needs to know that $\frac{a-ib}{a-ib} = 1$, which is what one is trying to prove. Of course, at third year university level, proofs of basic algebraic facts, such as $\frac{a+ib}{a+ib} = 1$ can be quite challenging. One cannot take Student $L$'s inability to supply a proof as strong evidence that $L$ did not understand the cancellation rule. The main evidence that $L$ was not sure is that Student $L$ said during the interview, "I'd have to work out a general form...Intuitively, I'd say yes."

If students are alerted to the possibility of cancellation they will use it. This can be seen in the increased use of this method among participants from Class 3 between Questions 3.a and 3.c. One student commented that cancellation did not seem very likely in most cases, so one would not normally look for it. However, this student decided that cancellation would be a good idea when it was available.

As they worked through these questions, most students revealed various misunderstandings. Student $C$'s interview gave an indication of how student confusion about vectors and scalars may appear in a practical problem. $C$'s contention that one can cancel a common complex factor even if there are other complex factors in the numerator or denominator was certainly correct. However it appeared that $C$ did not distinguish between "complex scalars" and real scalars, as evidenced by the response to the question: "Can you tell me what the allowed scalars are?" Though not a serious problem, it does indicate some confusion about the formalism.

In addition, other students speculated that the cancellation method would save time and be more accurate. All six of the students in Class 3 whom I questioned closely on the validity of the cancellation method were confident that the method was correct, although just two students attempted to prove it for themselves.

It appears that students are not learning the cancellation rule. Almost no one in Classes 1 and 2 attempted to simplify $\frac{2+2i}{1+i}$ by cancellation. However, once they were instructed, they found it useful and time saving. There is no conceptual problem here. Students simply are not learning that they can cancel and mostly do not think of it themselves. Teaching this might involve more than simply telling them. One student emphatically declared that cancellation definitely would be what the student would have done in Question 3.b.iii, but this same student did both Question 3.c.i and Question 3.c.ii by realizing the denominator.

**3.4. Polar Calculations.** Most of the questions I asked about the polar representation were intended to study other topics, such as continuity and analyticity, or shifting between Cartesian vectors and polar vectors (see Section 3.1). Nevertheless, I did ask two sets of questions intended to directly study the students' understanding of the polar representation.

**Polar Calculations: Interview Questions and Results.** The first set of questions comes from the first interview with Class 3 students. I asked three questions to test their ability to do basic computations in polar form.

> *Class 3, Interview 1, Question 4.b*
>   4.b)  Simplify the following given that $z = re^{i\theta}$.
>
>   i) $3e^{i\pi} \cdot 5e^{i\frac{\pi}{2}}$          ii) $3e^{i\pi} \div 5e^{i\frac{\pi}{2}}$          iii) $3e^{i\pi} + 5e^{i\frac{\pi}{2}}$

The second set of questions was discussed in Section 3.1 in regard to shifting representations. Class 1 and Class 2 students were asked a series of questions designed to see how complicated a question had to be before they would switch from using Cartesian methods to polar methods. It is interesting, however, to study what difficulties students had with the polar form in those cases where they used it, on Questions 6, 7, 8 (see Section 3.1). Note that there was a slight difference in Question 8 between the two classes: in Class 1 the argument in the denominator was written as $15°$, rather than $\frac{\pi}{12}$.

**Polar Calculations: Analysis.** The overall picture that emerged from studying the results of these two sets of questions was that the argument is the problem. Prior to studying complex analysis, most students may have been able to ignore questions of multivaluedness. No doubt, many students view the problems presented by the multivaluedness of the polar form as a nuisance and simply try to disregard them, rather than embracing multivaluedness as an essential and interesting aspect of the theory. In any case, I chose to analyze these questions in terms of the errors students made and their understanding of multivaluedness.

In the interviews with Class 3 about Question 4.b.iii, four students tried to put the answer in polar form. These students found the correct answer $(-3 + 5i)$ and then tried to convert $-3 + 5i = \sqrt{34}e^{i\left(\mathrm{Arctan}\left(\frac{-5}{3}\right)+\pi\right)}$. All four students got the modulus correct but made errors in the argument. For example, they used $\arctan x$ instead of $\mathrm{Arctan}\,x$, dropped the minus sign in front of $-5/3$, or dropped the phase angle $\pi$. These kinds of mistakes in themselves are not enough to draw conclusions of a general nature, but they are part of an overall picture. The difference between $\arctan x$ and $\mathrm{Arctan}\,x$ is very significant, and complex analysis is perhaps the first undergraduate course in which failure to understand this difference would lead to significant errors.

TABLE 5. Number of Attempts and Number of Correct Simplifications for Classes 1 and 2, Interview 1, Questions 6, 7, and 8

| Question | Class 1 Attempts | Class 1 Correct | Class 2 Attempts | Class 2 Correct |
|----------|------------------|-----------------|------------------|-----------------|
| 6 | 4 | 3 | 3 | 2 |
| 7 | 3 | 1 | 3 | 3 |
| 8 | 3 | 0 | 5 | 1 |

The results from the questions asked of Classes 1 and 2 are summarized in Table 5. Almost all of the difficulty with these questions was in computing the arguments. One student acknowledged that not knowing how to find the arguments, apparently not realizing (or forgetting) that one could use basic trigonometry.

Of the two incorrect answers to Question 7 using polar methods, there were two errors in the modulus and one student also miscalculated the argument. In addition, one of the students who is counted as having answered correctly, did not simplify $e^{i\pi}$ to $-1$. Nevertheless, the primary difficulty students had was finding the argument. Two used their calculators.

Finally, the results for Question 8 were certainly the worst, with only one correct answer out of eight attempts. Although some errors were arithmetic, several fit into the theme for this section. For example, one student was completely confused by $15°$: it did not occur to the student to convert degrees to radians.

Two students found Question 8 too complicated, because of the arguments, and gave up. Another student applied DeMoivre's theorem to the denominator (it is not correct to do this directly, but in this case it happens to work), without justification, but could not make further progress. This student explained that the purpose was to convert sine into cosine and vice versa using the addition of angles identities for sine and cosine, so that Euler's theorem could be applied. It clearly did not occur to this student to factor out $i$, to achieve the same objective.

Thus, there were several types of errors and difficulties on Question 8. Students could not find the argument or modulus correctly for one or more expressions, misused DeMoivre's theorem, did not realize that they could factor out $i$ in the denominator, or were intimidated by a question that looked hard, but is straightforward if one breaks it into parts. The main difficulty students appeared to have in finding the argument was in not being comfortable or fluent with $\arctan x$. In particular, they were not fluent with its multivalued qualities.

**Polar Calculations: Conclusion.** The main difficulties that students had with the polar representation were not having facility with basic trigonometry, particularly $\arctan x$; little or no grasp of its multivaluedness; and a tendency to assume that $\cos\theta$ and $\sin\theta$ can be interchanged. Although some students had trouble calculating the modulus correctly, they made mostly arithmetic mistakes. There did not appear to be any conceptual difficulties with the modulus.

**3.5. Symbolic Calculations.** This section analyzes data about how well students did calculations involving $z$ by using just the properties of $|z|$, $\bar{z}$, $\mathrm{Re}(z)$, $\mathrm{Im}(z)$, and the field properties of the complex numbers. The questions in this section can be done very readily without substituting $z = x + iy$, or $z = re^{i\theta}$, or without using a geometric approach. Obtaining some skill at symbolic methods is very useful for more advanced subjects, such as power series and contour integrals. The main result is that students were not obtaining any fluency in symbolic methods. In most cases, they did not attempt a symbolic solution.

**Symbolic Calculations: Interview Questions and Results.** I asked a number of direct questions, and questions that had some other primary focus, to look for evidence that students were obtaining symbolic skill. I have summarized the results in Table 6. The questions were as follows.

*Class 2, Interview 1, Question 11*

11. Show that if $|z| = 1$, then $|1 - z\bar{w}| = |z - w|$.

*Class 2, Interview 2, Question 1*

1. Find all solutions to $|z - i| = |z + 1|$.

*Class 2, Interview 3, Question 10*

10. If $|z| = 2$, show that $|2 - \bar{z}\bar{w}| = |zw - 2|$, for all $w$.

*Class 2, Interview 5, Question 3*

3. Show that for $\gamma(t) = e^{it}$ , $0 \le t \le 2\pi$, $\int_{\gamma} \left( \frac{1}{z} \bar{z} \right) dz$ is zero.

Is $\frac{1}{z} - \bar{z}$ analytic?

*Class 2, Interview 5, Question 8*

8. If $\gamma$ is a circle of radius 2 centered at $i$, what is $\int_{\gamma} \frac{|z + i|}{|\bar{z} - i|} dz$ ?

*Class 3, Interview 1, Question 7.a*

7.a. Find all solutions of the following equations:

i) $z^2 + 2iz - 1 = 0$     ii) $z^2 + (1 + i)z + i = 0$ .

*Class 3, Interview 2, Question 4*

4.a. For which $z$ is $|\bar{z}| = |z|$ ? Explain.

4.b. For which $z$ is $|\bar{z} + 1| = |z + 1|$ ? Explain.

4.c. For which $z$ is $|\bar{z} - i| = |z + i|$ ?

4.d. For which $z$ is $|iz + 1| = |\bar{z} + i|$ ?

4.d. Where is $f(z) = \frac{|\bar{z} - 1|}{|2z - 2|}$ analytic?

In Table 6, questions are identified by the essence of their content rather than by class and interview number. The column headed "Symbolic Correct" gives the total number of correct answers using symbolic methods. The column "Symbolic Attempt" gives the number of students who attempted to use symbolic methods but gave up, or tried another method. The "Other Correct" column gives the number of students who got the question correct using some other method; usually substitution of Cartesian or polar vector form, or occasionally geometric methods.

**Symbolic Calculations: Analysis.** Examination of Table 6 shows that students were acquiring very little proficiency in symbolic techniques. The results suggest that students did not have symbolic methods in their repertoire, as opposed to the possibility that they simply preferred other methods, because reducing many of the items was dramatically simpler using symbolic methods.

TABLE 6. Attempted and Correct Responses to Questions Asked
to Test Knowledge of Symbolic Methods in Classes 2 and 3

| Question Content | Attempted Correct | Symbolic Attempt | Symbolic Correct | Other Correct |
|---|---|---|---|---|
| $\lvert 1 - z\bar{w} \rvert = \lvert z - w \rvert$ | 4 | 0 | 1 | 0 |
| $\lvert z - i \rvert = \lvert z + 1 \rvert$ | 5 | 0 | 1 | 3 |
| $\lvert 2 - z\bar{w} \rvert = \lvert zw - 2 \rvert$ | 1 | 0 | 0 | 1 |
| $\displaystyle\int_\gamma \left( \frac{1}{z} - \bar{z} \right) dz$ | 4 | 0 | 0 | 4 |
| $\displaystyle\int_\gamma \frac{\lvert z + i \rvert}{\lvert \bar{z} - i \rvert} dz$ | 4 | 0 | 0 | 0 |
| $z^2 + 2iz - 1 = 0$ | 7 | 6 | 0 | 1 |
| $z^2 + (1 + i)z + i = 0$ | 4 | 2 | 2 | 0 |
| $\lvert \bar{z} \rvert = \lvert z \rvert$ | 6 | 1 | 0 | 5 |
| $\lvert \bar{z} + 1 \rvert = \lvert z + 1 \rvert$ | 5 | 1 | 1 | 3 |
| $\lvert \bar{z} - i \rvert = \lvert z + i \rvert$ | 4 | 1 | 0 | 3 |
| $\lvert iz + 1 \rvert = \lvert \bar{z} + i \rvert$ | 3 | 0 | 1 | 2 |
| $f(z) = \dfrac{\lvert \bar{z} - 1 \rvert}{\lvert 2z - 2 \rvert}$ | 1 | 0 | 0 | 0 |

For example, to show that $f(z) = \frac{\lvert \bar{z}-1 \rvert}{\lvert 2z-2 \rvert}$ is analytic everywhere except at $z = 1$ using the $z = x + iy$ representation requires calculation of the modulus of the numerator and denominator, and then further simplification. Using symbolic methods, one can see, almost by inspection, that $f(z) = 1/2$ if $z$ is not equal to 1. In addition, sometimes students would ask if there were an easier way and be very impressed by the symbolic method. It may be that many students at third-year level would prefer to use symbolic methods if they knew about them.

Students did use symbolic methods when asked to solve quadratic equations, but this is not an especially good indicator because of the tendency to "think real, do complex." If students are thinking of $z$ as just the $x$ of real polynomial theory, then it is not particularly encouraging to see them readily apply the quadratic formula. During interviews, I did not explore this point, so it is unclear whether the results of questions about quadratic equations indicate fluency in symbolic methods.

Students did not seem to have learned symbolic methods. For example, only one student out of six tried to use symbolic methods to solve an equation as elementary as $\lvert \bar{z} \rvert = \lvert z \rvert$. Also, I tutored one student who was not aware that symbolic methods existed. I examined the textbooks and lecture notes for the three classes to see how much instruction students were receiving in symbolic approaches. The question was: Is the topic so difficult that even very good students are not getting it, or is there too little instruction or emphasis?

In Class 1 the textbook (Priestley, 1990) has two sections, 1.3 and 1.4, on pages 3 and 4, that contain instruction on symbolic methods. Section 1.3 describes properties of complex conjugates, and Section 1.4 uses symbolic methods to prove three inequalities. There are no other direct examples, and just two problems, Numbers 5 and 6 on pages 11 and 12. These are complicated problems that were assigned for homework, but that can be done by substituting one of the vector forms. Priestley is quite terse, and perhaps too difficult for an introductory course at the

300 level. However, in contrast to the symbolic treatment, there are several pages of discussion of elementary geometric results, such as lines, circles, and inequalities. In addition, although logical, it is probably not good pedagogy to present symbolic methods so early in the text, since students are unlikely to have acquired an object understanding of the algebraic or vector representations so early in the course.

The instructor in Class 1 proposed to prove that $\overline{z_1 z_2} = \bar{z}_1 \, \bar{z}_2$, and asked for help from the class. Substituting $z = x + iy$ was suggested, and after commenting that a polar substitution would be better for multiplication, the instructor proceeded with the proof using the Cartesian substitution. To prove the triangle inequality in Section 1.4 of Priestley, the instructor used geometric arguments. For homework, students were assigned Problems 4 and 5, both of which can be done most conveniently with the help of symbolic methods. However, I found no examples of symbolic calculations done in class.

In Class 2, Sections 2, 3, 4 of the textbook (Churchill & Brown, 1990) cover the basic facts about complex numbers under the headings "Algebraic Properties," "Geometric Interpretation," and "Triangle Inequality." These sections establish many useful facts that can be used for symbolic manipulations. Although Churchill and Brown do not actually prove any of the results using symbolic methods, sketches are given that indicate how to prove several facts, with details left to the exercises. For example, Churchill and Brown suggest that one can use $\frac{1}{z} = z^{-1}$ to show that $(z_1 z_2)^{-1} = z_1^{-1} z_2^{-1}$. Some of the details are the subject of Exercise 11 on page 6. After establishing all these results, either directly or with the help of exercises, there are a number of applications of using symbolic methods, for example, Exercise 16 on page 6, and Exercises 4, 7 through 13, 15, and 16 on page 11.

Class 3 students got slightly more instruction in geometric and symbolic techniques. The instructor and textbook (Saff & Snider, 1993) established basic facts using Cartesian substitution, but the Class 3 instructor did two elementary proofs using symbolic methods: i) $\frac{1}{z} = \frac{z}{|z|^2}$, and ii) $|zw| = |z||w|$. Saff and Snider have four exercises that can be done with symbolic methods: 13, 14, 15, and 16 on pages 11 to 12. In addition, some parts of Exercise 7, page 11 could be done symbolically, but in examples, Saff and Snider use a geometric approach, so it is unlikely that many students would use symbolic methods on Exercise 7.

**Symbolic Calculations: Conclusion.** Students in all three classes received some instruction on symbolic methods, but not nearly enough for these methods to be accessible to them. This was a serious shortcoming of the courses studied, since symbolic methods are very useful, and I believe that practice with symbolic methods would improve students' skills with other methods, such as the Cartesian or polar methods. Pedagogically, it might be a good idea to defer instruction on symbolic methods until the third to fifth week of the course to give students a chance to gain fluency with the other representations.

## 4. Summary and Conclusions

In this section I give brief overviews of the results on shifting representation and computation methods presented Section 3. This followed by suggestions for teaching and directions for future research.

**4.1. Shifting representations.** The students studied had a good grasp of the Cartesian vector forms and polar forms, but slightly less skill translating from one form to the other. At least half the students did not have good judgment about when to shift from one form to another. All shifting was between the algebraic extension and polar vector forms, as students had almost no skill with other forms, such as the symbolic form.

The data support the three LPB criteria identified for understanding single representations and translating between representations. In addition, the data support inclusion of a further criterion requiring good judgment of when to shift from one representation to another. The expanded criteria can be briefly expressed as:

(1) Ability to use a single representation.
(2) Ability to represent a problem in different representations.
(3) Ability to translate between representations.
(4) Ability to judge when to shift from one representation to another.

Finally, most students had an object understanding of the algebraic and Cartesian vector forms, but only a process understanding of the polar vector form. The evidence for this conclusion is based largely on the fact that most students applied prior knowledge when using the simplification algorithm to simplify complex fractions; however, further research is needed. In particular, it would be useful to have more information about how an object understanding of each representation manifests itself when working with problems at an elementary level.

**4.2. Basic Facts and Calculations.** Many students had a difficult time consistently applying the fact that, in the Cartesian representations, $x$ and $y$ are real. In addition, several of the interviewed students were unsure whether the usual cancellation rule for simplifying a fraction applied for the complex numbers.

Students had two primary difficulties in applying polar methods: little flexibility with basic trigonometry, particularly $\arctan x$, and little or no understanding of multivaluedness. Generally, the modulus did not present difficulties.

Almost all interviewed students did not have access to symbolic methods, with the sole exception that they were able to solve quadratic equations, where they may well have been thinking real, rather than using a symbolic approach. I believe that this is a reasonable reflection of the way they had been instructed both in classes and textbooks. Symbolic methods were simply not emphasized in their classes or textbooks. I think this was a mistake because substitution of a vector form is a method of last resort for anybody experienced in complex analysis.

**4.3. Suggestions for Teaching.** First of all, especially for those readers who have not taught an introductory course in complex analysis, it needs to be emphasized that this course (at least the versions I observed) is a mad rush to get to residue theory. Many of the topics covered, such as line integrals, Cauchy's theorem, power series and Laurent series support this goal. The only application of residue theory at this level is to solve difficult, real, definite integrals. Since there are now powerful computer algebra systems that will easily evaluate such integrals, is the goal of covering residue theory still appropriate? For example, *Maple 7* will easily calculate all of the integrals given in Problem 3, page 161, of Alhfors (1979), at least if specific values of parameters are selected. This question arose during my data collection as I watched students struggle with basic calculations. A student who needs an advanced application of residue theory would probably have to take an advanced course in any event. Furthermore, Class 1 had time for just two examples using residue theory on the last day of classes, and Class 2 did not reach residue theory. So the goal is at best barely attainable. If the goal of covering residue theory were abandoned, then some of the supporting material would not need to be covered, freeing time to spend two or perhaps three weeks on basic computations.

Assuming one is working within the traditional course content, there is a general strategy that instructors of complex analysis could use to help students with some of the observed difficulties:

(1) Leave the usual first week survey of the complex numbers as it is. My data suggest that the treatment of symbolic calculations to do proofs within the first few classes is not being absorbed, so it is unclear how helpful this coverage is. Students generally get better with basics as the course progresses (for example, as shown by the data in Table 1), so it is probably a good idea to delay the more detailed instruction suggested below for a few weeks, so that students can gain some confidence and fluency (i.e., begin the processes of interiorization and encapsulation in APOS).

(2) Sometime in weeks three to five, a whole class (or more) should be devoted to introducing more advanced techniques of computation. By week five, the classes I studied had covered the geometry of the plane, functions of a complex variable, derivatives (Cauchy-Riemann conditions) and mappings, so that students had some experience with at least two or three of the representations. Particular care needs to be taken to ensure that students realize that the symbolic representation is a practical approach and not merely a theoretical tool used to do proofs. For example, instructors could do several problems using one or more of the techniques, but with the problems chosen so that one technique is much easier. Instruction also needs to include explanations of why and when to shift representations. If time permits, these kinds of problems would be particularly well suited for discovery exercises or group discussions.

(3) Towards the end of the course, between weeks 8 to 10 (in a 13 week semester course like those studied), one class should be devoted to further examples of basic calculations. By this time (week 10) in the course, power series and Laurent series have been covered, so there are many new problems that could be used as examples. My data give some indication that students at third year level are very impressed by and try to learn techniques that save work. I think it is a reasonable assumption that if instructors reviewed basic calculations and shifting representations late in the course, then most students would have enough experience and confidence with one or two representations to be thoroughly impressed and motivated to learn how to use geometric and symbolic methods.

(4) Generally, instructors need to recognize that the basic skills discussed in this paper are much harder for students to acquire than is being assumed (at least as was assumed in the classes I observed). Hence, all through the course instructors should try to give examples that illustrate basic techniques. For example, when studying contour integrals, give an example in which the integrand can be simpflified using symbolic methods. Then proceed with a parameterization.

(5) Finally, the data seem to suggest that student understanding suffers because complex analysis is largely a study of what works. Functions that are not analytic at more than a countable number of points in their domain

are of little interest. Hence, it is not easy for students to get computational practice with a wide variety of functions. How will students learn about $\bar{z}$ if instructors do not use $\bar{z}$ in examples, because the function $f(z) = \bar{z}$ is nowhere analytic?

**4.4. Directions for Future Research.** Very little work in mathematics education has been done in complex analysis, so any research at the university level opens a new area. There are many directions that research in this field could take, but several directions for future research are suggested by the data and by the analysis in Section 3. In particular, there are many problems of understanding the basic representations that need to be better understood. Some of these are:

(1) To what extent does not shifting representations indicate lack of understanding as opposed to personal preference?
(2) Can trigonometric difficulties be separated from actual difficulties with the polar representation?
(3) How well do students understand basic epistemological questions about the complex numbers, such as what is the meaning of "$+$" in $x + iy$?
(4) The analysis presented here of what constitutes action, process, and object understandings within the reification and APOS frameworks of the four representations of complex numbers is limited. A great deal of work needs to be done to further clarify how action, process and object understandings manifest themselves.

In addition, there are many questions about why students are not acquiring proficiency with symbolic methods and geometric methods. The data suggest that most students do not realize that symbolic methods exist even though there was some coverage of these methods in all three classes studied. Was there simply not enough coverage or are there more fundamental obstacles? If so, what are these obstacles and how can teachers help students overcome them?

I also collected considerable data on the question, "How do problems that students have with basic material affect more advanced work?" This is a big subject and my data only scratch the surface. For example, I did not attempt to correlate the level of understanding within the reification or APOS frameworks with proficiency with more advanced material.

I collected some data that suggest that students gain proficiency with basic representations as the course progresses, but there are many interesting questions that involve studying how understanding changes, or does not change, over the duration of the course. For example, I found that all six of the students studied in Class 2 had a very robust belief that every polynomial in $x$ and $y$ is analytic. This belief persisted despite the many counterexamples I offered throughout the course (e.g., $f(z) = x$ is nowhere analytic). Why was this belief so entrenched?

Aside from the many interesting research questions on basic material, the data indicate that the theme "thinking real, doing complex" is a very significant learning obstacle in complex analysis. This is a major area for further research efforts, since this theme seems to manifest itself in so many ways.

Another major area for research is the question of multivaluedness. This is a crucial aspect of complex analysis that is also relatively complicated. My data suggest that students are absorbing this material with some proficiency, but that there are many questions of understanding to be researched.

Finally, I identified many specific themes that warrant further investigation. Some of these are: treating $|z-a|$ as a circle rather than a distance; underestimating how important the $c_{-1}$ term is in a Laurent expansion about a pole; "Is every function analytic?", the difference between the domain of an analytic function and the disk of convergence of the power series for the function expanded about a point in the domain; and the interplay between the path of integration and the integrand of a path integral.

# References

Ahlfors, L. (1979). *Complex analysis* (3rd ed.). New York: McGraw-Hill.

Asiala, M., Brown, A., DeVries, D. J., Dubinsky, E., Mathews, D., & Thomas, K. (1996). A framework for research and curriculum development in undergraduate mathematics education. In J. Kaput, A. H. Schoenfeld, & E. Dubinsky (Eds.), *Research in collegiate mathematics education. II.* (pp. 1–32). Providence, RI: American Mathematical Society.

Breidenbach, D., Dubinsky, E., Hawks, J., & Nichols, D. (1992). Development of the process conception of function. *Educational Studies in Mathematics*, *23*, 247–285.

Churchill, R. V., & Brown, J. W. (1990). *Complex variables and applications* (5th ed.). New York: McGraw-Hill.

Confrey, J., & Smith, E. (1994). Exponential functions, rates of change, and the multiplicative unit. *Educational Studies in Mathematics*, *26*, 135–164.

Danenhower, P. (2000). *Teaching and learning complex analysis at two British Columbia universities*. Unpublished doctoral dissertation. Simon Fraser University, Vancouver, British Columbia, Canada.

Dreyfus, T. (1991). Advanced mathematical thinking processes. In D. Tall (Ed.), *Advanced mathematical thinking* (pp. 25–41). Dordrecht, The Netherlands: Kluwer.

Dubinsky, E., Dautermann, J., Leron, U., & Zazkis, R. (1994). On learning fundamental concepts of group theory. *Educational Studies in Mathematics*, *27*, 267–305.

Ginsburg, H. (1981). The clinical interview in psychological research on mathematical thinking: Aims, rationales, techniques. *For the Learning of Mathematics*, *1*, 4–11.

Glaser, B., & Strauss, A. L. (1967). *The discovery of grounded theory: Strategies for qualitative research*. Chicago: Aldine.

Hazzan, O. (1999). Reducing abstraction level when learning abstract algebra concepts. *Educational Studies in Mathematics*, *40*, 71–90.

Howe, K., & Eisenhart, M. (1990). Standards for qualitative (and quantitative) research: A prolegomenon. *Educational Researcher*, *19*(4), 2–9.

Hunting, R. P. (1997). Clinical interview methods in mathematics education research and practice. *Journal of Mathematical Behavior*, *16*, 145–165.

Kieran, C. (1992). The learning and teaching of school algebra. In D. A. Grouws (Ed.), *Handbook of research on mathematics teaching and learning* (pp. 390–419). New York: Macmillan.

Lesh, R., Post, T., & Behr, M. (1987). Representations and translations among representations in mathematics learning and problem solving. In C. Janvier (Ed.), *Problems of representation in teaching and learning mathematics* (pp.33–40). Hillsdale, NJ: Erlbaum.

Lincoln, Y., & Guba, E. (1985). *Naturalistic inquiry*. Beverly Hills, CA: Sage.

Meel, D. E. (2002). Models and theories of mathematical understanding: Comparing Pirie and Kieren's model of the growth of mathematical understanding and APOS theory. *Research in collegiate mathematics education. V.* (pp. 132–181). Providence, RI: American Mathematical Society.

Nehari, Z. (1968). *Introduction to complex analysis* (2nd ed.). Boston: Allyn and Bacon.

Priestley, H. A. (1990). *Introduction to complex analysis*. Oxford, England: Oxford University Press.

Romberg, T. A. (1992). Perspectives on scholarship and research methods. In D. A. Grouws (Ed.), *Handbook of research on mathematics teaching and learning* (pp.49–64). New York: Macmillian.

Rudin, W. (1976). *Principles of mathematical analysis* (3rd ed.). New York: McGraw-Hill.

Saff, E. B., & Snider, A. D. (1993), *Fundamentals of complex analysis for mathematics, science, and engineering* (2nd ed.). Englewood Cliffs, NJ: Prentice Hall.

Sfard, A. (1994). Reification as the birth of metaphor. *For the Learning of Mathematics, 14*, 44-55.

Sfard, A., & Linchevski, L. (1994). The gains and pitfalls of reification: The case of algebra. *Educational Studies in Mathematics, 26*, 191–228.

Thompson, P. W. (1994). Images of rate and operational understanding of the fundamental theorem of calculus. *Educational Studies in Mathematics, 26*, 229–274.

Tirosh, D., & Almog, N. (1989). Conceptual adjustments in progressing from real to complex numbers. In G. Vergnaud, J. Rogalski, & M. Artigue (Eds.), *Proceedings of the 13th annual conference of the International Group for the Psychology of Mathematics Education, Vol. 3* (pp. 221–227). Paris: CNRS - PARIS V.

White, P., & Mitchelmore, M. (1996). Conceptual knowledge in introductory calculus. *Journal for Research in Mathematics Education, 27*, 79–95.

Williams, S. R. (1991). Models of limit held by college calculus students. *Journal for Research in Mathematics Education, 22*, 219–236.

Zazkis, R., & Hazzan, O. (1999). Interviewing in mathematics eduaction research: Choosing the questions. *Journal of Mathematical Behavior, 17*, 429–439.

LANGARA COLLEGE, 100 W. 49TH AVENUE, VANCOUVER, B.C., CANADA, V5Y 2Z6
*E-mail address:* `pdanenho@langara.bc.ca`

CBMS Issues in Mathematics Education
Volume **13**, 2006

# Using Geometry to Teach and Learn Linear Algebra

## Ghislaine Gueudet-Chartier

ABSTRACT. Linear algebra is a difficult topic for undergraduate students. In France, the focus of beginning linear algebra courses is the study of abstract vector spaces, with or without an inner product, rather than matrix operations as is common in many other countries. This paper presents a study of the possible uses of geometry and "geometrical intuition" in the teaching and learning of linear algebra. Fischbein's work on intuition in science and mathematics is used to analyze the treatment and use of geometry in linear algebra textbooks as well as mathematicians' and students' uses of geometry in linear algebra. I indicate the possibilities and limitations of such uses of geometry and make suggestions for a linear algebra course that uses geometry to support learning.

## 1. Introduction

Prior to a description of the issues studied herein, it is necessary to clarify the use of the expression "linear algebra" in this paper. "Linear algebra" is used in accordance with the teaching context in France. This may be different from many other countries, where undergraduate students encounter mostly matrix-oriented courses. Though the issues I study are not specific to France, a brief description of the historical background of the teaching of linear algebra in France can help to clarify the context of the study.

Linear algebra was first taught in French universities at the graduate level, in 1939. The first courses were strongly connected with the study of Hilbert spaces. During the 1960s, the introduction of linear algebra into the secondary school curriculum led to many discussions among French mathematicians. These discussions included some contention about the presentation of geometry.

French mathematicians took two main opposing views. The first one, discussed in detail in Choquet's book *Teaching Geometry* (1964), recommended presenting geometry defined by axioms (independent of linear algebra), then using it for an intuitive presentation of linear algebra. Mathematicians like Dieudonné took a different view. They preferred to start directly with linear algebra because, as he said, geometry was a mere application of linear algebra. Understanding geometry was an immediate consequence of understanding linear algebra. During the "modern mathematics" reform, Dieudonné's position was adopted in France's national curriculum, and linear algebra started to be taught in secondary school. However, difficulties encountered by students led to the failure of this approach, and linear algebra disappeared from the secondary school curriculum during the 1980s. Since 1986, linear algebra – banished from secondary education – has been a requirement for undergraduate science students. It remains a difficult topic for them.

Many mathematicians have claimed that using geometry or "geometrical intuition" helps students in their understanding of linear algebra. This claim raises several questions. In the French teaching context, "linear algebra" is clearly identified as linear algebra in abstract spaces, with or without an inner product. I will refer to this later as "general linear algebra" or "the general theory." But what is meant by "geometry"? It can be geometry taught in secondary school[1] or Euclidean geometry (in its historic, axiomatic meaning). For some mathematicians, linear algebra itself (or at least parts of it) are a kind of geometry. Another question is: What is meant by "geometrical intuition"? It is certainly linked with the possibility of using drawings or mental pictures. But intuition does not mean only visualization, and there is no doubt that it has other aspects. Determining how geometrical intuition can help students in their learning of linear algebra, and whether mathematicians try to develop geometrical intuition in their linear algebra courses by specific choices, are additional issues.

Several published studies have investigated how geometry or geometrical aspects of linear algebra can be used to introduce the general theory. They have reported that teaching based on a geometrical approach can improve students' understanding, but have pointed out difficulties stemming from such a choice (see §2). These studies have confirmed that possible interventions of "geometrical intuition" in linear algebra requires a thorough study. This was the aim of my doctoral dissertation, which contains most of the results presented here. This paper has three main parts: a grounding of the study in related theory and research, analysis of the uses of drawings and geometry in teaching linear algebra, and particular results from interviews about the role of $\mathbb{R}^2$ and $\mathbb{R}^3$ in linear algebra teaching and learning.

*Overall presentation of the study (§2).* I start with the main theoretical framework for my research. Studying the question of intuition requires appropriate tools; I found them in Fischbein's (1987, 1993) work on intuition in science and mathematics. In this section, I present the notion of intuitive models, along with the research questions it allowed me to formulate. I also describe the setting of my study within the context of related studies. Though these constituted a starting point for my study, I present them at the end of the section in order to interpret their results in terms of the notion of geometrical intuition.

*Geometry in linear algebra courses (§3).* In this section, I present results about mathematicians' uses of geometry and drawings in linear algebra courses. These results come from the analysis of mathematicians' responses to a questionnaire. Since my research took place in France, the results of the questionnaire are influenced by French teaching of linear algebra and geometry. However, this is not the case for the results in other sections.

*Linear algebra in $\mathbb{R}^2$ and $\mathbb{R}^3$ as a source of geometrical intuition (§4 and §5).* In these two sections, I focus on a particular intuitive model: linear algebra in $\mathbb{R}^2$ and $\mathbb{R}^3$ with the dot product. I first establish some possibilities and limitations of that model through a textbook study (§4.) I then present a detailed analysis of the effects of the use of that model among students solving a problem in $\mathbb{R}^n$ (§5).

---

[1]In France, the geometry taught in secondary school is mostly plane and space vector geometry. Vectors are defined intuitively; there are no axioms presented.

## 2. Theoretical Framework and Research Questions

Many mathematicians have attempted to clarify the nature of intuition in mathematics, in students' learning as well as in their own research. Most of them have distinguished different kinds of intuition taking a psychological approach. But the writings of few provide a means for a precise analysis of intuition. Fischbein (1987) made a careful study of the nature of intuition, its function, and the factors that can shape it. His work is therefore the basis of my theoretical framework. Below I present those elements of Fischbein's work that are relevant for the present study.

### 2.1. Intuition in Mathematics: Fischbein's Theory.

2.1.1. *Intuition and the need for certitude.* According to Fischbein, every human being needs to act in accordance with a credible reality. Even within a conceptual structure, one's reasoning endeavors need a form of certitude. The role of intuition is to provide that kind of certitude. For Fischbein, intuition is synonymous with intuitive knowledge. It is a type of cognition, characterized by self evidence, immediacy, and certitude; it always exceeds the given facts. Productive reasoning requires intuitively acceptable cognition. Thus, when a notion is, for a given person, intuitively unacceptable, that person will probably produce (deliberately or unconsciously) a more acceptable substitute. Such a substitute is called by Fischbein an intuitive model. Models are a central factor of intuition in mathematics. A large part of Fischbein's book (1987) is devoted to them. In the following section, I present those aspects of the models used in my study.

2.1.2. *Intuition and models.* Also, according to Fischbein:

> A system $B$ represents a model of a system $A$ if, on the basis of a
> certain isomorphism, a description or a solution produced in terms
> of $A$ may be reflected consistently in terms of $B$ and vice versa.
> (1987, p. 121)

This definition is very general; the word "system" used in it can have several meanings. The following examples (all of them related to the present study), will be useful to make the definition precise.

*Example 1.* A "system" can be restricted to a single notion. For example, the system $A$ can be the notion of vector in the plane, and the model $B$ the drawing of arrows on a sheet of paper.

*Example 2.* A system can also be a whole theory: complex numbers (system $A$), associated with the vector geometry of the plane (system $B$).

*Example 3.* A system is not always a conceptual system: physical space (system $A$) can be associated with $\mathbb{R}^3$ considered as a vector space with an inner product (system $B$).

By Fischbein's definition, a property in $B$ may be "reflected consistently" in $A$. This means that the property can somehow be translated from one system to the other. Let us consider in Example 1 the relation $R_A$: "the vector $u$ is the sum of $v$ and $w$." $R_A$ can be associated with the drawing $R_B$ of a parallelogram, whose sides are the arrows associated with $v$ and $w$, and one of whose diagonals is the arrow associated with $u$. The relation $R_A$ in $A$ corresponds to a consistent relation $R_B$ in $B$.

But a model can also lead to misconceptions if it is wrongly used. In Example 1 again, students sometimes claim that two vectors in the plane have the same direction because the associated arrows are both pointing "up, on the right." In this case, the word "direction" exists in both systems. But the notion of the "direction" of a vector in the plane cannot be reflected consistently in the common notion of "direction" of an arrow.

In most cases, the word "isomorphism" used by Fischbein in his definition is not a mathematical isomorphism (but it can happen, as in Example 2). Rather, the word "isomorphism" is used to indicate a particular set of relations between some objects and properties of $A$ and some objects and properties of $B$. Extending this to additional relations is likely to be misleading.

The three examples given above correspond to three different kinds of models. Among the models distinguished by Fischbein are the following: figural models, abstract and intuitive models, and analogical and paradigmatic models. I discuss each in turn.

*Figural models.* Fischbein distinguishes between intramathematical and extramathematical models. In the case of an intramathematical analogy, the original and the model are both mathematical theories or objects. In contrast, an extramathematical object is something that does not lie strictly within mathematics. It is not a mathematical object, collection of objects, or theory. The extramathematical models that I will study here correspond to the use of drawings. I refer to such models as figural models. Here "drawings" means pictorial representations.

A figural model can be related to geometrical notions; for example, the calculation of the distance from a given point $p$ to a given plane $F$ in 3-space can be associated with a drawing of a parallelogram, a point, and a dotted line containing the point, perpendicular to the plane (Figure 1). But the same drawing can also be associated with a polynomial problem: calculation of the distance from $x^4$ to the set $F$ of polynomials of degree less than 3, in the space $\mathbb{R}_4[X]$ with the inner product defined by $< p|q >= \int_0^1 p(x)q(x)dx$ .

Three-dimensional objects, and computer-generated graphics are other kinds of extramathematical models that can be used in linear algebra. I do not discuss them here because they are used only sparsely in France.

*Abstract and intuitive models.* Models fall into two distinct categories: abstract and intuitive. Some mathematical objects are abstract models for concrete realities. In Example 3, $\mathbb{R}^3$ with the dot product is an abstract model of physical space.

In contrast, an intuitive model is one that seems concrete to the perceiver. Figural models are obviously intuitive. But a mathematical object can also be an intuitive model for someone who perceives it as a reality. In Example 2, vector geometry in the plane is a model for complex numbers. The existence of this intuitive model was very important in the emergence of the notion of complex numbers because it legitimated their existence. Complex numbers gained legitimate status with the work of Gauss (1831), who presented a geometric interpretation of imaginary quantities. Though other mathematicians like Wessel(1799), Buée (1805), and Warren (1828) had proposed such interpretations before, they probably lacked Gauss' influence on the mathematics community.

*Analogical and paradigmatic sub-categories.* Intuitive models themselves have two sub-categories: analogical and paradigmatic. An analogical model is external to the original object modeled; the model and the original belong to two distinct systems. An analogical model provides the reasoning process with a source of research hypotheses. According to Fischbein, two systems are said to be analogical if a partial similarity exists, that can lead a person to assume additional similarities.

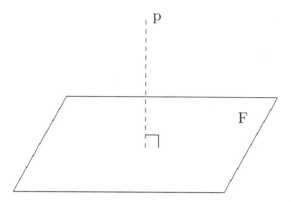

FIGURE 1. Distance from a point (or a polynomial) to a plane (or a subspace of polynomials).

Analogy justifies plausible inferences. Analogies become models if they can be productively used in reasoning. Figural models associated with mathematical models are analogical models.

A paradigmatic model is a subclass of the original that is used as a model of the original. It is a particular exemplar of the original. That is, Fischbein's concept of paradigmatic model is similar to the concept of prototype in cognitive semantics. One's understanding of the original system is influenced (correctly or incorrectly) by the paradigmatic model. The original mathematical object is represented in one's reasoning by that exemplar, and not by its abstract definition. For example, using $\mathbb{R}^3$ as a model for the concept of vector space can lead to attributing properties of $\mathbb{R}^3$ to all vector spaces. Someone using that model might claim that two two-dimensional subspaces can never be supplementary subspaces.

The above theoretical considerations led me to a definition of "geometrical intuition." It also provided me with an appropriate framework for formulating and investigating my research questions in a consistent way.

**2.2. Research Questions.** The first stage here is to clarify what can be called a geometry. Some mathematicians may consider linear algebra to be a geometry; this cannot be relevant in our case. Because my analysis mostly takes place in a teaching context, I am naturally referring to geometry as a mathematical domain. The definition of intuition given by Fischbein emphasizes credible reality. In the use of geometry in the teaching and learning of linear algebra, the link with reality is central. For this reason, I call "geometry" a mathematical theory whose main purpose is to provide an abstract model (using Fischbein's terminology) for physical space; it is notably restricted to three dimensions. The geometry taught at school is such an abstract model, as is axiomatic Euclidean geometry.

I will term "geometric intuition" the use of models stemming from a geometry. These models are intuitive models; they can be either analogical or paradigmatic. In the first case, this constitutes an intramathematical analogy.

Because of the nature of geometry, a geometric model will always be associated with an extramathematical, figural model. A geometric model can thus smuggle uncontrolled elements into the reasoning process. For example, when studying the general notion of quadratic form, students encounter self-orthogonal vectors.

That situation cannot be associated with anything in two-dimensional Euclidean geometry; it is contradictory to a drawing representing two orthogonal vectors in the plane.

In the following study, I will use geometric and figural models rather than the general expression "geometric intuition." The research questions can be formulated as follows:

(1) What are the possible uses of geometric models in linear algebra?
(2) How do mathematicians and students use geometric and figural models in linear algebra?
(3) What are the consequences of the observed uses of models on students' practices and thinking processes?

These are very general questions, for which my work only provides partial answers. However, some previous related research on linear algebra provides hints about the use of geometric and figural models, even if it was not formulated in those terms. I discuss this research in the following subsection, together with the setting of my study in the context of these works.

**2.3. Related Research.** Much research on linear algebra addresses the question of possible uses of geometry or geometric aspects of linear algebra. Here I only present research in which that question is central and whose results are meaningful in terms of intuitive models. I also discuss its connections to and differences with my own work.

2.3.1. *Modes of description.* Hillel (2000) identified three modes of description in linear algebra: the *abstract mode*, the *algebraic mode*, and the *geometric mode*. The abstract mode uses the language and concepts of the general theory (e.g., vector space, dimension, kernels). The algebraic mode uses the language and concepts of the theory in $\mathbb{R}^n$ (e.g., matrices, systems of equations). The geometric mode uses the language and concepts of 2- and 3-space (e.g., points, lines, planes). His approach was quite different from mine because Hillel studied the three modes and the mechanisms that enable one to move from one mode to another, but he did not examine the question of geometric intuition.

Nonetheless, his description of students' difficulties with the geometric mode can be interpreted as consequences of an irrelevant use, or of the use of irrelevant components, of a figural model. Difficulties attached to the point and arrow depictions of a vector are the most striking. Hillel observed that most mathematicians use both depictions and he described wrong interpretations of some representations by students. For example, it is well known that students may claim that it is possible for two 1-dimensional subspaces to have an empty intersection. This phenomenon can be interpreted as a misleading intervention of a figural model: 1-subspaces are represented as straight lines, and students believe that they can be parallel. In this case, the representation of a straight line that does not contain the origin is an irrelevant component of the model.

The question of how point and arrow depictions in linear algebra affect students' understanding could be studied from the point of view of geometric intuition. Although I encountered related difficulties in my work with mathematicians and students, I will not address them explicitly.

2.3.2. *Cabri-Geometry in linear algebra.* Sierpinska, Dreyfus, and Hillel (1999) designed a learning environment with Cabri-Geometre II software for the notions of vector space, linear transformation, and eigenvector. That environment was explicitly intended to use geometric intuitions. The intuitive model was provided by Cabri vectors. I focus here on the difficulties of students discussed by the authors. At one stage in the reported activities, students encountered the task: "Find the coordinates of a vector $v$ in the basis $v_1$, $v_2$." The vectors $v$, $v_1$, and $v_2$ were Cabri-vectors constructed on the screen by the students. They immediately reorganized their construction to obtain orthogonal vectors $v_1$ and $v_2$. They went on doing a mechanical calculation of coordinates, not attending to the initial meaning of the problem. That phenomenon can be interpreted in terms of intuitive models. The notion of "coordinates" is strongly associated with the drawing of two orthogonal axes. The students referred to that model because they did not have an appropriate figural model for the notion of basis vectors at their disposal. The authors had intended to avoid an explicit introduction of the notion of basis and expected the students to develop an intuition of it. It is interesting to observe here the emergence of a very familiar model, one that has the appearance of credible reality. It created a misleading geometric intuition.

Similar difficulties were observed by Sierpinska (2000). In a further analysis of the same teaching environment, she identified a phenomenon that she described as: "Thinking of mathematical concepts in terms of their prototypical examples rather than definitions." For example, some students, when asked to construct a linear transformation with given values on a basis, looked for a well-known geometric transformation (dilation, rotation, etc.) or for linear combinations of these transformations. These students were using a geometric model for linear applications: the model of well-known transformations of the plane. That model may have been derived from previous courses,[2] but it was insufficient for the given task. In this case as in the previous one, a familiar, misleading figural model emerged.

To have the appearance of credible reality, an intuitive model must be very familiar to students. The construction of a model (as a cognitive object) is a long process that requires regular and frequent rehearsal of the elements of the model. Teaching designed to help students form intuitive models must thus be long-term, regular teaching; otherwise more familiar models are likely to emerge.

2.3.3. *The concreteness principle and geometric models.* The *concreteness principle* was stated by Harel (2000) as follows:

> For students to abstract a mathematical structure from a given model of that structure the elements of that model must be conceptual entities in the students' eyes; that is to say, the student has mental procedures that can take these objects as inputs (p.177).

The definition of model given by Fischbein applies to the models mentioned here because there exists an isomorphism between a subclass of the model and a subclass of the corresponding mathematical structure. The concreteness principle is quite close to the following assumption, formulated in Fischbein's terminology: "For students to abstract a mathematical structure from a given model of that structure, that model must be an intuitive model for the student." However, establishing whether "conceptual entities" and "intuitive models" are equivalent would require a specific study.

---

[2]In France, the same kind of answers can be produced by students using transformations they studied in secondary school geometry.

Harel (2000) conducted a linear algebra teaching experiment using his concreteness principle and a geometric, paradigmatic model. That model can be considered as paradigmatic because it consisted of a geometric presentation of vector spaces; it was associated with a figural model. The studied teaching experiment had positive effects on the students' performances in linear algebra, including their ability to prove general linear algebra results. Students belonging to the group (Group B) that followed the experimental teaching seemed to have had a better control of the correctness of their answers than the students (Group A) who were taught linear algebra without geometric representations. The reason for that observation could be that the students of Group B formed an intuitive model that helped them to check the consistency of their reasoning. But Harel also observed difficulties attached to the use of a geometric model in linear algebra teaching. Some students could be captured by the model, and stay inside of it, instead of moving up to the general theory.

This brief review of previous research shows that several results relative to geometrical intuition in linear algebra have already been established. Namely, the use of a figural model can lead to inappropriate intuitions and a geometric model can be an obstacle if students stay captured in it. An intuitive model must consist of very familiar objects in order to have the appearance of credible reality.

Before starting with the study of mathematicians' choices and the presentation of my results, I need to mention an essential difference between the research presented above and my own study. Their authors elaborated and discussed linear algebra teaching experiments. I will consider the possibilities of elaborating teaching using geometric models, but my study deals with ordinary linear algebra courses taught at university.

## 3. Mathematicians and Geometric Models in Linear Algebra

Having observed the difficulties encountered by students in their learning of linear algebra, many mathematicians have recommended that geometry be taught before the general theory. But the content of such a geometry course, and the way it might be used to learn linear algebra, depends on mathematicians' views, and those can be very different from one teacher to another. In order to have a better idea of the different uses of geometrical (and associated figural) models by mathematicians in their linear algebra courses, I created a questionnaire for mathematicians (see Appendix A).

I first present my analysis of the mathematicians' answers to the part of the questionnaire devoted to the use of drawings in linear algebra, then I present the conclusions of the analysis of the entire questionnaire. The questionnaire was given to mathematicians who answered it outside of my presence. They took about one hour to complete the entire questionnaire. I collected 31 questionnaires, completed by mathematicians of various ages and research subjects, all of them having recently taught linear algebra.

**3.1. Mathematicians' use of drawings in linear algebra.** The two questions relevant to mathematicians' use of drawings in linear algebra in the questionnaire were:

**Question 3.1:** For each of the drawings in Table 1, indicate if you use it in your linear algebra courses; if the answer is "yes" indicate which notions

TABLE 1. Question 3.1

| Drawing | Used (yes/no) | That drawing illustrates |
|---|---|---|
| | | |
| | | |
| | | |

or properties you illustrate with it. (You can mention several uses of the same drawing.)

**Question 3.2:** If you use other drawings, draw them in the following table, and indicate the interpretation(s) you associate with them. (A blank table with five lines followed.)

Twenty-eight mathematicians answered Questions 3.1 and 3.2. For the analysis of the responses, I used the following criteria:

- Are the drawings in Question 3.1 used by the teacher?
- Does he (or she) mention other drawings (Question 3.2)?
- Does he (or she) mention interpretations of the drawings related to a general vector space or limited to dimension 3?

My analysis led to the following conclusions. In general, these mathematicians did not use many drawings in their linear algebra courses. Of the 28 responding, 16 mathematicians mentioned other drawings they might use in their courses. The average number of drawings mentioned by the 16 mathematicians was 2.25; this is very low considering the fact that there were five lines that could have been filled in the blank table given in the questionnaire. The average number of drawings per teacher, for both parts of the question, was only 3.2.

Moreover, most of the drawings were reported as being used to illustrate situations occuring in $\mathbb{R}^2$ or $\mathbb{R}^3$. Only 12 (43%) of the mathematicians proposed additional interpretations referring to an abstract vector space, rather than to $\mathbb{R}^2$ or $\mathbb{R}^3$. For example, for the first drawing proposed in Table 1, nine mathematicians gave the interpretation: "basis of the space,"[3] and three "orthogonal basis of the space," while only three of them mentioned the general notion "orthogonal basis," and only one the general notion of basis. For the second drawing, eight mathematicians mentioned using it in teaching as an intersection of planes, and five as an intersection of subspaces.

The drawings volunteered by the mathematicians were not very different from those provided in the questionnaire (Table 1). Except for two quadric surfaces, what the mathematicians offered were mostly combinations of planes, lines (plain or dotted) and vectors. Only five of them were drawings in the plane; the 31 others were perspective drawings evoking 3-space, even if they were used to illustrate situations in a general vector space. Space may have seemed more representative than the plane, a better candidate for a paradigmatic model. The notions illustrated included projections (3), orthogonal projections (4), symmetries (2), rotations (2), supplementary subspaces (3), and coordinates of a vector (2).

In fact, most of the notions and properties mentioned by the mathematicians already would have been encountered by students in secondary school geometry in France: lines, planes, symmetries, and projections. This was not the case for the few examples reported about supplementary subspaces and rotations around an axis. The three drawings proposed in Table 1 are used in secondary school textbooks. The first and the third occur frequently in the space geometry course; the second is used to illustrate the intersection of two planes in space, and sometimes the corresponding system of equations. So, the second drawing in Table 1 would be the least familiar of the three for secondary school students; it was also the least mentioned by the mathematicians. For the second drawing, 15 (54%) of the mathematicians declared that they used it in their linear algebra courses, whereas 23 (82%) used the first drawing and 20 (71%) used the third. These mathematicians did not have a well developed, specific, use of drawings for linear algebra. They reported drawing them mostly when presenting examples in a geometrical context.

## 3.2. Analysis and Discussion of the Mathematicians' Responses.
Considering the responses to the entire questionnaire, I was led to distinguish three groups of mathematicians, including 24 of the 26 mathematicians who answered all the questions.[4]

*Group A: Many drawings, geometry presented after linear algebra.* There were only four mathematicians in Group A. These mathematicians used many drawings in their linear algebra courses. The figural model corresponding to their reported use of drawings was associated with a part of linear algebra that was to be used as

---

[3]The term "space" refers here directly to geometry. In French, the word "space" used alone means "geometrical 3-space."

[4]I used statistical tools for that global analysis, but the small number of questionnaires prevented me from referring to the statistical results without explicitly examining the effective content of the questionnaire. So, the statistical results were only a way to identify possible connections and the conclusions result from a direct observation of the questionnaires.

a paradigm for the whole theory: the study of $\mathbb{R}^2$ and $\mathbb{R}^3$. Such use of drawings would be a geometric model, according to the definition of that concept I use here. Their use of a paradigmatic model can also be a wider part of linear algebra, not limited to dimension 3, and could refer to a general vector space.

*Group B: Almost no drawings, geometry presented after linear algebra.* The eight mathematicians in Group $B$ used no, or only a few, drawings. They did not refer to a geometric model to introduce linear algebra. They prefered to introduce the general theory directly and to present geometry afterwards as an application thereof.

*Group C: Many drawings, geometry presented before linear algebra.* The 12 mathematicians in Group $C$ referred to an analogical geometric model, stemming from a geometry independent of linear algebra. They used many drawings in the geometry course and also in linear algebra. The drawings involved were more or less the same in both cases.

The two last tendencies, of Groups $B$ and $C$, were very close to positions observed in France during the reform of modern mathematics. The structural choice of Group $B$ corresponds, more or less, to Dieudonné's views. In contrast, Choquet advocated a presentation of geometry preceding linear algebra like the mathematicians of Group $C$. Groups $B$ and $C$ included 20 of the 26 mathematicians who completed the entire questionnaire. This may have been an indication of the influence, still very strong, of the discussions held before and during the reform of modern mathematics on the choices of French mathematicians in their linear algebra courses.

Only the mathematicians of Group $A$ seemed to have escaped that influence. They proposed to students a geometric model *inside* linear algebra (and thus paradigmatic). That choice deserves a special study. I will now focus on this choice, and more precisely, on the use of linear algebra in $\mathbb{R}^2$ and $\mathbb{R}^3$ as a geometric model for general linear algebra.

## 4. The $\mathbb{R}^2$-$\mathbb{R}^3$ Model

The study of $\mathbb{R}^2$ and $\mathbb{R}^3$ as vector spaces with an inner product is a geometry. According to the definition given in Section 2, this is an abstract model for physical space. It seems to be a good candidate for a geometric model in the teaching of inner product spaces. I will call it "the $\mathbb{R}^2$-$\mathbb{R}^3$ model." I start this section with an overview of its possibilities and limitations. Then I present an example of the use of that model in a textbook.

**4.1. Possibilities and limitations of the $\mathbb{R}^2$-$\mathbb{R}^3$ model.** The $\mathbb{R}^2$-$\mathbb{R}^3$ model can be associated with a figural model, and coordinates offer a natural way to introduce $\mathbb{R}^n$. The link between $\mathbb{R}^n$ and other $n$-dimensional inner product spaces is evident for mathematicians.

Historical analysis (Dorier, 2000) of the development of linear algebra has suggested that axiomatic linear algebra finally emerged after several works about infinite dimensional spaces, as a way of unifying different mathematical domains. Linear algebra is a general theory designed to unify several branches of mathematics. Presenting linear algebra concepts only in $\mathbb{R}^2$ and $\mathbb{R}^3$ can appear very arbitrary to the students (see, for example, Robert's (1998) work about generalizing, unifying, and formalistic notions and Harel's (2000) Necessity Principle). When limited to $\mathbb{R}^2$ and $\mathbb{R}^3$, some concepts and properties of linear algebra may only seem to be geometrical tautologies to students. Below are some examples.

- The definition in $\mathbb{R}^2$ and $\mathbb{R}^3$ of a basis as a family of vectors that are linearly independent and spanning the whole space cannot appear as necessary to students. The notion of dimension is implicit and self-evident in that context; therefore a basis of $\mathbb{R}^2$ (or $\mathbb{R}^3$) is defined as a set of two (or three) linearly independent vectors. The notion of spanning the whole space does not seems to be required in that context.
- The property of existence of a basis for a given space is fundamental in general linear algebra. In $\mathbb{R}^2$ and $\mathbb{R}^3$, it appears to students as a observable fact. More generally, results stating the existence of a mathematical object are only needed in a theoretical context, where that existence cannot be directly observed.

Yet, there exist concepts and properties already relevant in $\mathbb{R}^2$ and $\mathbb{R}^3$ that can be generalized to any vector space (in fact, most of these properties occur in spaces with an inner product). For example, the Pythagorean Theorem, which is presented in the plane in secondary school and used then in several exercises, can be generalized to any space with an inner product: for a set $\{e_1, ..., e_k\}$ of orthogonal vectors, $|e_1 + ... + e_k|^2 = |e_1|^2 + ... + |e_k|^2$.

But there are obviously limitations to the use of the $\mathbb{R}^2$-$\mathbb{R}^3$ model. Some notions and properties of general linear algebra are not relevant in that context. And, the possibility of unification, central in linear algebra, is lacking in such a presentation. Moreover, the generalization to $\mathbb{R}^n$ and then to other vector spaces may not be natural for all students. For mathematicians, $\mathbb{R}^n$ is a natural model for any other real vector space of dimension $n$ because of the structural isomorphism. For students, considering a polynomial or a function as a vector is the result of a long process.[5] The use of drawings may aid students' understanding even when the vector space is different from $\mathbb{R}^2$ or $\mathbb{R}^3$.

I now will now make these general considerations precise by examining a university textbook that uses the $\mathbb{R}^2$-$\mathbb{R}^3$ model.

**4.2. A textbook using the $\mathbb{R}^2$-$\mathbb{R}^3$ model.** *Linear Algebra Through Geometry* is a textbook by Banchoff and Wermer (1991), designed for undergraduate students. The title clearly announces that the authors intend to use geometry to introduce and illustrate linear algebra. In the book's preface, the authors say:

> In this book we lead the student to an understanding of elementary linear algebra by emphasizing the geometrical significance of the subject. Our experience in teaching undergraduates over the years has convinced us that students learn the new ideas of linear algebra best when these ideas are grounded in the familiar geometry of two and three dimensions. Many important notions of linear algebra already occur in these dimensions in a non-trivial way, and a student with a confident grasp of the ideas will encounter little difficulty in extending them to higher dimensions and more abstract systems (Banchoff & Wermer, 1992).

---

[5]It is certainly linked with encapsulation; considering polynomials or functions as vectors means, in particular, considering them as objects instead of processes (Dubinsky, 1991).

The approach of the authors is clearly stated here: first build a geometric model limited to dimensions 2 and 3. They claim that the model will help students when learning linear algebra because these students will only have to extend now familiar notions.

Analysis of the book, which I do not give here in detail, made clear that the geometric model proposed by the authors is the $\mathbb{R}^2$-$\mathbb{R}^3$ model. Chapters 1, 2 and 3 are dedicated to it. Chapter 4 is a transition; it deals with $\mathbb{R}^n$, in fact, mostly $\mathbb{R}^4$. The remaining chapters are dedicated to general vector spaces. I present here a brief synthesis of the results of the whole analysis.

*Linear algebra notions presented within the $\mathbb{R}^2$-$\mathbb{R}^3$ model.* Many notions of elementary linear algebra, and of vector spaces with an inner product, appear in the model. Yet, there are important exceptions such as vector spaces, vector subspaces, spanned subspace, and basis. (The less general notion of coordinate basis vectors is already used for spaces of dimension 1, 2 or 3.)

*Use of drawings.* There are 92 drawings in the book. Two of them illustrate general situations (in an arbitrary vector space), and five illustrate situations in dimension 4. The other 85 (92%) of the drawings are associated with situations in $\mathbb{R}$, $\mathbb{R}^2$, and $\mathbb{R}^3$.

The elements of these vector spaces are sometimes represented as arrows, and sometimes as points. No explicit rationale is given for this, and possible confusions are not discussed.

*Moving from the $\mathbb{R}^2$-$\mathbb{R}^3$ model to the general theory.* The book displays two stages in generalizing from the $\mathbb{R}^2$-$\mathbb{R}^3$ model to general linear algebra.

*Stage 1. Introducing $\mathbb{R}^n$ (Chapter 4).* A special chapter is dedicated to generalizing from the $\mathbb{R}^2$-$\mathbb{R}^3$ model to $\mathbb{R}^n$. That chapter is very similar in structure to the previous ones, thanks to the use of coordinates. A specific choice is made by the authors; they emphasize $\mathbb{R}^4$. It appears as a first stage in the generalization, already outside of the "familiar geometry" but allowing an explicit description of vectors, thereby avoiding dots in their coordinate representations. Five drawings are given to illustrate situations in $\mathbb{R}^4$. Linear algebra in $\mathbb{R}^4$ is used as a first step towards $\mathbb{R}^n$, as an intermediate intuitive (paradigmatic) model.

*Stage 2. Abstract vector spaces (Chapter 5).* The introduction of abstract vector spaces (finite dimensional vector spaces over $\mathbb{R}$) is accompanied by a radical change in presentation. There are almost no drawings. Some very important notions, like vector space, vector subspace, spanned subspace and basis are introduced for the first time. The main link with the preceding chapters is $\mathbb{R}^n$, which is used as a new paradigmatic model.

No direct use is made of the $\mathbb{R}^2$-$\mathbb{R}^3$ model for abstract spaces. Rather, it is used as a paradigmatic model for $\mathbb{R}^n$; then $\mathbb{R}^n$ itself is used as a paradigmatic model for other vector spaces. The main link between the $\mathbb{R}^2$-$\mathbb{R}^3$ model and general linear algebra is that some of the terms used in the general context have already been encountered in the geometric context. However, the authors do not suggest referring to an associated drawing that might reinforce that link and help with the generalization process.

No geometric model is used for abstract vector spaces. Their introduction to abstract vector spaces does not appear as a natural generalization. Too many important notions are not included in the model.

Using the $\mathbb{R}^2$-$\mathbb{R}^3$ model as a geometric model for $\mathbb{R}^n$ seems worthwhile for shaping useful intuitions. Moreover, this can be an important stage in the generalization process. Studying the $\mathbb{R}^2$-$\mathbb{R}^3$ model can provide more general indications about possible thinking processes involved in moving from dimension 2 or 3 to dimension $n$, with $n > 3$. I will now examine a particular example of the way students use the $\mathbb{R}^2$-$\mathbb{R}^3$ model for $\mathbb{R}^n$.

## 5. Analysis of the use of the $\mathbb{R}^2$-$\mathbb{R}^3$ model by students

During the second semester of the academic year 2000–2001, I observed a six-week long linear algebra course for second-year university students. It focused on quadratic forms and vector spaces with an inner product. All the students had learned elementary linear algebra during their first year.

The course was taught by Proffesor Thomas[6], an experienced teacher and researcher at the university where the study took place. The teaching of the course consisted of two lectures ($1\frac{1}{4}$ hours each) and two tutorials (2 hours each) per week. During the lectures, all 110 of the students sat together in a large lecture hall. They copied down the lecture notes and mostly remained silent. During the tutorials, limited to groups of around 30, students attempted to solve exercises with the help of a teacher. The exercises were taken from a list given by the teacher $T$, who gave the lectures, one tutorial, and organized that particular teaching. After observing the class, I interviewed Professor Thomas and eight of his students individually. The student interviews were based on a questionnaire (see Appendix B) and included the following task: *"Find the length of a diagonal of a cube with edges of length 1 in $\mathbb{R}^n$."*

I present first a brief account of the teacher's interview. I discuss only the aspects that can be linked with students' solution attempts, which I analyze in the second subsection.

**5.1. The Teacher's Choices.** The course was supposed to offer an overall presentation of quadratic forms, inner products, and symmetric and orthogonal matrices. The teacher introduced all these notions, but he chose to emphasize the $\mathbb{R}^2$ and $\mathbb{R}^3$ case in the sense that, after stating general results, he often illustrated them in $\mathbb{R}^2$ or $\mathbb{R}^3$. Sometimes a result was only established in $\mathbb{R}^2$ and $\mathbb{R}^3$, and the students were asked to do the generalization as homework (only for results stated with coordinates). Among the 32 exercises proposed during the corresponding tutorials, 20 were set exclusively in $\mathbb{R}^2$ or $\mathbb{R}^3$.

Another typical choice of the teacher was the use of many drawings. He made drawings during the lectures (66 drawings during 15 hours of lecture). He also explicitly asked students to produce drawings in 10 of the 32 exercises. However, these drawings were exclusively used to illustrate situations in $\mathbb{R}^2$ or $\mathbb{R}^3$. Comparing these choices with the results of the mathematicians' questionnaire (§3) shows that they were not usual in the French teaching context.

For these reasons, I especially questioned the teacher about his use of drawings, the role of the $\mathbb{R}^2$-$\mathbb{R}^3$ model in his course, and its possible use by students. In summary, he answered these questions as follow:.

---

[6]This name for the teacher and the names for students are pseudonyms.

(1) *Drawings.* According to Professor Thomas, the drawings did not help understanding. They were just natural because quadratic forms were geometrical objects, stemming from physics. He said the drawings must be used only to illustrate situations in $\mathbb{R}^2$ or $\mathbb{R}^3$. Though he sometimes drew to illustrate a situation in $\mathbb{R}^n$, in such cases he asked the students to think in $\mathbb{R}^2$ or $\mathbb{R}^3$.

(2) *The $\mathbb{R}^2$-$\mathbb{R}^3$ model.* According to the teacher, the study of $\mathbb{R}^2$-$\mathbb{R}^3$ as vector spaces with the dot product as presented in his course was not intended to help with general spaces, or even with $\mathbb{R}^n$. Such study was interesting in itself and graduate students often had to manipulate general statements without being conscious of their meaning in small dimensional spaces. During the interview I insisted on a possible use of the $\mathbb{R}^2$-$\mathbb{R}^3$ model to learn, or understand, the general theory. Professor Thomas answered: "For quadratic forms, all the phenomena already happen in three-dimensional spaces. *It is necessary to understand how to move up from 2 to 3* (emphasis added). After that, there is nothing new."

The sentence emphasized above was a very important assumption of the teacher: a student who understands the underlying process when moving from $\mathbb{R}^2$ to $\mathbb{R}^3$ can easily carry out the generalization. I also asked Professor Thomas about the exercise used in student interviews and possible student answers. His assumption was: "If they are able to solve it in $\mathbb{R}^3$, they are able to solve it." The actual situation for students was more intricate.

### 5.2. Analysis of the students' interview responses.

The interview exercise was formulated as follows:

"Find the length of a diagonal of a cube with edges of length 1 in $\mathbb{R}^n$."

One student difficulty was linked to the geometrical vocabulary. These students had never been introduced to "cubes" and "diagonals" in $\mathbb{R}^n$. This might have embarrassed some of them and prevented them from solving the exercise, even if it were possible to answer without having clear insight into the $n$-cube's significance.

During the interviews, I chose to intervene as little as possible in order to avoid influencing the students' solving processes. The only hint I gave was to indicate where the diagonal of a cube lies in dimension 3, if they had drawn such a cube and needed such a hint. There were two main ways to solve the problem accessible to second-year students.

*Analytic solution method.* The first one possible solution method was analytic. One of the diagonals of the cube can be represented as a vector[7] $\overrightarrow{OA}$, where $O$ has coordinates $(0, 0, ..., 0)$ and $A$ has coordinates $(1, 1, ..., 1)$. Thus the length of the diagonal is $|\overrightarrow{OA}| = \sqrt{n}$. In this case, considering the problem for $n = 2$ or $n = 3$ can be helpful; it allows one to find the coordinates of $A$ for these values of $n$. That result can then be immediately generalized. The students who solved the exercise in this way were using the $\mathbb{R}^2$-$\mathbb{R}^3$ model as a paradigmatic model (in Fischbein's terminology). Then they moved up to $\mathbb{R}^n$ using coordinates. That move can be purely algebraic; but it can also be helped by the use of a figural model displaying the coordinates of $A$ for $n = 2$ or $n = 3$.

---

[7]It can also be interpreted as a segment.

*Inductive solution methods.* A second solution is by an induction process. For $n = 2$, the length of the diagonal is $\sqrt{2}$. One assumes that the length of the $(n-1)$-cube's diagonal is $\sqrt{n-1}$. The diagonal of the $n$-cube is then the sum of a diagonal of the $(n-1)$-cube and of a unit vector orthogonal to it. From the Pythagorean Theorem, the length of the diagonal is thus $\sqrt{n}$ (this is the vector version; the diagonals can also be interpreted as the sides of a right triangle). A related solution method requires consideration of the case $n = 2$ and also of the process leading from $n = 2$ to $n = 3$. That process provides the key to the induction; the associated drawing could display the diagonal of a side and a unit vector orthogonal to it.

*Students' solution attempts.* The eight students' answers fell into three groups.

*Group 1: The obstacle of geometrical vocabulary (Students: Ana, Barbara, Charles, Diane).* These students did not overcome the obstacle of vocabulary. They were not able to confer any meaning to the term "cube" in $\mathbb{R}^n$.

> *Ana:* For $n = 3$, it's as cube... For $n$? I do not see what a cube can be! For $n = 3$, the diagonal of a face is $\sqrt{2}$; it gives $\sqrt{3}$ ... But for dimension $n$...

Two students did not try to use a drawing; the other two drew only the 3-cube. They all solved the problem in dimension 3 and calculated $\sqrt{3}$. Barbara also mentioned a square as a cube in dimension 2 and calculated $\sqrt{2}$ in that case. But she did not link the two cases (dimension 2 and 3); she did not identify a generalizable process.

*Group 2: Using coordinates (Student: Edouard).* There was only one student in Group 2. I give here details about his reasoning process because it was very different from that of all the others. He said at the beginning: "I first do it in dimension 2." Then he made a drawing (Figure 2). He calculated the value $\sqrt{2}$ immediately after plotting the diagonal and wrote that on his drawing.

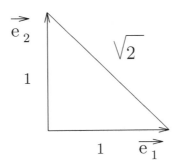

FIGURE 2. Student Edouard. Diagonal of the cube in $\mathbb{R}^2$.

He went on, speaking and drawing simultaneously:

> *Edouard:* I do the same now in dimension 3. $e_1$, $e_2$, $e_3$ is an orthonormal basis of $\mathbb{R}^3$. The diagonal of the cube is $\overrightarrow{OA}$, $A$ is there... So the coordinates of $A$ are all equal to 1, it gives $\sqrt{3}$... And it will always be the same thing, the coordinates of $A$ are 1, 1, 1 ... So $|\overrightarrow{OA}| = \sqrt{n}$.

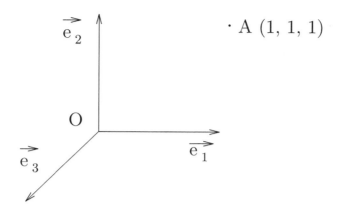

$$\cdot A\,(1,\,1,\,1)$$

FIGURE 3. Student Edouard. Diagonal of the cube in $\mathbb{R}^3$.

The corresponding drawing is shown in Figure 3.

Edouard was the only student who produced a drawing with vectors. He was also the only student who made a drawing but did not draw a complete 3-cube. On his first drawing, he plotted the diagonal; on the second, only its endpoints. He immediately thought in terms of vectors and arrows. Then he recognized an orthonormal basis on his first drawing, and was thus led to use coordinates. He identified a linear algebra context; then instead of using drawings stemming from secondary school geometry like the others, he used a figural model associated with linear algebra. It helped him answer the question for $n = 2$ and $n = 3$, and to formulate the problem with coordinates. Then the move from the $\mathbb{R}^2$-$\mathbb{R}^3$ model to the general case became obvious to him.

*Group 3: Induction process (Students: Fanny, Guy, Henri).* These three students first drew a square and then calculated $\sqrt{2}$ for $n = 2$; then they drew a cube and used the Pythagorean Theorem, explicitly or not, to compute $\sqrt{3}$ for $n = 3$. They all claimed that the general result was $\sqrt{n}$. They all used a kind of induction process, but none of them produced a rigorous proof. The most algebraic reasoning was produced by Fanny, who proposed no geometrical interpretation or justification for its generalization:

> *Fanny*: For $n = 2$, it gives $\sqrt{a^2 + a^2} = a\sqrt{2}$. Then for $n = 3$, I have $\sqrt{a^2 + 2a^2} = a\sqrt{3}$. Then it will go on the same way, there is always another $a^2$, and you get $\sqrt{a^2 + (n-1)a^2} = a\sqrt{n}$.

(rather than labeling them with 1, she labeled the edges $a$). Even if the calculations were not very different from those produced by thinking in terms of coordinates, the reasoning process was not the same. She was first dealing with lengths of segments and then with algebraic expressions, without any geometrical interpretation.

By contrast, the reasoning of Henri was based on geometrical statements. After calculating $\sqrt{2}$ for $n = 2$ and using it to deduce $\sqrt{3}$ for $n = 3$, he said:

> *Henri*: I would say then there are 1, and $\sqrt{3}$, and it is orthogonal, so it gives $\sqrt{4} = 2$, but I'm not sure, I do not see it clearly... But I think it works, because $\sqrt{3}$ is the diagonal, and the last edge is

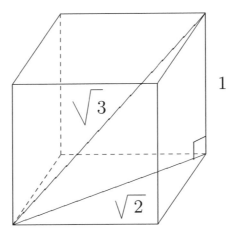

FIGURE 4. Student Henri. The diagonal of the cube in $\mathbb{R}^3$.

orthogonal to it. So it is actually 2 for $n = 4$. And it is always the same, so you obtain $\sqrt{\sqrt{\sqrt{1+1}^2 + \dots + 1}^2 + 1} = \sqrt{n}$.

At first he appeared to feel embarrassed because he lacked a picture depicting the $n$-cube. But his drawing of the 3-cube displayed a right triangle formed by the diagonal of a face, an edge of the cube, and the corresponding diagonal of the cube (as in Figure 4).

After producing this drawing, Henri focused on the orthogonality of the diagonal of a face and an edge of the cube. Even if he did not explicitly interpret the diagonal of the 3-cube as the diagonal of a "face" of the 4-cube, he used it that way; this allowed him to compute the result for $n = 4$. He did not try to provide any further geometrical interpretation, but immediately generalized his result for any value of $n$.

The observations described in this section came from a clinical study about a specific task, conducted with a small number of students. However, the behaviors described illustrate the more general phenomena discussed below.

**5.3. Conclusions.** The first issue I address here is the use of drawings. They played a central part in the interview exercise. The two students who did not make any drawing did not produce a solution. The one student who used vectors was immediately led to the solution. Representing the whole cube did not help. But the four students who represented a square, then a 3-cube, and interpreted the faces of the 3-cube as squares (one of them even drew the 3-cube *on* the square), found the solution.

5.3.1. *Using Coordinates.* Recall that the teacher said: "If they can do it in $\mathbb{R}^3$, they can do it." The interviews with students invalidated that claim. The teacher may have thought that the students were going to solve the exercise analytically, but only one student did so. The word "cube" placed all the other students in a geometrical context where they stayed captured. Moreover, during their tutorials, the students had never encountered tasks that requested analytic solutions where no coordinates were given in the task statement.

Using coordinates can be a good way to move from dimension 2 or 3 to dimension $n$. But students may not use such a process in their reasoning if coordinates are not a familiar tool for them. In particular, they must be able to formulate problems presented in the abstract, or geometric, mode in terms of coordinates (Hillel, 2000). Fostering this ability takes a specially designed course.

5.3.2. *Increasing the dimension.* The teacher also made a more general claim about quadratic forms and inner product spaces: "The important point is to be able to move from dimension 2 to dimension 3."[8] The observations in the particular interview exercise studied here are consistent with that claim. The four students who did not identify the process that led from $n = 2$ to $n = 3$ did not solve the exercise. The four others found that the length of the diagonal was $\sqrt{n}$.

Moving up from dimension $n - 1$ to dimension $n$ is a complex process. In particular, it is necessary to interpret the space of dimension $n - 1$ as a hyperplane of the space of dimension $n$. Mathematicians are quite used to such processes; they readily consider the first space as an hyperplane and then add a supplementary line to it, to obtain the whole $n$-space. This is a difficult process for students and requires familiarity with the notion of subspaces. Such a supplementing process could be explicitly addressed in linear algebra tutorials. The $\mathbb{R}^2$-$\mathbb{R}^3$ model could be used to study how to move from dimension 2 to dimension 3.

5.3.3. *Decreasing the dimension.* For the task discussed here, it was sufficient at each stage of the induction to reason in a plane that contained the $(n-1)$-diagonal and the $n$-diagonal. This is a familiar process for mathematicians. Many general results can be established by reasoning in a well-chosen 2-dimensional space, and it is then possible to help the reasoning with a drawing. The actual $n$-dimensional object cannot be pictured, but it is always possible to cut it along a plane and represent the obtained section. Henri was led to a solution by the use of such a process in $\mathbb{R}^4$. That possibility (cutting and drawing) could be explicitly emphasized in a linear algebra course. In that case, there would be no intervention of a geometric model. The figural model would be directly associated with the pertinent part of the general situation.

## 6. Conclusion

In Sections 3, 4, and 5, I reported on different aspects of my work: a mathematicians' questionnaire, results from a textbook study, and interviews with teacher and students. Some of the results obtained are general statements about the use of geometric or figural models in linear algebra; others are relative to the use of the $\mathbb{R}^2$-$\mathbb{R}^3$ model. I will now synthesize the answers provided by these approaches under each of the research questions presented in Section 2.

### 6.1. What are the possible uses of geometric models in linear algebra? The first answer is: it appears that linear algebra cannot be taught nor learned as a mere generalization of a geometry. The historical development of linear algebra (Dorier, 2000) has indicated that the modern theory emerged from the necessity of unification of several mathematical domains. The intellectual need for linear algebra (I refer here to Harel's (2000) Necessity Principle) is grounded in the unification of several mathematical domains.

---

[8]Banchoff and Wermer make a similar claim in their book. It can be formulated as, "If students can move up from $\mathbb{R}^3$ to $\mathbb{R}^4$, then they will have no problem with $\mathbb{R}^n$."

Yet, a geometric model can be helpful, especially because the associated figural model confers on the geometric model the appearance of concreteness. For example, the $\mathbb{R}^2$-$\mathbb{R}^3$ model allows one to present some notions and results of linear algebra before introducing the general theory as some properties appear as self-evident in a geometric context. The use of coordinates allows one to move up to $\mathbb{R}^n$. However, there is no evidence that this model can be used effectively to introduce abstract vector spaces.

**6.2. How do mathematicians and students use geometrical and figural models in linear algebra?** Most of the mathematicians in France advocate one of two opposite approaches. The first group advocates a structural approach to linear algebra, without geometrical or figural models. The second group recommends a geometry course before linear algebra so that geometry can then provide models and the associated drawings. In both cases, however, the mathematicians do not develop a figural model specifically for linear algebra; their drawings are only used in a geometrical context.

I did not discuss in this paper the general use of geometric models by students. Yet, the importance of familiarity with models must be emphasized. It was a direct consequence of the Necessity Principle (Harel, 2000) and was also observed by Sierpinska (2000). Students may use familiar models in their reasoning processes, even when a teacher proposes a geometric model for linear algebra. These familiar models can be inappropriate for linear algebra.

**6.3. What are the consequences of the observed uses of models on students' practices and thinking processes?** I studied this question in a particular context: the use of the $\mathbb{R}^2$-$\mathbb{R}^3$ model by students in their solving of a problem stated in geometric language in $\mathbb{R}^n$. The model can help students find an algebraic description of the problem that can be generalized to higher dimensions, provided they use coordinates. That possibility is strongly linked with the existence of an appropriate figural model that allows one to derive an algebraic description.

Understanding the process leading from $\mathbb{R}^2$ to $\mathbb{R}^3$ provides another possibility, if the student extends that process so as to use it in going from $\mathbb{R}^{n-1}$ to $\mathbb{R}^n$. The figural model is fundamental in that case as well, to help in understanding the generalization mechanism. But the model can also have negative effects for some students who stay captured in the geometrical context.

These results indicate that geometric models must be used carefully in linear algebra courses. Geometry cannot be the only starting point for linear algebra; other domains must intervene to justify the need for a general theory.

The geometric model requires long-term teaching so that it will become very familiar to students. A figural model, specially intended for linear algebra, should be presented. However, the uses of such a model for general vector spaces requires additional research.

Moreover, because geometric models belong to dimension 2 or 3, it might be useful to include in a linear algebra course the study of processes used by experts as they move from dimension $n$ to dimension $n+1$. And to discuss how they recognize, in an $n$-dimensional problem, that the main phenomenon occurs in a well-chosen 2 or 3-dimensional space. Linear algebra teaching could integrate these possibilities in order to try explicitly to develop students' geometric intuition. For the moment, this does not seem to be done by mathematicians, at least in France.

# References

Banchoff, T., & Wermer, J. (1991). *Linear algebra through geometry*. New York: Springer Verlag.

Brousseau, G. (1998). *La théorie des situations didactiques*. Grenoble, France: La Pensée Sauvage.

Choquet, G. (1964). *L'enseignement de la géométrie*. Paris: Hermann.

Dieudonné, J. (1964). *Algèbre linéaire et géométrie élémentaire*. Paris: Hermann.

Dorier, J-L. (Ed.). (2000). *On the teaching of linear algebra*. Dordrecht, The Netherlands: Kluwer.

Dubinsky, E. (1991). Reflective abstraction in advanced mathematical thinking. In D. Tall (Ed.), *Advanced mathematical thinking* (pp. 95–123). Dordrecht, The Netherlands: Kluwer.

Fischbein, E. (1987). *Intuition in science and mathematics: An educational approach*. Dordrecht, The Netherlands: D. Reidel.

Fischbein, E. (1993). The theory of figural concepts. *Educational Studies in Mathematics, 24*, 139–162.

Gueudet-Chartier, G. (2000). *Rôle du géométrique en algèbre linéaire*. Thèse de doctorat, Laboratoire Leibniz, Université Joseph Fourier, Grenoble.

Harel, G. (2000). Three principles of learning and teaching mathematics. In J-L. Dorier (Ed.), *On the teaching of linear algebra* (pp. 177–189). Dordrecht, The Netherlands: Kluwer.

Hillel, J. (2000). Modes of description and the problem of representation in linear algebra. In J-L. Dorier (Ed.), *On the teaching of linear algebra* (pp. 191–207). Dordrecht, The Netherlands: Kluwer.

Sierpinska, A., Defence, A., Khatcherian, T., & Saldanha, L. (1997). A propos de trois modes de raisonnement en algèbre linéaire. In J-L. Dorier (Ed.), *L'Enseignement de l'algèbre linéaire en question* (pp. 249–268.). Grenoble, France: La Pensée Sauvage.

Sierpinska, A., Dreyfus, T., & Hillel, J. (1999). Evaluation of a teaching design in linear algebra: The case of linear transformations. *Recherches en didactique des mathématiques, 19*, 7–40.

Sierpinska, A. (2000). On some aspects of students' thinking in linear algebra. In J-L. Dorier (Ed.), *On the teaching of linear algebra* (pp. 209–246). Dordrecht, The Netherlands: Kluwer.

## Appendix A: Mathematicians' Questionnaire

### 1. Use of geometry in the exercises.

**Question 1.1:** The following exercise is often proposed to second year students.

*Let $E$ be an inner product space, and $p$ a projection of $E$. Prove that $p$ is an orthogonal projection if and only if, for all $x$ in $E$, $|x| \geq |p(x)|$.*

- What solution would you present to the students?
- Are there geometrical aspects in the solution you propose, and for what purpose do you use them?
- Would you use a drawing in a solution presented to the students? If the answer is positive, which drawing would you use, and what do you expect from the use of that drawing?

**Question 1.2:** The following exercise is often proposed to first year students.

*Let $E$ be a vector space, and $x$, $y$, $z$ three vectors in $E$, linearly independent by pairs. Is the set of the three vectors $(x, y, z)$ linearly independent?*

- If you observe during a tutorial a student who says that (s)he is sure that the answer is positive, but (s)he can not find a proof, what do you tell him (her) to help? (Give a precise answer, and explain the reason for your choice).
- The same exercise is proposed in an examination. A student proposes the following solution:
  *No, the vectors drawn hereby provide a counter-example.*
  With the drawing of Figure 5.

FIGURE 5

- What mark, between 0 and 5, do you attribute to this answer?
- Which comments do you write on the student's sheet?
- Explain your mark and comments.

### 2. Use of the geometry taught in secondary school.

**Question 2.1:** In secondary school, the students encounter the words "basis" and "orthonormal basis" in the geometry courses. Do you think that some of the properties, techniques, results...presented in secondary school can be used at university in the linear algebra courses? If your answer is negative, explain why. If your answer is positive, present the results you consider useful and explain how they can be used.

**Question 2.2:** In secondary school, the students encounter the words "projection" and "orthogonal projection" in the geometry courses. Do you

think that some of the properties, techniques, results...presented in secondary school can be used at university in the linear algebra courses? If
your answer is negative, explain why. If your answer is positive, present
the results you consider useful and explain how they can be used.

## 3. Use of drawings in linear algebra

**Question 3.1:** For each of the drawings in the table, indicate if you use it in
your linear algebra courses; if the answer is "yes" indicate which notions
or properties you illustrate with it. (You can mention several uses of the
same drawing.)

| Drawing | Used (yes/no) | That drawing illustrates |
|---------|---------------|--------------------------|
| 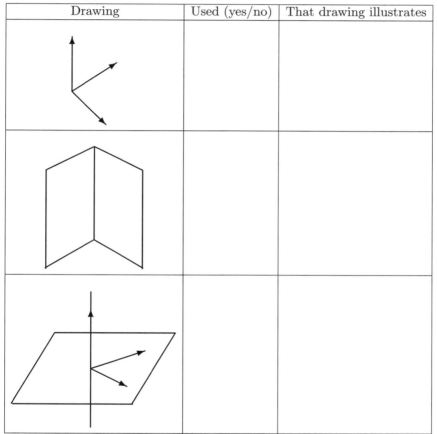 | | |

**Question 3.2:** If you use other drawings, draw them in the following table,
and indicate the interpretation(s) you associate with them. (A blank table
with five lines followed.)

## Appendix B: Students' Questionnaire

**Drawings on the exam** This section refers to the text of an exam passed by the students two weeks before the interviews. I also used their own exam sheets. Here are the two exercises of the exam that I used in the interviews.

### Exercise 1 of the exam
*Let $q$ be the quadratic form defined by: $q(x) = (2x_1 + 3x_2)^2$. Give its matrix, its rank, and draw its isotropic cone.*

### Exercise 2 of the exam
*Let $q$ be the quadratic form defined on $\mathbb{R}^3$ by:*
*$q(x) = 3x_1^2 + 2x_2^2 - x_3^2$. Give its rank, and draw its isotropic cone. Is $q$ positive ?*

About Exercise 1 of the exam, I asked the following question:
The isotropic cone is a straight line that can be illustrated by the following figure. How would you represent an element of that cone?

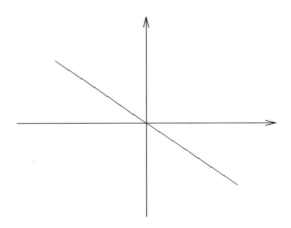

About Exercise 2 of the exam, I asked the following questions:
If the student drew axes, does he or she think that these axes were part of the requirement? Was the drawing useful to answer the question: *Is $q$ positive?*

**Other Exercises**

### Exercise 1
*Each of the drawings in the figure represents a subset $E_i$ of $\mathbb{R}^2$. In each case, indicate if it is possible to find a quadratic form $q$ of $\mathbb{R}^2$ such that:*
    *— $E_i$ is the kernel of $q$?*
    *— $E_i$ is the isotropic cone of $q$?*

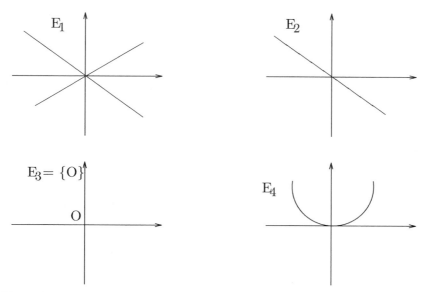

## Exercise 2

*Let $E = \mathbb{R}_3[X]$ be the inner product space of degree 3 polynomials and let $P$ and $Q$ be two orthogonal elements of $E$ whose length is 1. Can you determine the length of $P + Q$ ?*

## Exercise 3

*Find the length of a diagonal of a cube with edges of length 1 in $\mathbb{R}^n$.*

DidMaR, Université de Rennes 1, Campus de Beaulieu, 35042 RENNES CEDEX, France

*Current address*: IUFM de Bretagne, 153, rue de Saint Malo, 35000 RENNES, France

*E-mail address*: `Ghislaine.Gueudet@bretagne.iufm.fr`

CBMS Issues in Mathematics Education
Volume **13**, 2006

# Investigating and Teaching the Processes Used to Construct Proofs

## Keith Weber

ABSTRACT. Even if students possess an accurate conception of proof and an understanding of the facts and theorems about a mathematical concept, they still may be unable to prove theorems about that concept. One reason that students cannot construct proofs in these cases may be that they do not have appropriate decision-making strategies. In this paper, I address this issue by describing a prescriptive procedure that a student can use to construct proofs about group homomorphisms. I argue that this procedure is effective, mathematically appropriate, and indicative of mathematical thinking. I report two studies in defense of these claims. In the first study, students could prove a large number of theorems by employing this procedure. I describe instruction designed to teach students to apply this procedure and report on a second study in which five students received this instruction. Students were able to prove significantly more theorems after receiving this instruction. The results and significance of the two studies are discussed.

## 1. Introduction

The ability to prove theorems is a valuable skill for any student studying mathematics. Indeed, this ability is often a primary goal in most advanced mathematics courses and is typically the sole means of assessing students' performance. Despite its importance, there is a large body of evidence indicating that students at all levels have difficulty constructing proofs (Harel & Sowder, 1998; Hart, 1994; Moore, 1994; Schoenfeld, 1985b; Senk, 1985; Weber, 2001b, 2002b). Further illustrating this difficulty is the emphasis put on reforming proof instruction in the National Council of Teachers of Mathematics standards (National Council of Teachers of Mathematics, 2000) and by the appearance of "how-to" textbooks on proofs (Cupillari, 1989; Franklin & Daoud, 1988; Solow, 1990; Watson, 1978).

As most students cannot construct proofs effectively in advanced mathematical domains, it is natural to try to locate the cause of their difficulties. There has been a great deal of research on this issue, much of which can be broadly categorized into three classes. The first cause of students' difficulties is that they often possess an inaccurate conception of what constitutes a mathematical proof (Harel & Sowder, 1998; Knuth & Elliot, 1998; Martin & Harel, 1989). For instance, some students believe an argument using intuitive reasoning or empirical arguments constitutes a proof. Others believe that a proof of a theorem is valid if and only if it follows a traditional or "ritual" proof format, such as the two-column proofs taught in some high school geometry courses (Harel & Sowder, 1998; Martin & Harel, 1989).

A second cause of students' difficulties is that they may not have an understanding of a theorem or a concept and may systematically misapply it (Harel, 1998; Hazzan & Leron, 1994; Moore, 1994). For example, in abstract algebra many students use Lagrange's theorem to infer incorrectly that any three elements of a group of order six necessarily form a subgroup (Hazzan & Leron, 1994). Other students, in linear algebra, may be so intent on proving theorems by manipulating formulae that they do not reason about what the manipulated symbols represent and may execute inappropriate operations such as dividing by a vector (Harel, 1998).

Though an accurate conception of proof and an ability to derive logical inferences are clearly necessary for proof-writing competence, these skills are not sufficient. A third reason that students cannot construct proofs is that they do not have the decision-making strategies to do so (Schoenfeld, 1985b; Weber, 2001b). In this paper, I focus on students' decision-making processes while constructing proofs. How one chooses a proof technique or a theorem to apply is a complicated decision that is poorly understood (Weber, 2001b). Reif (1995) claimed that in order to teach an ability effectively, "one needs to understand the underlying cognitive mechanisms (knowledge and thought processes) required to achieve [that] ability." In his review of research on proof-writing, Hart (1994) remarks on the paucity of cognitively-based studies on proof processes and concludes more research needs to be done.

In this paper, I explore the issue of proof construction within the context of group homomorphisms, a central concept of group theory. While there has been some educational research within the domain of group theory (Asiala, Dubinsky, Mathews, Morics, & Oktac, 1997; Dubinsky, Dautermann, Leron, & Zazkis, 1994; Hart, 1994; Hazzan, 1999; Leron & Dubinsky, 1995; Weber, 2001b), more work is needed. In particular, there has been very little research investigating the topic of homomorphisms. Also, while there has been a great deal of educational research on mathematical proof, much of this research has either focused on proof in the abstract or at the pre-college level. There has been limited research on proof within the context of advanced mathematical domains. The existing research of this type has primarily investigated students' inability to construct proofs (Alcock & Simpson, 2002; Bills & Tall, 1998). More research is needed on how students can effectively construct proofs within advanced mathematical domains and on how one can successfully teach this ability.

The purpose of this paper is to address the following questions:

(1) How can one describe a set of cognitive processes that an undergraduate can use to prove statements about group homomorphisms?
(2) How can one teach undergraduates to use these thought processes to construct proofs?

To address the first question, I describe a prescriptive procedure that one can use to construct proofs and give evidence of the effectiveness of this procedure. To address the second question, I present one way that this procedure can be taught using an apprenticeship model for instruction (Collins, Brown, & Newman, 1989) and report on a study in which undergraduates received this instruction. Although there are important issues that must be addressed before making definitive claims, the results were positive: The undergraduates in the study were able to prove significantly more theorems after receiving this instruction.

**1.1. Theoretical Perspective.** To address the questions investigated in this paper, I adopted an information-processing (IP) theoretical perspective. The central hypothesis used in IP research is that all of human thought can be understood in terms of symbol structures and the processes that manipulate these symbols. A corollary of this hypothesis is that intelligent human behavior can be modeled by systems that have the capacity to store and manipulate symbolic structures, such as computers (Newell & Simon, 1972). In fact, many IP researchers metaphorically view the human mind as a computer and much of IP research consists of creating computer programs that model human problem-solving behavior.

A critical assumption used in much IP research is that human behavior can be decomposed into a relatively small set of basic cognitive processes. Of course some aspects of human behavior, such as mathematical problem solving or constructing proofs, are incredibly complex. However this complexity is due to the complicated composition of basic processes just as a computer can produce complex aggregate behavior from a small set of primitive commands (Anderson & Schunn, 2001). While there is no single methodology that is used in all IP research, the following approach is common in IP research with an educational orientation. After selecting a type of problem that one would like students to be able to solve, the researcher attempts to specify the knowledge and behavior that produce competent performance at solving these problems. Such a specification, sometimes called an "idealized model," is constructed with enough precision and detail that an individual could use this model to solve problems that he or she had been previously unable to solve. This approach has been used to reveal a great deal of information about mathematical problem solving (Schoenfeld, 1985b) and has served as the basis for the design of effective pedagogy (Anderson, Boyle, Corbett, & Lewis, 1990; Bagno, Eylon, & Ganiel, 2000; Heller & Reif, 1984; Palinscar & Brown, 1984; Reif & Scott, 1999; Schoenfeld, 1985b, 1998). Some of this pedagogy will be discussed in Section 1.3. A more thorough overview of this approach can be found in Schoenfeld (1987, p. 9-20) and Anderson and Schunn (2001).

In the 1980s, problem-solving was a main focus of mathematics educators (National Council of Teachers of Mathematics, 1980) and the IP theoretical perspective was commonly used to investigate this topic. However, by the end of the decade, mathematics education research had shifted away from determining how students solve difficult mathematical problems and toward investigating how students' understanding of mathematical concepts develops, often within the context of a social setting. For a description of this change, see Davis' (1992) article describing the goals of the mathematics education community at that time. Due to this shift in attention, and to perceived weaknesses in the IP approach, the IP perspective is not frequently used in current mathematics education research.[1] Research using an IP perspective tends to focus on performance, but often ignores understandings that support it (Schoenfeld, 1994). Further, some have argued that IP theories of mind cannot account for the cognitive mechanisms required for a student to develop mathematical understanding (Cobb, 1990).[2] A general consensus among

---

[1]The IP approach is still widely used in cognitive psychology, as well as in other branches of educational research, such as physics (Bagno et al., 2000; Reif, 1995; Reif & Scott, 1999).

[2]Some cognitive psychologists strongly disagree with these claims (Anderson & Schunn, 2001).

mathematics educators is not that this type of research is wrong, but rather incomplete; it ignores (and perhaps is not equipped to address) the questions that
these researchers believe are most important (Davis, 1996; Ohlsson, 1990; Ohlsson, Ernst, & Rees, 1992; Schoenfeld, 1992). In the last decade, research using an
IP perspective has rarely been published in the mainstream mathematics education
journals.[3] In 1994, Schoenfeld reported that there had been "virtually no such work
at the college level" (Schoenfeld, 1994), an assertion that still holds true today.

**1.2. Theoretical Assumptions.** The theoretical assumptions in which this
work was oriented can be summarized as follows:

(1) Proof construction is a highly complex problem-solving task in which one
    wishes to derive a particular piece of information in a logically permissible
    way.
(2) The complex task of constructing proofs within a particular mathematical
    domain can be decomposed into a set of more basic facts, algorithms, and
    decision-making processes.
(3) A primary reason that students have difficulty with writing proofs is that
    they lack effective decision-making processes. (This assumption has em-
    pirical support (Weber, 2001a).) One reason for this shortcoming is that
    instruction often is not aimed at fostering students' development of effec-
    tive proof-constructing strategies.
(4) By outlining a procedure (i.e. a set of processes) that is effective at con-
    structing proofs, one can design instruction to help students apply this
    procedure. This can lead students to become significantly more capable
    at constructing proofs.
(5) Students can and should learn to apply this procedure in a mathematically
    meaningful way. By mathematically meaningful, I mean that:
    (a) Students will be behaving in a manner that is indicative of mathe-
        matical reasoning (in the eyes of a teacher or mathematician);
    (b) The procedure will not be divorced from their conceptual understand-
        ing;
    (c) The students will know why the procedure is a useful method for
        constructing proofs.

As previously noted, research using such a paradigm is uncommon in mainstream
mathematics education research. Nonetheless, I believe that research using an IP
approach can contribute to collegiate mathematics education research. First, this
approach can be used to investigate aspects of advanced mathematics that currently
receive little attention in the mathematics education literature. Most educational
research in advanced mathematical courses focuses on how students can develop
their understanding of mathematical concepts. For instance, Hazzan (1999) claimed
that educational research in abstract algebra has emphasized concept acquisition
and pedagogical techniques. There has been very little research on how one can use
their understanding of group theoretic concepts to construct proofs. The research
in this paper focuses on this important issue. Second, many researchers in mathematics education stress that looking at the same issue from different theoretical

---

[3]John Anderson and his colleagues at Carnegie Mellon (Anderson et al., 1990; Anderson &
Schunn, 2001) as well as others (Hecht, 1998) continue to publish mathematics education research
within this framework. However the target audience for this research consists primarily of cognitive
psychologists and the artificial intelligence community.

perspectives is critically important (Schoenfeld, 1994). In particular, influential researchers, such as Robert Davis (1996), have encouraged mathematics educators to give more attention to research using psychological paradigms that may have fallen out of favor. Much of the educational research on abstract algebra has been conducted using constructivist paradigms. Research using an IP theoretical perspective can compliment this existing research by investigating the issues from a different point of view.

**1.3. What can be gained from this type of research?** Based on Reif's (1995) paradigm of pedagogy, teaching the skill of proof construction can be viewed in the following way. Prior to instruction, a student has an initial set of cognitive processes (as well as factual knowledge, conceptual understanding, beliefs) that he or she will invoke when trying to construct a proof. Instruction can be viewed as a deliberate attempt to lead students to transform their initial processes to an improved set of processes (as well as improved factual knowledge, conceptual understanding, beliefs) that allow them to construct substantially more proofs. As educators, we are trying to lead students to acquire an improved set of processes and it would be helpful to have a clear and explicit conception of what these processes are.[4] Thus, a description of a set of processes sufficient to construct the types of proofs that an undergraduate student would be expected to construct would be beneficial in the sense that it would specify an explicit model of desired student performance.

Once the desired student performance has been specified, the task of the researcher or instructor is to lead students to achieve this performance. This research paradigm has been used to develop many successful instructional units. For example, John Anderson and his colleagues developed a "production system" (i.e., a highly specified procedure in the form of if-then rules) to prove high school geometry theorems. They taught students to apply their procedure by first having them apply the most basic components, and then having them apply successively more complicated components until students had mastered the entire procedure. Instruction took place by way of a computer tutor that provided immediate feedback when the student performed an action inconsistent with the desired production system. Empirical studies demonstrated that students could prove far more geometry theorems with less total study time when traditional instruction was supplemented with the computer tutoring for procedures (Anderson et al., 1990).

Palinscar and Brown (1984) delineated a set of specific metacognitive questions that successful readers ask themselves to assess whether they comprehended what they have just read. Using a modeling-scaffolding-fading approach, Palinscar and Brown created instruction that they labeled "reciprocal teaching" designed to teach at-risk students to monitor their understanding by asking these questions. Students' reading comprehension test scores after receiving this instruction showed dramatic gains, and Palinscar and Brown's reciprocal teaching instruction has been widely cited as a paradigm of successful educational research.

Schoenfeld outlined specific heuristics and metacognitive behaviors that he believed would lead to effective mathematical problem solving (Schoenfeld, 1985b).

---

[4]Focusing on processes is characteristic of research using an information processing approach. However, it is worth noting that improving students' conceptual knowledge, internal representations of group theoretic concepts, or beliefs can influence the proof strategies that they will invoke (Moore, 1994).

Using a variety of pedagogical techniques, Schoenfeld (1998) created his now famous problem-solving course. His techniques included renegotiating the classroom's didactical contract to organize his students as a community of learners, as well using an "apprenticeship" model of instruction (Collins et al., 1989). In summary, these researchers all designed effective instruction once they had an explicit model of desired student performance in mind. How they went about developing that instruction varied, depending upon, among other factors, the philosophies of the researchers.

**1.4. Criteria for an Effective Procedure.** The procedure described in this paper was created with the goal of prescribing a set of processes that an undergraduate can use to construct proofs about group homomorphisms. There are several desirable features that such a procedure should have. First, the procedure should be sufficient to construct the types of proofs that it was designed to do. While it would be unreasonable to expect that this procedure would be capable of proving every proposition about homomorphisms that a student might encounter, a student applying this procedure should exhibit substantial improvement and be able to prove a reasonable percentage of such propositions. Second, this procedure should be compatible with students' abilities. It should not, for instance, overload their working memory, ask them to execute an unreasonable number of steps, or require students to apply factual or procedural knowledge they do not possess. Third, the procedure should be precise and specific. While a certain degree of flexibility is necessary in any method designed to accomplish a wide range of tasks, it is also true that vague and imprecise methods are very difficult to teach effectively. There is a large body of research literature indicating that instruction designed to teach heuristic methods, without an explicit and precise understanding of how those methods are applied, is usually ineffective (Begle, 1979; Schoenfeld, 1985b). Finally, this procedure should be one that students can apply in a mathematically meaningful way (as defined in section 1.1). Lester and Garafalo (1982) have described effective but conceptually barren methods a child can use to solve arithmetic word problems. (e.g., the word "left" in a problem is a cue find two numbers in the problem and subtract the smaller from the larger). Schoenfeld (1988) reported a case study in which one mathematics teacher taught strategies that were quite effective for achieving high scores on formal assessments, but probably did the students more harm than good. Simply because a procedure can accomplish a task effectively does not imply that it is desirable to teach.

Thus, the procedure used in this work was evaluated using the following criteria. First, could this procedure accomplish what it was designed to do? That is, could it be used to construct proofs about group homomorphisms? Second, would students who were taught to apply the procedure be able to prove more statements in a meaningful way after they received this instruction?

In Section 2, I describe a prescriptive procedure that one can use to prove the types of statements about group homomorphisms that an undergraduate would encounter in a first abstract algebra course. In Section 3, I demonstrate that both a computer and students can use this procedure to prove these statements. In Section 4, I describe one way that students can be taught to apply the procedure. I then report the results of a study in which students received this instruction. Finally, in Section 5, I offer suggestions for implementing similar types of teaching in advanced mathematical classrooms.

## 2. A Prescriptive Procedure to Construct Proofs

**2.1. Experts and Novices Constructing Proofs in Group Theory.** The procedure discussed in this section is largely based on two previous studies in which I observed students constructing proofs (Weber, 2001b, 2002b; Weber & Alcock, 2004). In each of these studies, I observed two groups of students. The first group consisted of undergraduates who had recently completed an introductory course in abstract algebra. All of these undergraduates could write basic proofs. The second group was made up of advanced graduate students who were completing dissertations in an algebraic topic. I observed participants, individually, as they proved either a non-trivial statement about group homomorphisms (Weber, 2001b) or about group isomorphisms (Weber, 2002b; Weber & Alcock, 2004). After their proof attempts, I asked all participants to describe why they had tried to prove the theorem in the way they had and then tested them to see if they were aware of the relevant facts and theorems needed to construct the proof. If they had been unable to write a proof but were aware of the necessary facts and theorems needed to construct the proof, I asked them to try to construct the proof again using these facts and theorems.

The undergraduates were not able to prove many of the statements. The eight undergraduates in the two studies could construct proofs 20% of the time. Due to the design of the studies, I was able to restrict analysis to statements where the undergraduates were aware of the facts and theorems necessary to construct the proofs, and the undergraduates could potentially write a proof if I told them specifically which facts to use. Even in these cases, without my assistance the undergraduates were still able to successfully construct a proof only 32% of the time. Analysis of their proving behavior indicated that primary causes of the undergraduates' difficulties included how they viewed the task of proving and the nature of the strategies that they employed. The undergraduates rarely considered the structural properties of the concepts that they were working with, instead they focused on logically manipulating these concepts' definitions. For instance, when trying to prove or disprove that two groups were isomorphic, students did not use the fact that isomorphic groups have the same structure, even though most were aware that isomorphic groups must share the same group theoretic properties. Their proof strategies generally consisted of manipulating logical symbols. For example, to prove a statement B, some undergraduates would search for a theorem of the form "A implies B." They would then try to prove the antecedent A without considering whether A might be true.[5]

The doctoral students' proof attempts differed significantly from those of the undergraduates. Before deriving inferences, the doctoral students would first try to understand the statement that they had been asked to prove. Specifically, they would tend to view these statements as requiring them to find relationships between structures (e.g. disproving that two groups were isomorphic would involve showing that they were structurally different). Their proof strategies included invoking powerful theorems that illustrated relationships between mathematical structures (e.g.,

---

[5]For a dramatic illustration of this, one undergraduate tried to prove a group G is abelian by showing that the square of every element in G is the identity. Although this would certainly be sufficient to show that G is abelian, there was no reason to suppose that such a strong condition would necessarily hold.

the First Isomorphism Theorem). In both previous studies the doctoral students achieved near perfect performance.

These results from earlier work suggest that if undergraduates viewed statements about homomorphisms as exhibiting relationships between mathematical structures and employed effective proof strategies, then their performance in proof-writing might improve substantially. Based on this conclusion, I developed a procedure that was designed to accomplish this. The goal was not to model the thought processes of experts' constructing proofs. The doctoral students' decision-making processes were based on their extensive experience with algebraic concepts, experience that undergraduates do not share. Rather, the purpose of the procedure was to specify behavior that would be sufficient to construct non-trivial proofs about group homomorphisms. Using the doctoral students' actions and the undergraduates' approaches as a guide, as well as relying on the general problem-solving literature in cognitive psychology (Mayer, 1992), literature on mathematical theorem proving (Leron, 1985), and my own experiences working in abstract algebra, I developed a procedure that an undergraduate could use to prove statements about group homomorphisms. This procedure was then presented to several algebraists. They were asked whether they believed such a procedure could be effective and whether they would be pleased if their students were reasoning in the way that the procedure indicated. Their responses were generally positive; indeed one algebraist claimed that when he taught undergraduate group theory, he explicitly tried to lead students to construct proofs in a similar way, but had been unable to do so. The algebraists' comments were used to modify the procedure.

To refine the procedure further, I asked two students to apply the procedure to prove statements about group homomorphisms selected from an undergraduate textbook (Dummit & Foote, 1991). One student was a doctoral mathematics student who had completed several graduate courses in abstract algebra. The other student was a computer science major who had some experience with writing proofs but had never taken a course in abstract algebra. These students' failures to construct proofs were used to identify deficiencies in the procedure. If the students were following the procedure but could not construct a proof, I would compare my own thought processes when proving this statement to the actions that the procedure dictated and revise the procedure accordingly. If the students were behaving in a way that I did not intend, I concluded that my procedure was lending itself to misinterpretation and therefore was not sufficiently precise. I would then specify in greater detail what actions a student using the procedure should take. This process of successive refinement continued until a precise procedure was developed that could lead to proofs from most of the relevant exercises related to homomorphisms that were found in undergraduate group theory textbooks.

## 2.2. Proof Construction Procedure.

The task of constructing a proof can be characterized as the problem of deriving information in a mathematically acceptable way. In abstract algebra proofs, one is often asked to find information about a particular structure. Although this is certainly not the only way to view the situation, I view the would-be prover's greatest obstacle as a lack of information about the particular structure. With this interpretation, I designed a recursive procedure in which students first would determine what structure they needed information about, then choose a method to find that information, then determine what type of information would be needed to apply that method, and so on.

*Procedure for proof construction.*

*Step 1.* First, determine what mathematical structure the conclusion of the theorem is asking you to derive information about.

*Step 2.* Choose a fact or theorem that one can use to derive information about this structure. To choose this fact or theorem:
- First identify the main methods to derive information about that structure. These methods largely consist of definitions and theorems that relate other structures to this structure.
- When choosing a fact or theorem, consider the following questions. Is it likely that the hypotheses of this theorem can be satisfied? Is it likely that the conclusion of this theorem will be useful in finding the desired information?

*Step 3.* If attempting to apply a theorem, check to see that the hypotheses of the theorem are satisfied. If the hypotheses are not satisfied, start at Step 1 with the goal to derive the information necessary to satisfy the hypotheses of this theorem. If this fails, return to Step 2 and choose another fact or theorem.

*Step 4.* Apply this fact or theorem.

*Step 5.* See if you now have enough to attain your goal. If you do, achieve it. If not, return to Step 1.

Interpreting this procedure within the particular domain of group homomorphisms, I hypothesize that there are four main structures concerning homomorphisms: the domain of the homomorphism, its image, its kernel, and the mapping (i.e., treating the homomorphism *qua* function). The main methods for deriving information about structures are the theorems that relate facts about one structure to another. For instance, the fact that homomorphisms preserve structural properties allows one to infer properties about the homomorphism's image from properties about its domain. A schematic description of this procedure is given in Figure 1.

This proof construction procedure relies heavily on the user's knowledge of group theory. For a student to use this procedure effectively, he or she would need to understand how theorems and definitions represent relationships between different mathematical structures. Likewise, the student would also need to be able to assess the likelihood that the hypotheses of theorems could be satisfied and that the conclusion of theorems would be useful in particular proof attempts. An example of an application of this procedure is given below.

> Suppose $G$ and $H$ are groups and $G$ is a group of order $pq$, where $p$ and $q$ are prime, and $f$ is a surjective homomorphism from $G$ to $H$. Prove that either $H$ is abelian or isomorphic to $G$.

*Step 1.* Since we are asked about $H$ and $f$ is surjective, what we desire is information about the image.

*Step 2.* As we have information about the size of the domain, we will attempt to use the fact that $|G|/|ker f| = |H|$. (This fact allows one to derive information about the image from information about the domain and the kernel.)

*Step 3.* To use this fact, we need information about the size of the domain and the size of the kernel. We have the information about the domain (it has

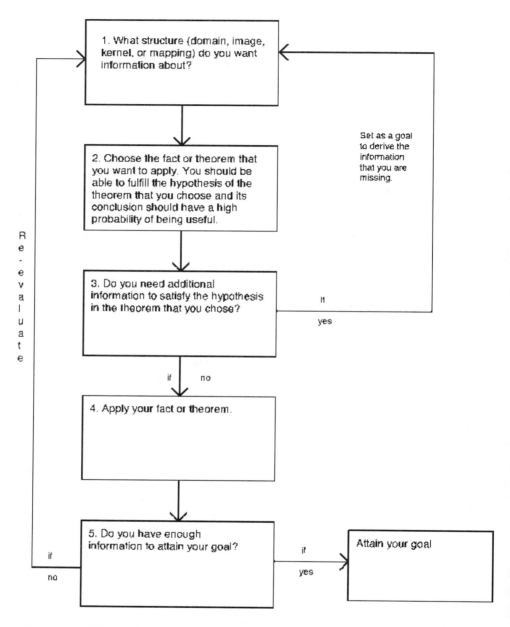

FIGURE 1. Schematic of a procedure to construct proofs about homomorphisms.

order $pq$), but we need information about the size of the kernel. We set as
a sub-goal to derive this information.

  *Step 3.1.* We desire information about the kernel.

  *Step 3.2.* We will use the fact that the kernel is a normal subgroup
        of the domain. (This fact allows one to derive information about
        the kernel from information about the domain.)

*Step 3.3.* This theorem has no hypotheses that need to be satisfied.

*Step 3.4.* We apply our fact to derive that the kernel of $f$ is a normal subgroup of $G$, a group of order $pq$.

*Step 3.5.* We derived that the kernel of $f$ is a normal subgroup of a group of order $pq$. Can we use basic facts about group theory to derive information about its size? The answer is yes. By Lagrange's theorem, we know $|ker f| = 1, p, q$, or $pq$.

*Step 4.* Since $|ker f| = 1, p, q$, or $pq$, $|H| = pq, p, q$, or 1.

*Step 5.* We derived that $|H| = pq, p, q$, or 1. Can we use basic facts about group theory to prove $H$ is abelian or isomorphic to $G$? The answer is yes. If $|H| = p, q$, or 1, then it is abelian (since groups of prime order and order 1 are cyclic). If $|H|$ has order $pq$, $f$ is bijective since $G$ and $f(G)$ have the same cardinality, and therefore an isomorphism. So $H$ is isomorphic to $G$.

## 3. Effectiveness of the Procedure

In an unpublished study, I tested the effectiveness of the procedure by creating a LISP computer program called PROCON (PROof CONstructor) designed to simulate the proof constructing procedure reported in the previous section. To make certain that this procedure could prove the types of statements an undergraduate in a typical group theory course would encounter, I created a list of such problems from the exercises in a pair of popular undergraduate algebra textbooks (Dummit & Foote, 1991; Rotman, 1996). Every exercise in these textbooks that required the student to prove a statement about homomorphisms was selected. This search produced nineteen statements regarding homomorphisms. PROCON effectively sketched proofs for 13 of the 19 statements. The simulation program was unable to construct proofs for three of the statements because these statements involved composite and two-valued functions, which PROCON was not equipped to represent. Of the statements the simulation PROCON was able to represent, it was able to prove 13 out of 16. The simulation was also able to construct other challenging proofs that were beyond the ability of most undergraduate abstract algebra students (see Weber, 2001b, for the PROCON code and for details about the study). Thus, one can model the described procedure in a form precise enough to be simulated by a computer and this procedure is capable of proving many of the theorems about homomorphisms that undergraduates are asked to prove.

In this section, I report a study in which I tested the effectiveness of the procedure with students enrolled in an introductory abstract algebra course. The students were briefly taught the procedure and then asked to apply it to prove eight statements about group homomorphisms. The students in this study did not learn how to apply this procedure in a very meaningful way; one could argue that these students only learned a sequence of steps for manipulating logical formulae. The main purpose of the study was to assess the utility of the procedure. I do not advocate teaching students using the methodology in the experiment.

### 3.1. Experimental Methods.

*Participants.* Six undergraduates enrolled in an introductory abstract algebra course at a technical university in the eastern United States participated in this study. They received a small fee in exchange for their participation. All participants expressed familiarity with group homomorphisms and their associated theorems.

*Problems.* Students were asked to prove eight of the problems presented below. Each problem was attempted by at least two students.

A. The following will be used for *A1-A3*: $f$ is a mapping from $\mathbb{Q}$ to $\mathbb{Q}$ (where $\mathbb{Q}$ is the group of rational numbers under addition) such that $f(q) = kq$ (where $k$ is a non-zero constant rational).

  *A1.* Show $f$ is a homomorphism.

  *A2.* Show $f$ is surjective

  *A3.* Show $f$ is injective.

  *A4.* $G$ is a group of order $pq$, where $p$ and $q$ are primes. $H$ is a group. $f$ is a surjective homomorphism from $G$ to $H$. Prove that either $H$ is abelian or $H$ is isomorphic to $G$.

B. The following is used for *B1* and *B2*: $G$ is a group, $f$ is a mapping from $G$ to $G$ given by $f(g) = g^2$.

  *B1.* If $G$ is abelian, show $f$ is a homomorphism.

  *B2.* If $f$ is a homomorphism, show $G$ is abelian.

  *B3.* $G$ is a finite group of odd order. $f$ is a homomorphism from $G$ to $G$. $f(g) = g^2$. Show $f$ is an injective mapping.

  *B4.* $H$ is a group. $f$ is a surjective homomorphism from $S_n$ to $H$ with kernel $A_n$. ($|S_n| = n!, |A_n| = n!/2$). Show that $f(S_n)$ is isomorphic to $\mathbb{Z}_2$. (This helps explain the idea of even and odd permutations. Ask if you are interested).

C.

  *C1.* Let $R$ be a group. Let $G$ be the group $R \times R$. Let $f$ be the mapping from $G$ to $R$ such that $f(x,y) = y$. Then $f$ is a homomorphism from $G$ to $R$.

  *C2.* $G$ is a finite group of order $n$. $H$ is a group. $f$ is a homomorphism from $G$ to $H$. Show $|f(G)|$ divides $n$.

  *C3.* Let $G$ and $H$ be groups. Let $f$ be a homomorphism from $G$ to $H$. Let $h$ be a member of $H$. $f^{-1}(h) = \{g_1, g_2\}$. (That is, if $f(x) = h$, then $x = g_1$ or $x = g_2$ and $g_1$ is not equal to $g_2$). What is the kernel of $f$?

  *C4.* Show the composition of two homomorphisms is a homomorphism. (That is, show $f(g)$ is a homomorphism).

D. The following is used for *D1*, *D2*, and *D3*. $G$ is the group of complex numbers with absolute value 1 under multiplication. $f$ is the mapping from $G$ to $G$ such that $f(g) = g^n$ (where $n$ is an integer greater than 1).

  *D1.* Show $f$ is a homomorphism.

  *D2.* Show $f$ is a surjection.

  *D3.* Show $f$ is not an injection.

  *D4.* $G$ and $H$ are groups. $G$ is a finite group of order $n$. $G$ is a simple group (This means there are no normal subgroups in $G$ besides $G$ and the trivial group.) Let $f$ be a surjective homomorphism from $G$ to $H$. Prove either $H$ has order $n$ or $H$ has order 1.

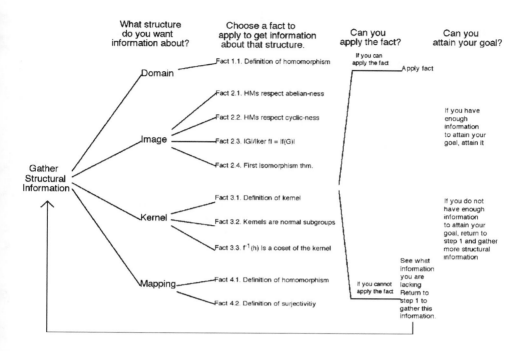

FIGURE 2. Handout describing procedure.

*Materials and procedure.* Students were asked to apply a highly specific form of the procedure presented above in Section 2.2. Students were given two handouts describing an outline of the procedure.[6] The first outline looked very similar to Figure 1. The second outline is given in Figure 2. Students were shown an example of the procedure's application. They were then given more in-depth instruction on how to apply each step of the procedure. In addition, students were given a "fact sheet" listing all the important facts that were associated with each structure (domain, image, kernel, and mapping). Part of this fact sheet is given in Appendix A. Each fact on the list was accompanied with the conditions specifying when it should be chosen, the information required to apply the fact, and an example of the fact's application.

The fifth step of the procedure asks students to use their knowledge of group theory to produce a finished proof. When students arrive at this step, they should have derived enough information to construct a proof using reasoning that is independent of the concept of group homomorphisms. For example, to prove statement *B3*, that if $G$ is a finite group of odd order and $f(g) = g^2$ is a homomorphism from $G$ to $G$, then $f$ is an injective mapping, the first four steps of the procedure allow the user of the procedure to construct a proof if the user can prove that $g^2 = e$ implies $g = e$. Proving this for groups of odd order is what the user would be asked to do in the fifth step, and this has nothing to do with group homomorphisms.

---

[6]All materials for this study can be found in Weber (2001a).

After each step of the procedure had been described in detail, students were tested to make sure that they could apply the step correctly. Any errors were discussed. The total time preparing students ranged between 30 and 40 minutes.

Students were then asked to prove theorems regarding homomorphisms by applying the procedure. Before a student could perform an action, the student was asked to state aloud the conditions that permitted them this action. Students typically required less than half an hour to prove all eight theorems; all students were finished within 40 minutes. Details and the rationale for this procedure are discussed in Weber (2001b).

*Coding.* Each of the students' proof attempts was coded as follows:

> *Correct proof:* A valid proof was constructed.
>
> *Failure in executing the first four steps of the procedure:* The student executed one of the first four steps of the procedure erroneously and thus could not construct a proof.
>
> *Failure to execute Step 5:* The students executed the first four steps correctly, but could not execute the fifth step. The fifth step requires students to apply their own group theoretic knowledge, independent of homomorphisms, to finish the proof.

**3.2. Results of the First Study.** In total, 41 out of 48 (85%) of the students' proof attempts produced a valid proof. There were two errors classified as "Failure to execute the first four steps of the procedure." One student interpreted the condition "You need to know the nature of the mapping" to be satisfied because he knew that the mapping $f$ was surjective; another implemented a fact incorrectly, not realizing $\mathbb{Q}$ denoted the group of rational numbers under addition.

There were five errors implementing Step 5. In these cases, students could not use their own group theoretic reasoning independent of homomorphisms to finish their proofs. For instance, both students who attempted problem *B3* were unable to demonstrate that in finite groups of odd order, $g^2 = e$ implies $g = e$. It would be very difficult to prove proposition *B3* without being able to derive this.

In summary, students were able to use the procedure to construct proofs effectively and efficiently. Collectively, the six students in this study were able to construct proofs for 85% of the propositions that they encountered and they averaged less than five minutes per proof attempt. The undergraduates' efficiency and rate of success were much better in this study than in other reported studies when students were asked to prove similar theorems without guidance (Weber, 2001a, 2001b).

**3.3. What do these results tell us about pedagogy?** To address this important question, consider what Schoenfeld (1985a) offered in answering a similar question about SAINT, a classic artificial intelligence program that could perform integration tasks (Slagle, 1963):

> Written in the early days of AI, SAINT solved indefinite integral problems with the proficiency of a better than average MIT freshman. SAINT worked "somewhat" like a mathematician. It was not, for instance, a "mechanized table of integrals." SAINT had roughly two dozen "basic forms" in memory, and had to derive answers to integrands not on that short list. The eight basic

transformational rules known to SAINT were [less] powerful [and less] numerous than those a student would know. SAINT did not rely on "brute force" computation for its solutions... So in many ways SAINT was handicapped, limited to performance on a human scale...

SAINT solved 84 of the 86 problems it worked, 54 of which were chosen from the MIT final exams in freshmen calculus. There is no doubt that the strategies used by SAINT "work" ... the question, of course, is what the machine implementation of these strategies say about human problem solving. In the case of SAINT, they said a great deal: Using a small number of strategies, a problem solver with fewer "basic facts" committed to memory than most students, fewer techniques than they might use, and none of the other skills that they bring to problem solving, managed to do far better than most students. The moral is *not* that students should be trained to do integrals, but a few powerful strategies – chosen for human implementation rather than machine implementation – might yield comparable results for humans. (Schoenfeld, 1985b)

The parallels between SAINT and the proof constructing procedure are obvious, and my argument is essentially the same as the one above. I make no claims that students should be taught how to construct proofs using the heavy-handed, overly symbolic, conceptually barren approach that I employed in this experiment. Rather, the moral here is that if we could teach students to execute similar strategies, strategies that are both mathematically meaningful and appropriate for undergraduate mathematics majors, they might achieve results comparable to the students in this study.

## 4. Instruction Designed to Teach this Procedure

In this section, I investigate the questions of how, and to what extent, one can teach students to apply the procedure described in this paper. My goal in this section is to offer an "existence proof" that it is possible for undergraduates to use the thought processes described here and that undergraduates who use them can show significant improvement in their proof-writing ability. However, no claims are made that this is the best or only way to teach students this type of reasoning.

**4.1. Development of Instruction.** I designed the instruction to achieve two specific goals. First, students should be able to prove significantly more statements about homomorphisms after receiving the instruction. One might argue that this goal was already achieved in the experiment described in the Section 3, but this would be misleading. Those students did demonstrate a degree of proof-writing competence, but this was only when they were given written materials explicitly telling them what to do. Also, those students were *forced* to use the procedure in a highly specified way. I would like students receiving my instruction to choose to use the procedure to construct proofs and be successful in doing so without the aid of external materials.

My second goal in designing the instruction was for students to use the procedure in a meaningful way. In Section 1.1, I specified that this entailed that: the thought processes and actions of the students would be indicative of mathematical reasoning (in the eyes of a mathematician), the procedure would not be divorced from the students' conceptual understanding, and the students would know why this procedure was useful for constructing proofs. The experimental methods described in Section 3 failed to meet this goal. In particular, the if-then rules did not require conceptual understanding and no effort was made to have students understand why they did what they did.

To accomplish these two goals, I used a cognitive apprentiship model with the approach of modeling-scaffolding-fading (Collins et al., 1989). This teaching approach suggests a specific instructional sequence. To teach a particular skill, the instructor illustrates to the students how that skill can be implemented. The students are then asked to practice that skill and the instructor offers guidance and support as needed. Gradually the instructor's support is removed to the point where students are applying the skill independently. This sequence was used in my instruction. For each step in the procedure, I described what the step was trying to accomplish and why it was useful. I then asked students to complete the step themselves with successively less guidance until they could apply the step independently. Students were never explicitly told how to implement a skill; they were not given a set of specific instructions for applying each step of the procedure.

A second principle of this teaching approach is that local skills are learned within the context of realistic global tasks. Each skill was practiced by students within the context of writing proofs for actual statements about group homomorphisms. Collins et al. (1989) argue that this style of instruction is more likely to lead to meaningful and flexible understanding than the algorithmic instruction students receive in traditional classrooms. By attempting to model the instructor's performance rather than learning how to mechanically apply a sequence of manipulations, students construct their own methods for accomplishing tasks that are likely to be meaningful to them. By learning skills in context, students are likely to apply them more flexibly, understand their purpose, and appreciate their value.

In pilot studies, I found that after learning the procedure, students would frequently revert back to their own strategies, which were often not effective for constructing proofs. To address this problem, I attempted to illustrate to students that some of their proof strategies were inadequate. I did this at a general and specific level. At a general level, during my second meeting with the students, I informed them that they did poorly on the pre-test. I then described several student strategies and showed why they were not effective. I hoped that this would create a student need to learn more effective strategies. At a specific level, when a student made a poor decision in constructing a proof, I would illustrate to the student why that decision was poor within the context of the proof. For instance, if a student tried to write a proof using a theorem whose hypotheses could not be satisfied, I would first ask the student to attempt to satisfy these hypotheses. When he or she could not, I would then ask if the hypotheses were necessarily true. I hoped that in this way, the students would understand why one should not let their proof attempts hinge on using theorems that are not likely to be applicable.

**4.2. Description of Instruction.** In this sub-section, I describe the instruction that I presented to five students in a controlled experiment that was designed to teach students to use the procedure. This instruction contained four one-hour components. Students spent the first forty minutes of the first hour and the last forty minutes of the final hour taking a pre-test and a post-test respectively. If one removes the hour and twenty minutes that the students spent completing the pre-test and the post-test, the total time of actual instruction was two hours and forty minutes. The instruction that students received is described below.

*Day 1.* The students spent the first 40 minutes of this lesson completing the pre-test (described in detail in Section 4.3). Afterwards, they were informed that critical facts and theorems about homomorphisms were the ones that demonstrated relationships between structures and they were asked to look over a sheet entitled, "Key facts about homomorphisms" (given in Appendix B).

*Day 2.* I first read the text from Appendix C, which motivated the value of the instruction by first informing the students that they performed poorly on the pre-test, illustrating how students often employed poor strategies in their proof attempts, and stressing that a team of scientists had developed a better way to construct proofs. I described the first step of the procedure and modeled how this step was carried out with two examples. Students were then given propositions about group homomorphisms and asked to apply the first step of the procedure. If a student's response was correct, I told the student why the response was correct. Otherwise, I would illustrate, within the context of a proof, why the response was incorrect.

*Day 3.* I described how to choose which fact to apply, stressing that students should choose a fact that was both applicable – that is, its hypotheses were satisfiable – and useful for proving the proposition. For each homomorphism structure (domain, image, kernel, and mapping), I first modeled how one would choose a fact to apply with a particular proposition. The students were then given several propositions and asked to choose which fact they would like to apply, first with leading prompts, and then without. If a student chose correctly, I explained why the student's choice was correct and asked the student to prove the proposition. If not, I demonstrated why the choice was incorrect by attempting to do what the student wanted and showing how this led to an impasse. After this, I explained how the third step of the procedure was applied and discussed the recursive nature of the procedure (all students who received this instruction were familiar with computer programming.) I modeled how this step of the procedure was applied, and then asked the students to prove a proposition where this step was required. After their proof attempts were completed, I evaluated their correctness. If a proof was incorrect, we discussed why.

*Day 4.* I asked the students to prove four practice problems (different from the pre- and post-test problems). These problems contained suggestive prompts reminding students to employ the procedure that we had discussed during our previous sessions. After fifteen minutes, I evaluated the students' proof attempts. Incorrect proof attempts were discussed. The students then completed the post-test, whose details are described in the next section. Further details of the instruction, including scripts, materials, and practice problems, are given in Weber (2001b).

### 4.3. Experimental Methodology.

*Participants.* This study was conducted twice with a total of five participants. The first time, there were three students who had recently completed a first undergraduate algebra course at a university in the eastern U.S. The second time, there were two students who were enrolled in a similar course at a nearby university that had covered the topic of homomorphisms about a month prior to the study. The

abilities and performance of the participants in this study were varied but, on average, typical. Of the five students, one student earned an A in his abstract algebra
course, three students earned a B (or went on to earn a B), and one student earned
a D. All participants expressed familiarity with the concept of homomorphisms and
its related facts and theorems. The students were paid a fee for their participation.
Difficulty recruiting students was the reason why there were so few participants.

   *Problems.* The critical measures in this study were the students' performance
on a pre-test given prior to instruction and a post-test given after instruction. The
pre-test and post-test problems are presented below.

   *Proposition Set A.*

   **1A:** Let $\mathbb{Q}$ be the group of rational numbers under addition. Let $f$ be the
   mapping from $\mathbb{Q}$ to $\mathbb{Q}$ such that $f(q) = 3q$ for all $q$ in $\mathbb{Q}$. Prove that $f$ is
   a homomorphism from $\mathbb{Q}$ to $\mathbb{Q}$.

   **2A:** Let $G$ be a group with 13 elements. Let $H$ be a group. Let $f$ be a
   homomorphism from $G$ to $H$. Show that $f(G)$ is abelian.

   **3A:** Let $S_4$ be the symmetric group of four elements. Let $H$ be a group.
   Let $f$ be a surjective (onto) homomorphism from $G$ to $H$. Let the kernel
   of the homomorphism be the subgroup of 2 x 2 cycles in $S_4$, that is, the
   set $\{e, (12)(34), (13)(24), (14)(23)\}$. Prove that for all $h$ in $H$, $h^6 = e_H$.

   **4A:** Let $G$ be an abelian group with an odd number of elements. Let $f$ be
   the homomorphism from $G$ to $G$ such that $f(g) = g^2$ for all $g$ in $G$. Show
   that $f$ is injective (one-to-one).

   *Proposition set B.*

   **1B:** Let $G$ be an abelian group. Let $f$ be the mapping from $G$ to $G$ such
   that $f(g) = g^2$ for all $g$ in $G$. Show that $f$ is a homomorphism.

   **2B:** Let $G$ be an abelian group and let $H$ be a group with 6 elements. Let
   $f$ be a surjective (onto) homomorphism from $G$ to $H$. Show that $H$ is
   isomorphic to $\mathbb{Z}_6$ (the cyclic group of six elements).

   **3B:** Let $G$ be a group of order $pq$, where $p$ and $q$ are primes. Let $H$ be a
   group. Let $f$ be a surjective (onto) homomorphism from $G$ to $H$. Show
   that either $H$ is abelian or isomorphic to $G$.

   **4B:** Let $G$ be a group of order 25. Let $f$ be the mapping from $G$ to $G$
   such that $f(g) = g^3$. Assume that $f$ is a homomorphism. Show that $f$ is
   injective (one-to-one).

   Three students used proposition set $A$ as the pre-test and proposition set $B$ as
the post-test; the other two used set $B$ as the pre-test and set $A$ as the post-test.
The propositions on these tests were superficially different, but were designed to
be analogous to each other. For example, the first problem on each test requires
students to verify that a particular mapping is a group homomorphism and the
third problem requires students to use a weak version of the First Isomorphism
Theorem to derive a fact about the image of a homomorphism.

   *Procedure.* The participants received the instruction described in Section 4.1 in
groups of two or three. The four one-hour instruction units were held over a period
of two weeks. Practice problems that the students worked were varied, and for
the most part were significantly different from the problems used on the pre- and
post-test. The list of practice problems is given in Appendix D. When completing
the pre-test and post-test, participants were not allowed to use supporting material
such as their course textbook or other materials.

*Coding.* Two graders (the author of this paper and a mathematics professor) independently scored each proof attempt using the scale of Malone et. al. (1980) that was used by Hart (1994) in his analysis of students' proofs about group theoretic topics. Each proof attempt was coded as completely correct, mostly correct (i.e., correct except for minor and insignificant errors), incorrect with substantial progress, or incorrect with no substantial progress. The mathematics professor was blind to which proofs were written before and after instruction.

**4.4. Quantitative Results.** The graders achieved a high level of inter-rater reliability. They assigned each proof the same score in 85% (34 out of 40) of the cases (85% of the time both for the pre-test and the post-test questions). Using the Spearman ranked correlation coefficient, their inter-rater correlation was 0.944. Further, they agreed whether a proof attempt was correct or mostly correct 97.5% (39 out of 40) of the time. Using the Spearman ranked correlation coefficient, their inter-rater correlation was 0.950. Finally, they achieved 100% agreement on whether a proof attempt represented a fully correct and valid proof. In all cases of disagreement, the mathematician's scoring was used because he was blind to which proofs were written before and after instruction.

Figure 3 presents aggregate students' performance on the pre-test and the post-test. Students' performance on the post-test was much better than their performance on the pre-test. 65% (13 out of 20) of the students' proof attempts on the post-test were correct or mostly correct, as opposed to only 30% (6 out of 20) on the pre-test. Students' proof attempts were correct, mostly correct or they could make substantial progress 75% of the time on the post-test, compared with only 50% of the time on the pre-test. Finally, each student who participated in this study was able to produce more correct or mostly correct proofs on the post-test than on the pre-test.

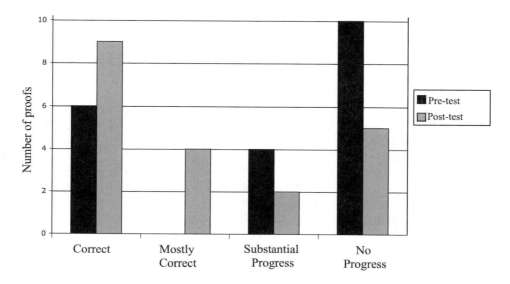

FIGURE 3. Pre-test and post-test performance.

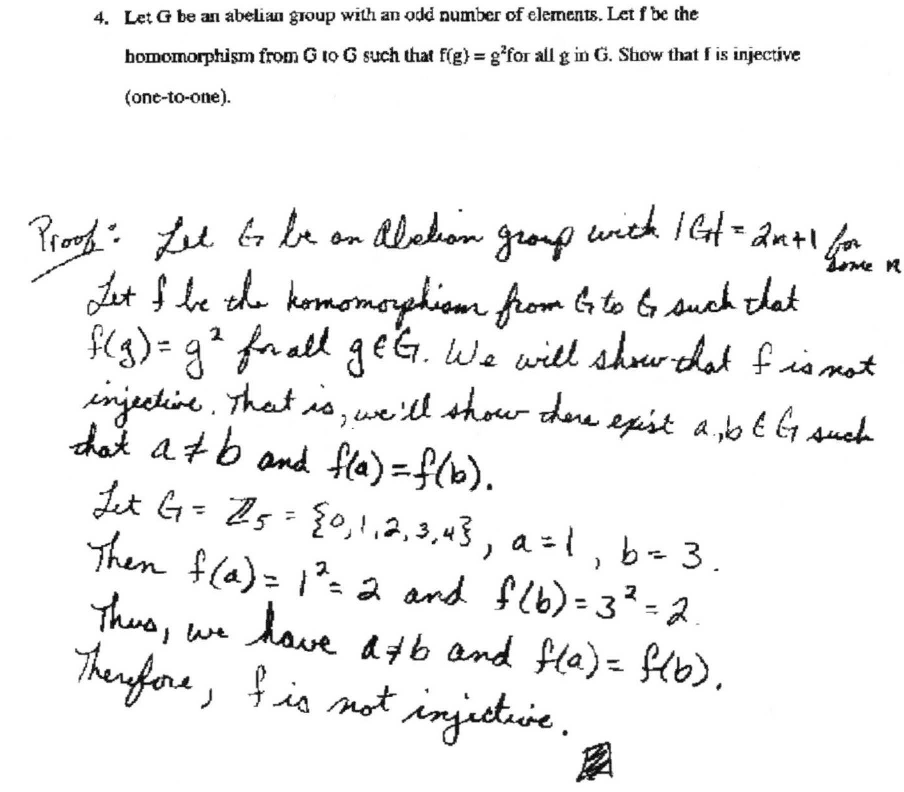

4. Let G be an abelian group with an odd number of elements. Let f be the homomorphism from G to G such that $f(g) = g^2$ for all g in G. Show that f is injective (one-to-one).

Proof: Let $G$ be an Abelian group with $|G| = 2n+1$ for some $n \in \mathbb{Z}$. Let $f$ be the homomorphism from $G$ to $G$ such that $f(g) = g^2$ for all $g \in G$. We will show that $f$ is not injective. That is, we'll show there exist $a, b \in G$ such that $a \neq b$ and $f(a) = f(b)$.

Let $G = \mathbb{Z}_5 = \{0, 1, 2, 3, 4\}$, $a = 1$, $b = 3$.

Then $f(a) = 1^2 = 2$ and $f(b) = 3^2 = 2$.

Thus, we have $a \neq b$ and $f(a) = f(b)$.

Therefore, $f$ is not injective. ∎

FIGURE 4. JJ's work on the pre-test.

**4.5. Analysis of students' work.** The data presented in Section 4.4 demonstrate that the students were able to construct more proofs on the post-test than on the pre-test, but do not indicate why. An analysis of students' written work on these tests reveals that some of the students' difficulties on the pre-test were due to their use of strategies that relied strongly on symbol manipulation and that many of the students were explicitly making use of the procedure when writing proofs on the post-test. This suggests that the students' use of the procedure was helpful in alleviating their strategic difficulties.

Two of the students showed little scratch work on any of their valid proofs. As a result, it was difficult to determine whether they were using the taught procedure on the post-test, or what aspect of the instruction led them to perform better on the post-test than on the pre-test. This difficulty can be illustrated with JJ, who may have been the strongest student to receive this instruction. On the pre-test, he answered three of the four questions correctly and he obtained a perfect score on the post-test. Figure 4 presents the one question on the pre-test that JJ answered incorrectly and Figure 5 presents JJ's valid proof for the corresponding question on the post-test.

JJ's difficulty in Figure 4 was due in part to an erroneous calculation, specifically computing $f(3) = 2$ when in fact $f(3) = 1$. One cannot conclude whether JJ

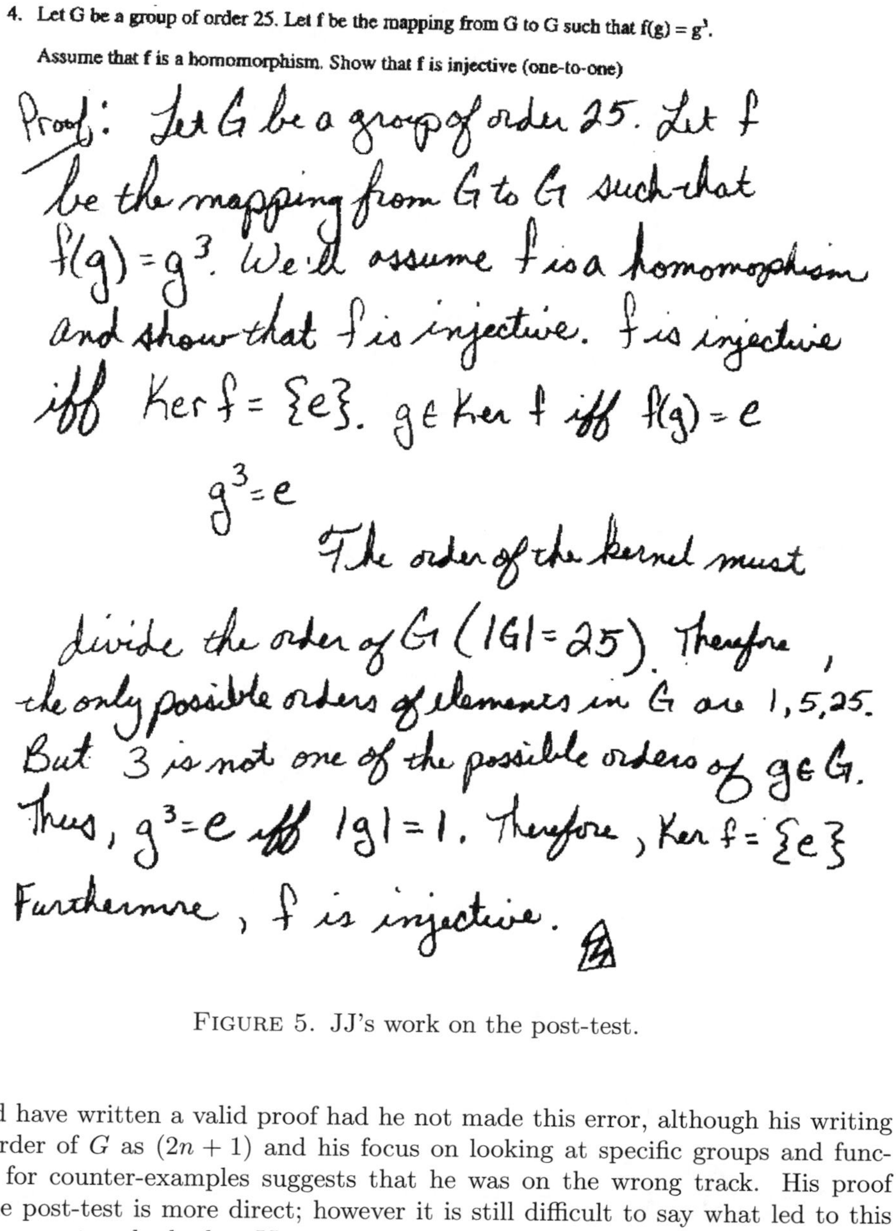

4. Let G be a group of order 25. Let f be the mapping from G to G such that $f(g) = g^3$.

Assume that f is a homomorphism. Show that f is injective (one-to-one)

Proof: Let G be a group of order 25. Let f be the mapping from G to G such that $f(g) = g^3$. We'll assume f is a homomorphism and show that f is injective. f is injective iff $\text{Ker } f = \{e\}$. $g \in \text{Ker } f$ iff $f(g) = e$

$$g^3 = e$$

The order of the kernel must divide the order of G ($|G| = 25$). Therefore, the only possible orders of elements in G are 1, 5, 25. But 3 is not one of the possible orders of $g \in G$. Thus, $g^3 = e$ iff $|g| = 1$. Therefore, $\text{Ker } f = \{e\}$ Furthermore, f is injective.

FIGURE 5. JJ's work on the post-test.

would have written a valid proof had he not made this error, although his writing the order of $G$ as $(2n + 1)$ and his focus on looking at specific groups and functions for counter-examples suggests that he was on the wrong track. His proof on the post-test is more direct; however it is still difficult to say what led to this improvement and whether JJ was using the procedure that he learned during my instruction.

On the other hand, the scratch work of two other students suggests that they were using poor strategies on the pre-test and were explicitly using the procedure on the post-test. Figures 6 and 7 present SF's work on a pre-test and post-test problem respectively. SF's work in Figure 6 is similar to students' behavior observed in previously reported studies (Weber, 2001a) and described in Section 2.1 of this paper; the inferences that SF drew were logically valid, but the approach that he took to his proof appeared to consist primarily of manipulating symbols.

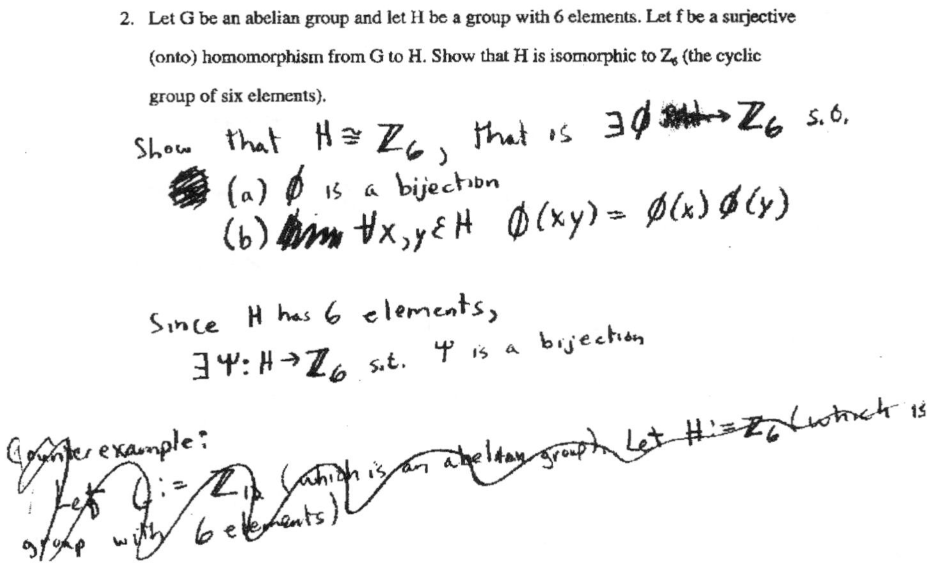

2. Let G be an abelian group and let H be a group with 6 elements. Let f be a surjective (onto) homomorphism from G to H. Show that H is isomorphic to $Z_6$ (the cyclic group of six elements).

Show that $H \cong Z_6$, that is $\exists \phi \text{ : } H \to Z_6$ s.o.

(a) $\phi$ is a bijection

(b) $\forall x, y \in H$  $\phi(xy) = \phi(x)\phi(y)$

Since H has 6 elements,

$\exists \Psi : H \to Z_6$ s.t. $\Psi$ is a bijection

Counter example:

Let $G := Z_{12}$ (which is an abelian group) Let $H := Z_6$ (which is group with 6 elements)

FIGURE 6. SF's work on the pre-test.

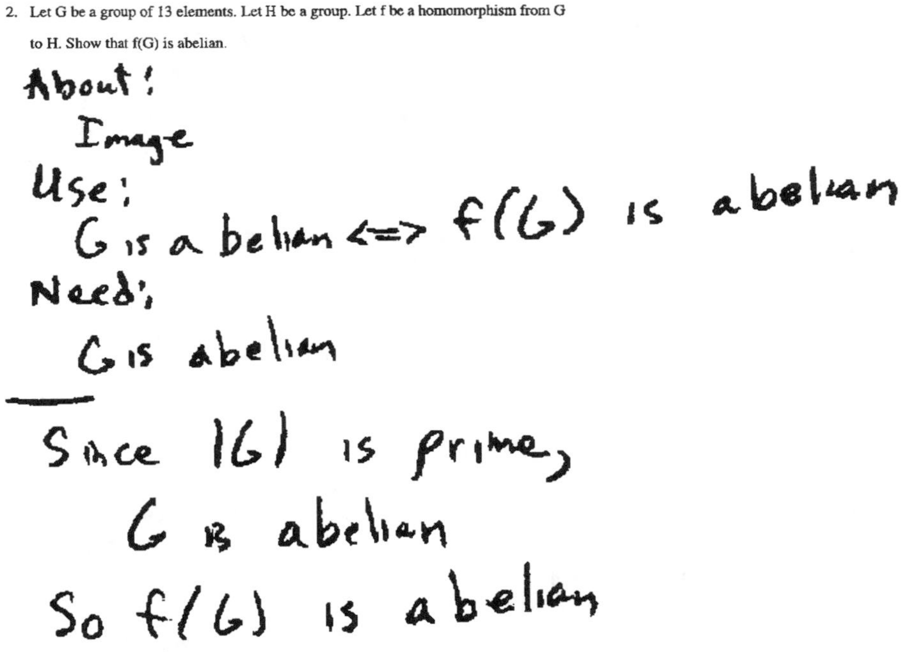

2. Let G be a group of 13 elements. Let H be a group. Let f be a homomorphism from G to H. Show that f(G) is abelian.

About!

Image

Use:

G is abelian $\iff$ f(G) is abelian

Need:

G is abelian

___

Since $|G|$ is prime,

G is abelian

So f(G) is abelian

FIGURE 7. SF's work on the post-test.

In his scratch work on the post-test, SF gave a clear indication that he was using the taught procedure to construct the proof. The proof was coded as "mostly correct" since it lacked details.

3. Let G be a group of order pq, where p and q are primes. Let H be a group. Let f be a surjective (onto) homomorphism from G to H. Show that either H is abelian or isomorphic to G.

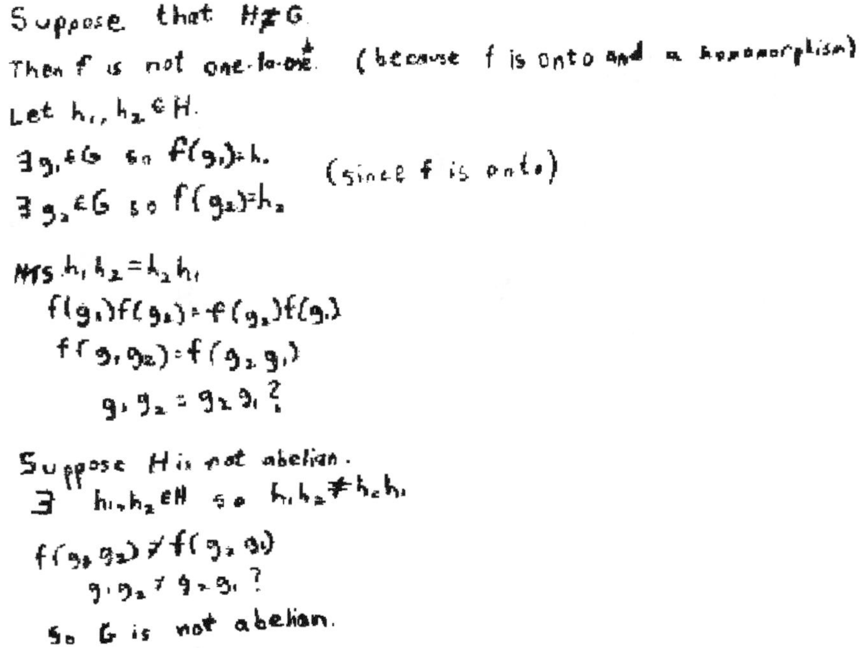

Suppose that $H \neq G$.

Then f is not one-to-one. (because f is onto and a homomorphism)

Let $h_1, h_2 \in H$.

$\exists g_1 \in G$ so $f(g_1) = h_1$.       (since f is onto)

$\exists g_2 \in G$ so $f(g_2) = h_2$

WTS $h_1 h_2 = h_2 h_1$

   $f(g_1) f(g_2) = f(g_2) f(g_1)$

   $f(g_1 g_2) = f(g_2 g_1)$

      $g_1 g_2 = g_2 g_1$ ?

Suppose H is not abelian.

   $\exists h_1, h_2 \in H$ so $h_1 h_2 \neq h_2 h_1$

$f(g_1 g_2) \neq f(g_2 g_1)$

   $g_1 g_2 \neq g_2 g_1$ ?

So G is not abelian.

FIGURE 8. HL's work on the pre-test.

Figures 8 and 9 present HL's solutions to similar problems on the pre-test and the post-test. Her work in both figures suggests that HL is a logically capable student with at least a symbolic understanding of group theory. Her proof approaches on the pre-test and post-test were quite different. In Figure 8, HL primarily approached the proof by trying to unpack definitions and manipulate symbols. Her scratch work and her subsequent valid proof in Figure 9 clearly indicate that she is using the taught procedure. Also, it is worth mentioning that HL used the phrases "similar structure" and "similar size." These phrases were not used during instruction, suggesting that HL may have given personal meaning to some of the steps in the procedure.

Finally, although the instruction was generally successful for most students, there was still one student (EB) who performed very poorly on both the pre-test and the post-test. On her post-test, EB produced no valid proofs and only one proof that was coded as mostly correct. Every other student produced at least one completely valid proof and two or more proofs that were coded as mostly correct or better. On the other three proofs, such as the one presented in Figure 10, EB made no substantial progress.

From my observations in teaching EB, I suspect that her difficulties stemmed from a deep lack of conceptual understanding of group theoretic concepts. EB was required to generate her own methods for modeling my behavior; I would not tell her exactly what she should do. I noticed that EB had a very difficult time doing

3. Let $S_4$ be the symmetric group of four elements. Let H be a group. Let f be a surjective (onto) homomorphism from $S_4$ to H. Let the kernel of the homomorphism be the subgroup of $2 \times 2$ cycles in $S_4$, that is the set $\{e, (12)(34), (13)(24), (14)(23)\}$. Prove that for all h in H, $h^6 = e_H$.

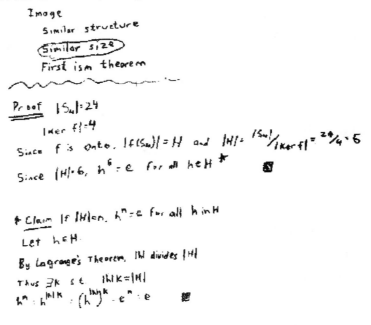

FIGURE 9. HL's work on the post-test.

this as she did not seem to perceive the semantic meaning of the statements that she was proving. As a result, she was often reduced to searching for symbolic features of statements that appeared to yield correct performance. For instance, in the problem presented in Figure 10, EB indicated that she believed that this statement asked her to find information about the kernel, presumably because the symbolic expression "$= e_H$" appeared in the conclusion. Based on EB's performance, I tentatively hypothesize that this type of instruction might not be effective for students with limited conceptual understanding.

**4.6. Limitations of This Study.** The data support the hypothesis that the students' performance improved because they had learned to apply a powerful procedure to prove these types of statements. However, several other factors may have contributed to students' improvement. For instance, perhaps improvement was due to practice applying powerful theorems or exposure to complex theorem proving.

Another troubling issue is that the most variable component of this instruction was the instructor, namely the author of this paper. The instructor had full confidence in his method and had a vested interest in demonstrating its effectiveness. It is possible that a less motivated instructor, one more skeptical of this teaching paradigm, would not achieve equivalent results. Also, only five students participated in this study. Clearly research incorporating more participants is necessary to determine how robust these results are. For these reasons, the results should be interpreted as illustrative and suggestive, and any conclusions drawn from this study should be tentative.

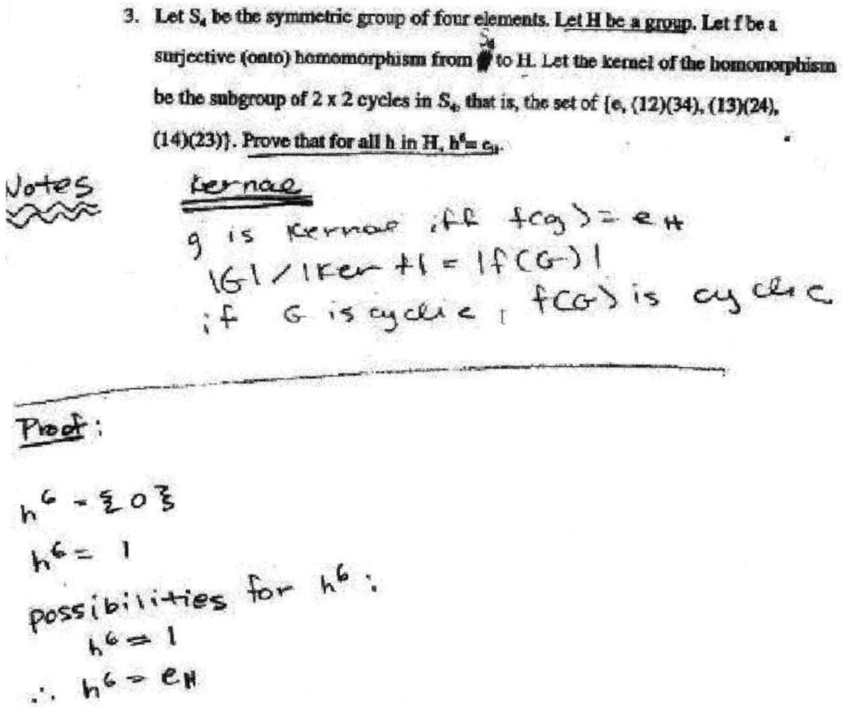

3. Let $S_4$ be the symmetric group of four elements. Let H be a group. Let f be a surjective (onto) homomorphism from $S_4$ to H. Let the kernel of the homomorphism be the subgroup of 2 x 2 cycles in $S_4$, that is, the set of {e, (12)(34), (13)(24), (14)(23)}. Prove that for all h in H, $h^6 = e_H$.

Notes

Kernae

g is Kernae iff $f(g) = e_H$

$|G| / |Ker f| = |f(G)|$

if G is cyclic, $f(G)$ is cyclic

Proof:

$h^6 = \{0\}$

$h^6 = 1$

possibilities for $h^6$:

$h^6 = 1$

∴ $h^6 = e_H$

FIGURE 10. EB's work on the post-test.

One goal of this instruction was to teach the students to apply the procedure in a meaningful way, and not just as a set of arbitrary rules. Toward this end, the instruction was designed so students were not given rules, but were led to construct how to execute each step of the procedure themselves. Also, if students favored a less efficient approach to constructing proofs, the instructor attempted to lead the student to see why their approach was inefficient. This was done so that students would acquire an internally meaningful procedure. Unfortunately, the analysis in Sections 4.4 and 4.5 fails to assess whether this goal was met. The analysis primarily examined students' written work on the post-test to address two questions: To what extent was their work correct? What evidence was there that the students used the procedure that was taught to them? From their written work, it was not possible to determine if they were using the procedure in a mathematically meaningful way – one cannot determine if they knew why the procedure was effective or whether the procedure was linked to their conceptual knowledge.

Despite these reservations and shortcomings, I regard the results as promising. The procedure that I attempted to teach the students was primarily heuristic, and heuristic knowledge is notoriously difficult to teach. Meta-analysises of research on the teaching of problem-solving have revealed that almost all studies designed to teach students heuristic knowledge were failures (Begle, 1979; Schoenfeld, 1985a). Thus, the success in this study is encouraging. While studies employing more students and a suitable control group are required before making conclusive claims,

I hope these results will convince the reader that it may be plausible to teach students this procedure.

## 5. Discussion

A common criticism of psychologically-based educational research is that this type of research involves analyzing a small number of subjects in an idealized, sterilized environment while actual classroom teaching is a messy interaction between the teacher and many students. Hence the connections between psychologically-based research and teaching practice are not always clear (Davis, 1996). The research described in this paper is certainly subject to this critique. In the study reported in Section 4, I attempted to teach a small number of students specific proof strategies within the narrow domain of group homomorphisms. One goal of my study was to explore whether it was possible to improve students' proof-writing abilities within a relatively short period of time. In an actual classroom, the instructor attempts to teach a large number of students a complex collection of skills, knowledge, and beliefs across a wide range of concepts over the span of several months. One critical goal of such instruction is to lead students to develop ways of thinking that can be used in subsequent mathematics courses. In this section, I discuss ways that my results might be incorporated into the classroom.

**5.1. Retention and Transfer.** Teaching students proof strategies that they cannot use flexibly and will forget quickly is of limited value. The important issues of retention and transfer were not addressed directly in this study. However, there is reason to suspect that the students who received my instruction might not retain what they learned or be able to apply these strategies to different types of proofs. Both educational and psychological research send clear messages on the issues of retention and transfer: For strategies to be used flexibly, students must have practice applying them in a variety of settings. The extent to which a skill is retained is highly dependent upon the extent to which it is practiced (Anderson & Schunn, 2001; Cooper, 1991).

To illustrate the problem of obtaining long-term benefits from novel classroom instruction, consider Yusof and Tall's problem-solving course. After completing their problem-solving course, students expressed desirable beliefs about mathematics, such as mathematics is a meaningful, pleasurable activity in which one solves problems and relates ideas. However, after completing a subsequent advanced mathematics course taught in a traditional manner, students once again viewed formal mathematics as a fearful activity that largely involved memorization and obtaining correct answers. These students' approaches to learning reverted back to memorizing procedures (Yusof & Tall, 1999).

The educational literature can be used to make a very clear prediction: If the students in the studies reported here returned to environments where effective proof strategies were ignored, these students would likely regress to using ineffective strategies. For students to develop flexible strategies that they continue to use, proof strategies must receive attention throughout their abstract algebra course (and more generally, throughout their advanced mathematics curriculum). Below, I give some suggestions on how this might be done.

**5.2. Scope of the procedure.** I applied the general proof constructing procedure described in Section 2.2 to the specific domain of group homomorphisms.

However, I believe that this procedure can be applied to other domains as well, and can be effective for constructing proofs under the following conditions. The statements to be proven must be provable by a direct proof in which the conclusion can be linked to the hypotheses by a series of important facts and theorems. As the analysis at the beginning of Section 3 illustrates, the majority of theorems pertaining to group homomorphisms that appear in undergraduate textbooks are of this type. I suspect that many of the other statements that undergraduates are asked to prove also fall into this category.

It is important to note the types of statements that cannot be proved using this procedure. First, statements that are proved by contradiction, induction, or contraposition cannot be proved by the procedure. Second, proofs that require the student to use new or uncommon proof techniques, or require the student to create new mathematical constructs, are also beyond the scope of the procedure.

**5.3. Proof presentation.** In advanced mathematics classrooms, teachers often present their proofs in a linear fashion, thereby hiding the reasoning used to create the proof. Although commonly used, such presentations are widely maligned by both mathematics educators and mathematicians (Davis & Hersh, 1981; Kline, 1977; Leron, 1985; Leron & Dubinsky, 1995; Thurston, 1994). Many mathematics educators (Alibert & Thomas, 1991) have endorsed using alternative methods for presenting proofs, such as generic proofs (Rowland, 2002; Tall, 1979) or structured proofs (Leron, 1985), because of their greater explanatory value. While explaining why a theorem is true is an important and sometimes overlooked purpose of proof (Hanna, 1990), it is not the only reason that proofs are presented in advanced mathematical courses. One reason that proofs are presented to students is to illustrate proof techniques (Weber, 2002a, 2004). By observing one proof, an instructor hopes that students can produce similar proofs in the future.

It is often suggested that a teacher discuss the reasoning that could be used to create a proof. This suggestion has two shortcomings. First, much of mathematicians' decision-making processes are tacit – that is, when doing mathematics, mathematicians are often not consciously aware of the strategies that they are employing. Second, mathematicians' decision-making processes are often dependent upon their extensive experience, an experience that their students do not share. For these reasons, accurately articulating the thought processes used to construct a proof can be quite difficult.

The procedure discussed in this paper offers a framework that can be used to discuss such processes. When a statement is to be proved, the instructor can first explain how this statement is making a claim about a mathematical structure. An instructor can then delineate the main methods for deriving information about this structure. For some of the methods that are not used, the instructor can explain, or perhaps illustrate, why these methods are unlikely to be successful. For the chosen method, the instructor can explain why this was a reasonable choice. In doing this, the instructor is modeling a reasoning process in enough detail that students can emulate it.

**5.4. Proof writing.** It is almost tautological to state that in order for students to become capable at constructing proofs, they must have extensive experience at constructing proofs. However, it does not follow that experience alone will lead students to develop effective proof-constructing strategies. On the contrary,

Lester (1994) claims that the research literature on problem-solving demonstrates that practice and experience are not sufficient for students to develop effective problem-solving heuristics; not putting any effort into problem-solving heuristics will ensure that students will not acquire this knowledge. This has been illustrated recently by the fact that many good differential equations students cannot solve non-routine calculus problems (Selden, Selden, Hauk, & Mason, 2000) and some Ph.D. mathematicians do not use basic problem-solving heuristics (DeFranco, 1996). Hence, practice is a necessary, but not sufficient, condition for mathematical competence. If one attempts to solve problems using ineffective strategies and does not reflect on his or her work, all one is doing is practicing and reinforcing these ineffective strategies. This raises an important question: How can we ensure that when students are engaged in proof writing, they will be reasoning in a way that will lead them to develop effective proving strategies?

Again, the procedure described in this paper suggests a method for doing this. When students are engaged in proof-related activities, they should be encouraged to reason about the structures they are working with and the methods they are applying. Perhaps before students are asked to construct a proof, they can be asked to describe the statement that they are proving in terms of deriving information about mathematical structures and to explain why various methods of proof may or may not be effective in these circumstances. Addressing these questions prior to constructing a proof might lead students to view proving as deriving information about mathematical structures, to plan before engaging in derivations and symbolic manipulations, and to reflect on their decision-making.

I hope that the work in this paper will allow mathematicians and mathematics educators to more fully understand the decision-making processes used in constructing proofs. Further, I have described and analyzed a specific method that an instructor can implement to improve students' proof-writing ability.

## Acknowledgments

This article is based on the author's doctoral dissertation completed in 2001 at Carnegie Mellon University under the direction of John R. Hayes. Much of this article was written while the author was a faculty member in the Department of Mathematics and Statistics at Murray State University in Murray, KY. I would like to thank Dr. Hayes, James Cummings, and Fred Reif for their helpful guidance while completing this project. I am also indebted to Lara Alcock, James Cummings, David Gibson, and the reviewers for helpful comments on drafts of the manuscript.

## References

Alcock, L., & Simpson, A. (2002). Definitions: Dealing with categories mathematically. *For the Learning of Mathematics*, *22*, 28–34.

Alibert, D., & Thomas, M. (1991). Research on mathematical proof. In D. O. Tall (Ed.), *Advanced mathematical thinking* (pp. 215–230). Dordrecht: Kluwer.

Anderson, J. R., Boyle, C. F., Corbett, A. T., & Lewis, M. W. (1990). Cognitive modeling and intelligent tutoring. *Artificial Intelligence*, *42*, 7–49.

Anderson, J. R., & Schunn, C. D. (2001). Implications of ACT-R learning theory: No magic bullets. In R. Glaser (Ed.), *Advances in instructional learning theory: Educational design and cognitive science, Vol. 5* (pp. 1–34). Mahwah, NJ: Erlbaum.

Asiala, M., Dubinsky, E., Mathews, D. M., Morics, S., & Oktac, A. (1997). Development of students' understanding of cosets, normality, and quotient groups. *Journal of Mathematical Behavior, 16*, 241–309.

Bagno, E., Eylon, B.-S., & Ganiel, U. (2000). From fragmented knowledge to a knowledge structure: Linking the domains of mechanics and electromagnetism. *American Journal of Physics, 68*(educational supplement), S16–S26.

Begle, E. B. (1979). *Critical variables in mathematics education: Findings from a survey of the empirical literature.* Washington, DC: Mathematics Association of America and National Council of Teachers of Mathematics.

Bills, E., & Tall, D. O. (1998). Operational definitions in advanced mathematics: The case of least upper bound. In A. Olivier & K. Newstead (Eds.), *Proceedings of the 22nd conference of the International Group for the Psychology of Mathematics Education* (Vol. 2, pp. 104–111). Stellenbosch, South Africa.

Cobb, P. (1990). A constructivist perspective on information-processing theories of mathematical activity. *International Journal of Educational Research, 14*, 67–92.

Collins, A., Brown, J. S., & Newman, S. E. (1989). Cognitive apprenticeship: Teaching the crafts of reading, writing, and mathematics. In L. B. Resnick (Ed.), *Knowing, learning, and instruction: Essays in honor of Robert Glaser* (pp. 453–494). Hillsdale, NJ: Erlbaum.

Cooper, R. (1991). The role of mathematics transformation and practice in mathematical development. In L. P. Steffe (Ed.), *Epistemological foundations of mathematical experience* (pp. 102–123). New York: Springer-Verlag.

Cupillari, A. (1989). *The nuts and bolts of proof.* Belmont, CA: Wadsworth.

Davis, P. J., & Hersh, R. (1981). *The mathematical experience.* Boston, MA: Birkhauser.

Davis, R. B. (1992). Understanding "understanding". *Journal of Mathematical Behavior, 11*, 225–241.

Davis, R. B. (1996). One complete view (though only one) of how children learn mathematics. *Journal for Research in Mathematics Education, 27*, 100–106.

DeFranco, T. (1996). A perspective on mathematical problem-solving expertise based on the performances of male Ph.D. mathematicians. In J. Kaput, A. H. Schoenfeld, & E. Dubinsky (Eds.), *Research in collegiate mathematics education. II* (pp. 195–213). Providence, RI: American Mathematical Society.

Dubinsky, E., Dautermann, J., Leron, U., & Zazkis, R. (1994). On learning fundamental concepts of group theory. *Educational Studies in Mathematics, 27*, 267–305.

Dummit, D. S., & Foote, R. M. (1991). *Abstract algebra.* Englewood Cliffs, NJ: Prentice Hall.

Franklin, J., & Daoud, A. (1988). *Introduction to proofs in mathematics.* Sydney, Australia: Prentice Hall.

Hanna, G. (1990). Some pedagogical aspects of proof. *Interchange, 21*, 6–13.

Harel, G. (1998). Two dual assertions: The first on learning and the second on teaching (and vice versa). *American Mathematical Monthly, 105*, 497–507.

Harel, G., & Sowder, L. (1998). Students' proof schemes: Results from exploratory studies. In A. H. Schoenfeld, J. Kaput, & E. Dubinsky (Eds.), *Research in collegiate mathematics education. III* (pp. 234–283). Providence, RI: American Mathematical Society.

Hart, E. W. (1994). A conceptual analysis of the proof-writing performance of expert and novice students in elementary group theory. In J. Kaput & E. Dubinsky (Eds.), *Research issues in undergraduate mathematics learning: Preliminary analyses and results* (pp. 49–63). Washington, DC: Mathematical Association of America.

Hazzan, O. (1999). Reducing abstraction level when learning abstract algebra concepts. *Educational Studies in Mathematics, 40*, 71–90.

Hazzan, O., & Leron, U. (1994). Students' use and misuse of mathematics theorems: The case of Lagrange's theorem. *For the Learning of Mathematics, 16*, 23–26.

Hecht, S. A. (1998). Toward an information-processing account of individual differences in fraction skills. *Journal of Educational Psychology, 90*, 545–559.

Heller, J., & Reif, F. (1984). Prescribing effective human problem-solving processes: Problem description in physics. *Cognition and Instruction, 1*, 177–216.

Kline, M. (1977). *Why the professor can't teach: Mathematics and the dilemma of university education.* New York: St. Martin's.

Knuth, E., & Elliot, R. (1998). Characterizing students' understanding of mathematical proof. *Mathematics Teacher, 91*, 714–717.

Leron, U. (1985). Heuristic presentations: The role of structuring. *For the Learning of Mathematics, 5*, 7–13.

Leron, U., & Dubinsky, E. (1995). An abstract algebra story. *American Mathematical Monthly, 102*, 227–242.

Lester, F. K., & Garafalo, J. (1982). *Metacognitive aspects of elementary school students' performance on arithmetic tasks.* (Paper presented at the annual meeting of the American Educational Research Association, New York)

Malone, J., Douglas, G., Kissane, B., & Mortlock, R. (1980). Measuring problem solving ability. In S. Krulik (Ed.), *NCTM 1980 Yearbook: Problem solving in school mathematics* (pp. 204–214). Reston, VA: National Council of Teachers of Mathematics.

Martin, G. W., & Harel, G. (1989). Proof frames of preservice elementary teachers. *Journal for Research in Mathematics Education, 20*, 41–51.

Mayer, R. (1992). *Thinking, problem solving, and cognition.* New York: Freeman.

Moore, R. (1994). Making the transition to formal proof. *Educational Studies in Mathematics, 27*, 249–266.

National Council of Teachers of Mathematics. (1980). *An agenda for action: Recommendations for school mathematics of the 1980s.* Reston, VA: Author.

National Council of Teachers of Mathematics. (2000). *Principles and standards for school mathematics.* Reston, VA: Author.

Newell, A., & Simon, H. (1972). *Human problem solving.* Englewood Cliffs, NJ: Prentice Hall.

Ohlsson, S. (1990). Cognitive science and instruction: Why the revolution is not here (yet). In H. Mandl, E. DeCorte, N. S. Bennett, & H. F. Friedrich (Eds.), *Learning and instruction: European research in an international context. Vol. 2.1: Social and cognitive aspects of learning and instruction* (pp. 561–600). Oxford, England: Pergamon.

Ohlsson, S., Ernst, A., & Rees, E. (1992). The cognitive complexity of learning and doing arithmetic. *Journal for Research in Mathematics Education, 23*, 441–467.

Palinscar, A. S., & Brown, A. L. (1984). Reciprocal teaching of comprehension fostering and comprehension monitoring activities. *Cognition and Instruction*, *1*, 117–175.

Reif, F. (1995). Millikan Lecture 1994: Understanding and teaching important scientific thought processes. *American Journal of Physics*, *63*, 17–32.

Reif, F., & Scott, L. (1999). Teaching scientific thinking skills: Students and computers coaching each other. *American Journal of Physics*, *67*, 819–831.

Rotman, J. J. (1996). *A first course in abstract algebra.* Upper Saddle River, NJ: Prentice Hall.

Rowland, T. (2002). Generic proofs in number theory. In S. Campbell & R. Zazkis (Eds.), *Learning and teaching number theory: Research in cognition and instruction* (pp. 157–183). Westport, CT: Ablex.

Schoenfeld, A. H. (1985a). Artificial intelligence and mathematics education: A discussion of Rissland's paper. In E. Silver (Ed.), *Teaching and learning mathematical problem solving: Multiple research perspectives* (pp. 177–187). Hillsdale, NJ: Erlbaum.

Schoenfeld, A. H. (1985b). *Mathematical problem solving.* Orlando, FL: Academic.

Schoenfeld, A. H. (1988). When good teaching leads to bad results: The disasters of 'well-taught' mathematics courses. *Educational Psychologist*, *23*, 145–166.

Schoenfeld, A. H. (1992). Learning to think mathematically: Problem solving, metacognition, and sense-making in mathematics. In D. Grouws (Ed.), *Handbook of research on mathematics teaching and learning* (pp. 334–370). New York: Macmillan.

Schoenfeld, A. H. (1994). Some notes on the enterprise (research in collegiate mathematics education, that is). In E. Dubinsky, A. H. Schoenfeld, & J. Kaput (Eds.), *Research in collegiate mathematics education. I* (pp. 1–19). Providence, RI: American Mathematical Society.

Schoenfeld, A. H. (1998). Reflections on a course in mathematical problem solving. In A. H. Schoenfeld, J. Kaput, & E. Dubinsky (Eds.), *Research in collegiate mathematics education. III* (pp. 81–113). Providence, RI: American Mathematical Society.

Selden, A., Selden, J., Hauk, S., & Mason, A. (2000). Why can't calculus students access their knowledge to solve nonroutine problems? In E. Dubinsky, A. H. Schoenfeld, & J. Kaput (Eds.), *Research in collegiate mathematics education. IV* (pp. 128–153). Providence, RI: American Mathematical Society.

Senk, S. (1985). How well do students write geometry proofs? *Mathematics Teacher*, *78*, 448–456.

Slagle, J. (1963). A heuristic program that solves symbolic integration problems in freshman calculus. In E. Feigenbaum & J. Feldman (Eds.), *Computers and thought* (pp. 191–203). New York: McGraw-Hill.

Solow, D. (1990). *How to read and write proofs* (2nd ed.). New York: Wiley.

Tall, D. (1979). Cognitive aspects of proof, with special reference to the irrationality of $\sqrt{2}$. In D. O. Tall (Ed.), *Proceedings of the International Conference for the Psychology of Mathematics Education (3rd)* (pp. 206–207). Coventry, England: University of Warwick. (ERIC Citation Number ED226956)

Thurston, W. (1994). On proof and progress in mathematics. *Bulletin of the American Mathematical Society*, *30*, 161–177.

Watson, F. R. (1978). *Proof in mathematics.* Staffordshire, UK: University of Keele.

Weber, K. (2001a). *Investigating and teaching the strategic knowledge needed to construct proofs* (unpublished doctoral dissertation). Pittsburgh, PA: Carnegie Mellon University.

Weber, K. (2001b). Student difficulties in constructing proofs: The need for strategic knowledge. *Educational Studies in Mathematics, 48*, 101–119.

Weber, K. (2002a). Beyond proving and explaining: Proofs that justify the use of definitions and axiomatic systems and proofs that illustrate technique. *For the Learning of Mathematics, 22*, 14–17.

Weber, K. (2002b). Instrumental and relational understanding in proofs about group isomorphisms. In I. Vakalis, D. Hughes Hallett, C. Kourouniotis, D. Quinney, & T. Constantinos (Eds.), *Proceedings of the second International Conference on the Teaching of Mathematics (at the undergraduate level).* Hersonisoss, Crete, Greece: Wiley & Sons. (Electronic Proceedings, Paper Number 86 (pap86.pdf). Retrieved March 22, 2006 from http://www.math.uoc.gr/ ictm2/Proceedings/pap86.pdf.)

Weber, K. (2004). Traditional instruction in advanced mathematics. *Journal of Mathematical Behavior, 23*, 115–133.

Weber, K., & Alcock, L. (2004). Semantic and syntactic proof productions. *Educational Studies in Mathematics, 56*, 209–234.

Yusof, Y., & Tall, D. (1999). Changing attitudes to university mathematics through problem solving. *Educational Studies in Mathematics, 37*, 67–82.

## Appendix A. Part of the "Facts Sheet" Given to Students in the Experiment in Section 3

**IMAGE:** If you are asked to prove something about the image, here are the inferences that you can apply.

**2.1:** If $G$ is abelian, then $f(G)$ is abelian. (Note: abelian = commutative)

*When to use:* Use if you know that $G$ is abelian.

*Information needed to use this:* You need to know if $G$ is abelian.

*Example of application:* $\mathbb{Q}$ is the group of rational numbers, which is known to be an abelian group, $f$ is a homomorphism from $\mathbb{Q}$ to $H$.

Fact 2.1. If $G$ is abelian, $f(G)$ is abelian.

Application: Since $\mathbb{Q}$ is abelian, $f(\mathbb{Q})$ is abelian.

**2.2:** If $G$ is cyclic, then $f(G)$ is cyclic

*When to use:* Use if you know that $G$ is cyclic.

*Information needed to use this:* You need to know if $G$ is cyclic.

*Example of application:* $\mathbb{Z}$ is the group of integers, which is known to be a cyclic group, $f$ is a homomorphism from $\mathbb{Z}$ to $H$.

Fact 2.2. If $G$ is cyclic, $f(G)$ is cyclic.

Application: Since $\mathbb{Z}$ is cyclic, $f(\mathbb{Z})$ is cyclic.

**2.3:** $|f(G)| = |G|/|ker f|$ (Note: $|A|$ denotes the order or size of $A$)

*When to use:* If you know the order of $G$.

*Information needed to use this:* You need to know the order of $G$ and the possible orders of the kernel.

*Example of application:* $f$ is a homomorphism from $G$ to $H$. $G$ has order 20. The kernel of $f$ has size 5 or 10.

Fact 2.3. $|f(G)| = |G|/|ker f|$

Application: $|f(G)| = 20/5 = 4$ or $|f(G)| = 20/10 = 2$

**2.4:** First Isomorphism Theorem: $f(G)$ is isomorphic to $G/ker f$

*When to use:* If you have information about $G$ and the kernel, or the quotient group $G/ker f$

*Information needed to use this:* You need info about the domain and kernel.

*Example of application:* $f$ is a homomorphism from $G$ to $H$ with kernel $K$.

Fact 2.4. $f(G)$ is isomorphic to $G/ker f$

Application: $f(G)$ is isomorphic to $G/K$

## Appendix B. Key Facts about Homomorphisms

### I. DOMAIN.

- Definition of homomorphism: $f(x)f(y) = f(xy)$ (Relates information about the mapping to information about the domain)

### II. IMAGE.

- If a structural property (e.g. abelian, cyclic) is true in $G$, then it is true in $f(G)$. (Relates information about the domain with information about the image.)
- First isomorphism theorem: $G/ker f$ is isomorphic to $f(G)$. (Relates information about the domain and kernel with information about the image.)
- $|G|/|ker f| = |f(G)|$. (Relates information about the domain and kernel with information about the image.)

### III. KERNEL.

- Definition of kernel: $g$ is in the kernel iff $f(g) = e_H$. (Relates information about the mapping to information about the kernel.)
- The kernel is a normal subgroup of $G$. (Relates information about the domain to information about the kernel.)
- If $f(g) = h$, $f^{-1}(h) = gker(f)$. (Relates information about the mapping to information about the kernel.)

### IV. MAPPING.

- Definition of homomorphism: $f$ is a homomorphism iff $f(x)f(y) = f(xy)$. (Relates information about the domain to information about the mapping.)

## Appendix C. Introductory Text for Day 2 of the
## Instruction Described in Section 4

OK, I looked over the test you took on Tuesday and you guys didn't do very well. But that's okay. Our research team knew that students have difficulty with these types of questions. In fact, we made the test hard on purpose. We want to show that our lessons will make you better at proving theorems, so we wanted to use a set of theorems you could not do at first.

One reason students have difficulty proving theorems is because there are a lot of things you can do, or a lot a facts and theorems you can use. You can't possibly try them all, so you need some strategies telling you which facts to try. What our research has shown is that the strategies students use often are not very good.

Let me give you a couple of examples. Sometimes if a student tries to prove a statement B, they look for any theorem A implies B and try to prove A. I'll be more concrete. One student wanted to show a group $G$ was abelian, so he tried to show every element squared in G was the identity. Now if he could do this, he certainly could show $G$ is abelian. If every element in $G$ has order 1 or 2, then $G$ is abelian. [*Write on board* "Show G is abelian", "If for all $g$ in $G$, $g^2 = e$, then G is abelian" *and* "Show for all g in G, $g^2 = e$"]. But the student had no reason to assume $G$ would have this unusual property. Almost no groups, abelian or otherwise, have this property. The student might be trying to do the impossible!

Here is another example of a poor student strategy. Students have a tendency to go straight to the definition of the term. "Suppose $f$ is a homomorphism from $S_5$ to H. Show the kernel of $f$ is $S_5$ or the identity." Students might immediately go to the definition of the kernel. [*Write on board,* "$g$ is in the kernel of $f$ if and only if $f(g) = e_H$."]

But where does the student go from here? [Pause] Nowhere. He needs to know some more information about $f$ if this is to be useful. The student strategies sometimes work, but there are a lot of problems you can't solve if you use them. Our research team has developed more sophisticated strategies. We've shown that these strategies are effective and that students can use them to prove theorems. What we are doing now is testing instruction designed to teach these strategies. Since there are only four meetings, you can learn these strategies in a short amount of time. And they will make a big difference when you try to prove the problems I gave you on Tuesday so they really are in your interest to learn.

In mathematics, there are a lot of ways to try to prove statements or solve problems. Mathematicians often first start their proofs or problem solutions by finding out what the problem is essentially asking for information about. That is, what structure is the problem asking you to find information about. The importance of this is that there are usually only a few important ways that mathematicians find out information about a particular structure.

Let me show you some examples that illustrate the differences between how some students and mathematicians would approach a problem
[*Write* "Show $4x^3 - x^4 = 30$ has no solutions."] It has been shown that this problem is extremely difficult for students. They typically spend their time trying algebraic techniques, such as $4x^3 - x^4 = x^3(4 - x)$. A mathematician sees this as a question about functions. A mathematician knows a very important way to find information about a function is its derivatives. [*Instructor proceeds to show the maximum value of $4x^3 - x^4$ is 27, so the left side never attains a value of 30*].

## Appendix D. Practice Problems Used During
## Instruction Described in Section 4

- Suppose that $G$ is a group and $f$ is the mapping from $G$ to $G$ such that f$(g) = g^2$. Prove that if $f$ is a homomorphism, then $G$ is abelian.
- Suppose that $G$ is a group and $f$ is the mapping from $G$ to $G$ such that $f(g) = g^{-1}$. Prove that if $f$ is a homomorphism, then $G$ is abelian.
- Suppose that $H$ is a group and $f$ is a homomorphism mapping $\mathbb{Z}$ (the group of integers under addition) to $H$. Prove the image of $f$ is cyclic.
- Suppose that $f$ is a homomorphism that maps $\mathbb{Z}$ into $S_3$ (the symmetric group of three elements). Prove that $f(\mathbb{Z})$ has order 1, 2, or 3.
- Suppose that $H$ is a group and $f$ is a surjective homomorphism from $\mathbb{Q}$ (the group of rational numbers under addition) to $H$. Prove that $H$ is abelian.
- Suppose that $H$ is a group and $f$ is a surjective homomorphism that maps $S_3$ into $H$. If $f$ is not a trivial homomorphism, prove that $H$ is cyclic.
- Suppose $G$ is the group $\mathbb{Z}/9\mathbb{Z}$ and suppose that $f$ is the homomorphism from $G$ to $G$ such that $f(g) = 6g$ for all $g$ in $G$. What is the kernel of $f$?
- Suppose $G$ and $H$ are groups and $f$ is a homomorphism from $G$ to $H$. Suppose that for $h$ in $H$, $f^{-1}(h) = \{g_1, g_2\}$ (that is, the only elements of $G$ that are mapped by $f$ to $H$ are $g_1$ and $g_2$. Compute the kernel of $f$.
- Suppose that $G$ is a group and the inverse mapping from $G$ to $G$ ($f(g) = g^{-1}$ for all $g$ in $G$) is a homomorphism. Prove that $f(g)$ is an injective homomorphism.
- Suppose that $G$ is a simple group (that is, a group with no normal subgroups) and $H$ is a group. Prove that if $f$ is a non-trivial homomorphism from $G$ to $H$, then $f$ is injective.
- Suppose that $G$ and $H$ are groups and $f$ is a homomorphism from $G$ to $H$. Suppose that for an element $h$ of $H$, $f^{-1}(h) = g$. Prove that $f$ is injective.
- Suppose that $f$ is the homomorphism from $\mathbb{Q}$ to $\mathbb{Q}$ such that $f(q) = -2q$. Prove $f$ is a surjective homomorphism.
- Suppose that $f$ is the homomorphism from $\mathbb{Z}/12\mathbb{Z}$ to $\mathbb{Z}/12\mathbb{Z}$ such that $f(z) = 5z$. Prove that $f$ is a surjective homomorphism.
- Suppose that $G$ is a group and the inverse mapping from $G$ to $G$ ($f(g) = g^{-1}$ for all $g$ in $G$) is a homomorphism. Prove that $f(g)$ is a surjective homomorphism.
- Suppose that $M$ is the group of 2 by 2 non-singular matrices and $N$ is the group $\mathbb{R}/\{0\}$ (the group of real numbers without zero under multiplication). Prove that the mapping $f$ from $M$ to $N$ such that $f(m) = det(m)$ for all $m$ in $M$ is a homomorphism.

GRADUATE SCHOOL OF EDUCATION, 10 SEMINARY PLACE, RUTGERS UNIVERSITY, NEW BRUNSWICK, NJ 08901

*E-mail address*: `khweber@rci.rutgers.edu`

CBMS Issues in Mathematics Education
Volume **13**, 2006

# The Transition to Independent Graduate Studies in Mathematics

## Janet Duffin and Adrian Simpson

ABSTRACT. In this paper we explore some of the changes that students undergo in coming to resolve difficulties they perceive in coping with independent graduate study in mathematics. By interviewing current UK PhD students about their existing ways of learning new areas of mathematics and their personal learning histories, we developed an expanded theory of cognitive styles and explored the changes in those styles that the students experienced. The data suggest that, with few exceptions (each fascinating in its own right), alien cognitive styles are incompatible with independent graduate study. We discuss some of the possible sources of changes to cognitive style in the changed pedagogies experienced at this level.

## Introduction

Moving between educational phases can cause significant difficulties for learners. Not only does a new phase bring new content to be learned and new ways in which one is expected to engage with that content, but it often brings a new 'didactic contract' (Brousseau, 1997): the set of unwritten rules about the conduct of the teacher-learner role pair (in the sense of Skemp, 1976). In the area of advanced mathematical thinking, a considerable amount of effort has been expended in exploring the transition from school to university, in terms of content, engagement and the altered didactic contract (Daskalogianni & Simpson, 2000; Tall, 1991).

The undergraduate–graduate transition has been less well explored, though there are an increasing number of interesting papers that have investigated the thinking of doctoral and post-doctoral mathematicians. For example, Smith and Hungwe (1998), investigated the nature of conjecturing amongst young mathematics doctoral students and found contrasts among students who saw mathematics as very formal, as a 'box of tricks,' or as about mental objects. Burton (1999) interviewed a range of professional mathematicians about how they come to know mathematics, contrasting those who use an emotional sense of certainty as a sign of knowing with those who see their ability to use a piece of mathematics appropriately as a sign of knowing.

The authors wish to thank the anonymous reviewers for ensuring this article is considerably better than it would otherwise have been.

Herzig (2002) considered drop-out rates in US PhD programs and suggested that the issue of 'participation' is at the core of many students' decisions about continuing their studies. Put together, these three papers suggest that there is something crucial at the heart of becoming a mathematician that is related to what people believe mathematics to be, how they come to know it and what their level of participation in creating new knowledge.

Thus, this study considered how students are affected by the dramatic change in the didactic contract that occurs when they move from a traditionally taught undergraduate degree course to conducting independent research for a PhD (in the UK context). In particular, we examined how students who enter graduate research programs with different cognitive styles cope in different ways with the changes. To some extent, this study complements and contextualizes more general studies such as Pole, Sprokkereef, Burgess, and Lakin (1997) that explored the transition to independent graduate study across a range of disciplines focusing on the relationships between PhD supervisors and their students.

After discussing the context of this study and the notion of cognitive style as a sensible construct for investigation, we consider several case studies exemplifying three radically different responses to the new teaching/learning situation. We use these case studies to show that existing cognitive style dichotomies do not suffice to describe the differences we observed. We propose a broader set of styles that allows us to chart the changes the students went through, and finally, we suggest possible sources of the difficulties they met.

## Cognitive Style

In a wide-ranging survey of papers on cognitive and learning styles, Riding and Cheema (1991) noted the difficulty researchers have in pinning down these constructs, notably because the phrases 'cognitive style' and 'learning style' are used interchangeably and, across different papers, with markedly different meanings. Despite this, the majority of authors define 'style' to mean, in some sense, a general tendency to respond in particular ways to learning tasks. However, some theorists even challenge the very idea of learning style as a valid construct, suggesting that there are no such general tendencies and that each new problem faced by a learner may be responded to in quite different ways: that is, there are no styles, only strategies (Riding & Rayner, 1998).

Stanovich and West (2000) reviewed a large range of psychological literature in exploring the notion of rationality. They noted that the majority of researchers in the field, while often taking opposing views on some of the crucial rationality experiments, such as the classic Wason Selection Task (Wason, 1966), had developed so-called 'dual process theories' in which they distinguished two distinct ways of thinking. Within each of these psychological theories, Stanovich and West identified one cognitive process that is "automatic, largely unconscious and undemanding of computational capacity" (perhaps akin to a 'module' in the sense of Fodor (1983)) and that is "highly contextualised, personalised and socialised ... driven by considerations of relevance." On the other hand, within each theory, they also discerned a competing cognitive process that "serves to decontextualise and depersonalise problems ... is more adept at representing in terms of rules and underlying principles." These two distinct ways of thinking are seen to be relatively resistant to change and thus provide support from the psychological literature for something more fundamental than strategies.

Similarly, in a comprehensive review by Messick (1984), the majority of papers that developed theories of cognitive style described styles in terms of two qualitatively different approaches. They generally distinguished between, on the one hand, approaches that are about connecting the problem to a range of existing knowledge (using terms like 'holistic', 'deep', 'simultaneous' and 'global') and approaches that are about applying set methods or procedures to problems (using terms like 'serialist', 'surface', 'successive' and 'analytic'). As well as matching the dichotomies given in the psychological literature (such as Evans & Over, 1996; Sloman, 1996), theories of cognitive style fit closely with well known mathematical dichotomies like those of procedural/conceptual thinking (Hiebert & Lefevre, 1986) and relational/instrumental understanding (Skemp, 1976). While one may consider all such dichotomies to be considerable simplifications, they perhaps merely focus attention on two ends of a continuum. However, even as simplifications, they appear to have been adopted by many in the mathematics education community as a useful way of providing high contrast between approaches.

The initial model of cognitive style adopted here is one adapted first by Pinto (1998) from our earlier work (Duffin & Simpson, 1993), that new experiences faced by learners can be classified into three distinct categories: natural, conflicting, and alien. A *natural* experience is one that fits with a learner's existing internal mental structures and thus can be easily dealt with and used to strengthen an existing way of thinking. A *conflicting* experience is one that might elicit two distinct responses that would lead to contradictory results (or where the learner brings an existing internal mental structure to bear on the task, expecting it to help solve the problem, but finds it does not appear to do so). In this case, the learner may modify her or his thinking by destroying existing mental structures that fail to cope, restricting their scope or merging apparently conflicting structures once their contradictions are resolved. Finally, an *alien* experience is one which neither fits with nor jars existing structures and may be ignored by the learner or used as the seed for a new and initially separate mental structure. This theory has close links with Piaget's 'alpha', 'beta' and 'gamma' reactions to cognitive disequilibrations (Piaget, 1975).

Pinto (1998) adapted this theory of experience responses to develop a theory of cognitive styles by suggesting that learners may have relatively stable preferences for dealing with new experiences. On the one hand, natural learners seek to link new areas of mathematics to their existing mental structures and might find it difficult to move rapidly through new material without being able to see connections or analogies with their existing ways of thinking. Alien learners,[1] on the other hand, are happy to absorb new areas of mathematics, initially, as separate mental structures that might later become connected as a result of encountering conflicts.

In this study, we began with a view that the natural-alien cognitive style dichotomy would provide a context in which to study the transition to independent graduate research, and, in particular, would allow us to study how students with different cognitive styles respond to the changed didactic contract.

---

[1]In her description of students in a first course in analysis, Pinto used the phrase 'formal learners' for those who tended to 'extract meaning,' that is, who tended not to link new information to existing thinking and thus concentrated on the formal symbolic expressions. We use the phrase 'alien learners' to indicate what we see as having wider applicability.

## The Study

The study was conducted with 13 individuals who constituted the entire cohort of mathematics PhD students at a medium sized UK university. The nature of graduate research in mathematics is such that the population that can be drawn upon is small, and rather than restrict it further, it was decided to interview all the students regardless of the stage of PhD study they had reached or their research topic. Thus, the respondents ranged from those just starting their dissertations, through to those in the final stages of writing up, and varied across both pure and applied mathematics.

The traditional PhD program in the UK is markedly different from similarly titled programs in some other countries, such as the US. There is often no formal requirement to attend or pass taught courses during the period of doctoral study and almost all of a student's three years of study is spent in independent research towards the production of a single, substantial dissertation of no more than 100,000 words. The dissertation is required to be the result of original research, to show an awareness of the relationship of the research to a wider field of knowledge and, potentially, to be publishable. Thus, the change in the didactic contract and in the teacher-pupil role pair relationship from a taught undergraduate program is, at least superficially, dramatic.

Each of the students was invited to a one-on-one interview with one of the authors. The interviews focussed on their espoused cognitive styles (particularly in response to trying to learn a new area of mathematics) and their descriptions of their learning histories. None of the students declined the offer, though one gave such a taciturn performance that an analysis of his transcript yielded little and he plays no part in the subsequent development of this paper. Since almost all the students were also involved in helping to teach undergraduates, many touched on this as an issue and it also provided some insights into their cognitive styles. The interviews were conducted as what Burgess (1988) calls a 'conversation with a purpose': that is, they were ethnographic interviews designed around a given focus to draw ideas flexibly from the respondents. We began with focussing questions like 'How do you go about learning a new piece of mathematics?'

Each of the interviews was transcribed, checked, and then coded independently by each author. Those codes were then compared, categories drawn out and dimensions within the categories discerned, along lines suggested by Glaser and Strauss (1967). In particular, we were able to uncover three quite distinct ways in which the students appeared to be responding to the transition to independent graduate study that are best illustrated by examining three prototypical individuals.

**Natural Students.** After an analysis of the transcripts, from their descriptions of their approaches to learning as undergraduates, we found a number of students who appeared to be natural learners. That is, they spoke of deliberately seeking out ways to connect new knowledge to existing mental structures, increasing the richness and complexity of their understanding. Oliver showed the response of a natural learner to the experience of independent graduate study most typically.

Indeed, while Oliver said little about his early school learning, he was quite forthcoming about his later school (sixth-form) and undergraduate experiences. In describing his initial encounters with calculus, he seemed concerned that "a lot of people just seemed to recite the mantra 'well if $y = x^n$ then $\frac{dy}{dx}$ is $nx^{n-1}$,' " and it was a mantra and a notion that one learnt." He felt unable to learn the 'mantra,' apparently preferring to make sense of the work in graphical terms that he felt meant he "was one of the slow ones in catching on." His view of his cognitive style as an undergraduate seemed tinged with some frustration as he compared it to others who were apparently alien learners in their approach:

> With some subjects - analysis for example - standard questions "prove whatever theorem," ...there are some people who will just develop short term memory tricks just rather like an actor learning lines. They don't necessarily mean anything to the individuals they just learn it from an exam oriented point of view. It's a perfectly satisfactory approach: from a mathematics one it's (and indeed from just a general intellectual approach) it's not entirely satisfactory. So I would find myself going for a walk in the local park and reciting a proof to myself slowly and going over every single line in my head and interrogating it to death, so I suppose in some sense I emerged with a greater understanding for having done so. But the downfall of that I suppose is that I would spend too long on learning certain proofs and therefore not learn enough proofs and would then invariably be left in the dark on certain questions.

There is a sense in this quotation that Oliver is perturbed by peers who succeed by these 'memory tricks' and considers his apparently slower approach may be the cause of his lower than expected grade "perhaps reflected by the fact I didn't get a first." These concerns were echoed by Daphne, another natural student reflecting on her undergraduate study:

> I had some friends in my university who were like that: they were actually not interested in maths, they were just interested in getting the answers and passing their exams. They did pass ... but, you know, all the lecturers and teachers they feel it and they see it when you understand or when you just play the system, play with maths again.

As Oliver focussed on the issue of becoming an independent graduate student, he appeared to see differences in the structure of his study: the undergraduate syllabus was 'pre-ordained,' while at the postgraduate level "one has to decide for oneself what one is learning. So the first thing is, well, what do I want to learn?" He was able to discern the role of existing literature in determining the directions he should be taking, but where he felt he was unable to link this new material to his existing understanding, there was a role for "different people in the coffee room" who could give him what he felt were important examples of key concepts.

His cognitive style, however, appeared to have changed little. His emphasis was still on making connections between new material and existing knowledge and merely accepting new processes or theorems was still difficult:

> I dislike in mathematics, and in many other things – I am often reluctant to make a huge leap of faith and just accept something, and I find I take on the role of the devil's advocate and interrogate something to death.

For Oliver, as for the other students who showed evidence of having had a natural cognitive style as undergraduates, the transition to independent graduate study was relatively smooth. If anything, the freedoms and space given in the new situation appeared to be a relief – the speed with which new concepts were met was no longer dictated by a 'pre-ordained' syllabus and the university timetable. The ways in which one could compare oneself with peers who took a different approach to learning (a source of some of the frustration) seemed to have diminished. So the changed context seemed to make things easier for the natural students, and their cognitive style seemed to be unaffected.

**Alien Students - Part I.** Some of the alien learners who, as undergraduates, so frustrated Oliver and his fellow natural learners, also began to study for a PhD. We found a number of students in the study who appeared to have had an alien cognitive style as undergraduates: talking of the importance of memorizing, being prepared (at least initially) to accept new ideas without being overly concerned with their meanings and focussing on the procedural aspects of the material encountered.

Lucy was typical of one form of response to the new context provided by independent graduate study. She spoke clearly about her approach as an undergraduate:

> It's a bit difficult if you are trying to muddle through things for yourself. I like to be told. Once I've been told I can then understand it, I don't really like finding things out for myself very much.

Like Oliver, she was apparently aware of the changed context she was now in:

> That [an alien cognitive style] is an easy way to learn if you have a syllabus to learn from, if you have set aims, but in research you don't have ... you have a few aims, but you don't really know how you are going to go about meeting those aims, so no, it tends to be quite different.

As she, and some of her fellow alien learners, described how they tried to make sense of the transition to independent graduate study, we got a sense of uncertainty and of a movement to adopt a different approach to learning. Lucy's description of how she approaches learning a new piece of mathematics was a curious mixture of alien and natural cognitive styles:

> I think if I am approaching something for the first time, I always sit down and try to read the theory behind it, sort of memorize it first, before I actually try to go back and understand what's happening with it. So I suppose ... I don't know, I try and get everything into order in my head first, before I actually try understanding it. Like, for example, learning theorems and proofs and such, I learn them without really understanding what's going off, and then I go back later and think about it.

A more explicit sense of the way she adopts some of the traits of a natural cognitive style were clear when she was asked to expand upon her perception of the differences between her undergraduate and postgraduate approaches:

> ... as I say there's no set syllabus, so instead of learning things by rote and memorizing things, I'm having to much more go into the background and start with things I already know and, I don't

> know, learn by analogy or build them up, so anyway I found that
> a lot harder.

The ideas of 'analogy' and 'building up' seem to resonate well with what Oliver
said about his (unchanging) learning style.

So for Lucy, and some of the other alien learners, the transition to independent
graduate study was far from smooth. Instead of the relief we saw from those with
a natural cognitive style, we saw frustration at no longer 'being told.' Alan, in
a similar situation to Lucy, talked of his concerns at becoming independent: "I
do find it difficult being a researcher having to find things out for myself." The
'easy way of learning' that had worked so effectively for them as undergraduates
no longer appeared to be appropriate for the new context and they appeared to
be responding by trying to adopt aspects of a natural cognitive style: looking for
analogies, building new ideas upon old ones. Alan was even more explicit about his
new need to make links to existing ideas:

> If I just go away by myself and sit down, and I first read through
> the mathematics and see what I recognize, and get what I recognize
> and try and build on that.

However, as we explored the transcripts of the students who had apparently
had alien cognitive styles as undergraduates, we found that not all of them had the
same reaction. Lucy, Alan and some of their fellow students seemed to have moved
from an alien towards a natural cognitive style; others, it appeared, had not.

**Alien Students - Part II.** Sudeep was a typical alien learner. His discussions
of his school and undergraduate learning had the characteristic emphasis on memory, procedures and disjoint mental structures. Here he highlights the separation
of the teacher-led theory from the repetitive student work:

> Yes, as an undergraduate we used to have several courses that
> involved mathematical physics and mathematics that would back
> up all the physical calculations that we would go on to do in later
> years, and so typically a piece of science would be learned by going
> through the theory and it lasts for an hour or so followed by several
> questions, examples of problems based on the same piece. Then we
> were expected to understand this piece of science by solving these
> questions, problems and so on, so they would give us some time.

However, along with a number of other students who appeared previously to
have been alien learners, Sudeep did not seem to adopt directly some of the traits of
a natural cognitive style. Indeed, the adaptation he appeared to make in response
to the transition to independent graduate study seemed subtly but interestingly
different from that of Lucy and Alan described above. Sudeep seemed to look not
for ways of linking new mathematical ideas to existing ones, but to look for a sense
of structure within the new mathematics itself:

> As for me personally, I've been interested more in physics and
> applied physics as adapting physics rather than pure mathematics
> as such and therefore, each mathematical construct goes with a
> certain physical notion and so if one starts asking questions like
> why the mathematics actually works the way it does, then there is
> usually a physical sort of explanation to it all. I would go for that
> basically. ... I would to start with, I would try to find a certain

structure in the piece of mathematics that I'm looking at, try to
find some logical procedure that's involved and try to understand
how and why the result is obtained.

Some of the essence of the alien cognitive style is still evident – the emphasis is on
a procedure, but on a 'logical procedure' that is involved.

Rebecca, an unusually thoughtful and reflective student, described her previous
cognitive style as having many alien traits, and had a view similar to that of Sudeep:

> Well, say I come across a new proof or something: if I've got a
> theorem I understand the theorem and I'm trying to learn the
> proof or be able to explain it to somebody else, the way I usually
> do it, is to go through each of the steps and keep going over each of
> the steps until it tells the story properly. So I think I'm probably
> more looking at each of the steps on their own to start with, to
> work out this bit and then the next bit, and then the next bit,
> until I've got that, and then eventually it all comes together ...,
> the way that I look at each of the separate bits and then put it all
> together afterwards, and I think that's how I've been doing most
> of my work: concentrating on the little ... . You know, each of
> the bits on their own to start with, because I can't understand
> the next bit until I've thoroughly understood the first one. But of
> course at the same time I always want to know where it's going,
> so the first thing would be to see where it's going to, and to get
> there I know I've got to do each of the steps one after the other,
> after the other.

The emphasis here seems to be on a relatively procedural approach (and perhaps even the remnants of an attempt to memorize 'each of the separate bits') but
with an important focus on "where it is going." There may have been a sense, in
looking back at her later undergraduate courses, that she could see this emphasis
on the structure within mathematics from some of her lecturers:

> I used to like that because it kind of told me what was going to
> happen but then each of the bits I could kind of see, oh that's
> going to be fitting into here and that's what I like best about it.

Again here the issue is not quite the same as a natural cognitive style, which
is about fitting new knowledge in with existing mental structures. Neither does it
appear to be a transitional stage on a continuum *between* alien and natural styles.
For Sudeep, Rebecca and some of the other students in the study, the transition
to independent graduate study had encouraged them to focus on the mathematical
structure within new notions. We came to call this cognitive style *coherence*, as its
focus appeared to be on the logical and structural coherence of new material.

Thus, across the majority of the students in this study, we saw a number of consistent ways in which cognitive styles were affected by the transition to independent
graduate study.

- Natural cognitive styles appeared stable across the transition and were
  relatively unaffected by it.
- Frustrations felt at the undergraduate level by those with natural cognitive
  styles seemed to dissipate.

- Alien cognitive styles appeared to be unstable across the transition and students responded by adopting alternative styles.
- The attempts to adopt new cognitive styles may have been accompanied by a sense of uncertainty.
- Alien learners appeared either to
    - adopt natural learning styles, attempting to make analogical links between new mathematical ideas and existing ones (seeking *structure around mathematics*), or
    - adopt coherence learning styles, attempting to explore the logical and structural form of a new mathematical idea to enable them to fit new material, piecewise, into a new mental structure (seeking *structure within mathematics*).

**Alternative Reactions to the Transition.** Three of the students in the sample did not appear to fit this general pattern, but appeared to have adapted to the new context of independent graduate study in somewhat different ways. Sally, for example, appeared to have been a natural learner as an undergraduate, but, rather than having a smooth transition to her PhD studies, she appeared to have many of the feelings of pressure that the other natural learners had attributed to their *undergraduate* experiences and this led her to adopt some of the characteristics we would associate with alien cognitive styles:

> Well now I read things and I have to just accept what it says. Then I go off and apply it to whatever I need it for, but ... [as an undergraduate] ... I wouldn't be able to accept something unless I could work it all through, the complete proof myself.

There was some indication that the pressure on Sally came from her movement to an area of mathematics she had not excelled at earlier. Thus, the need to acclimatize quickly to the new context may have led her to adopt different characteristics.

Jansher, in contrast, seemed to have been an alien learner as an undergraduate, but appeared to cope relatively well with the transition without a need to alter his cognitive style substantially. Instead he 'retreated' (in the sense of Perry (1970)) into a radical form of an alien cognitive style, adopting a formalist view of mathematics that denied a significant role for 'meaning':

> Yeah, I think my basic ... I'm not an educationalist – I don't know much about these things, but my basic idea is to learn mathematics up to a certain standard one has to play it as a formal game. With a certain amount of rigidity, certain rigid rules which you follow. It's the same thing that always occurs with undergraduates here, for example I was teaching analysis last semester ... they [the students] would turn round and say 'what does this definition mean' and that's an absolutely wrong way of looking at it. The definitions are atomic statements they are not meaningful, they are only meaningful within what they can give you.

Such an approach, though philosophically stable, seemed psychologically a long way from all the other students in our sample.

The final student in our sample who did not fit the general pattern discussed above was, in many ways, the most interesting. Paul exhibited what we came to call a flexible cognitive style and appeared to have had this approach to learning for some time.

As mentioned earlier, a cognitive *style* is an approach to learning new material (in this case, new mathematical content) that is relatively stable over time. We contrast this with the definition of a cognitive *strategy*, that is, an approach adopted for a particular situation. All the students in our sample (including Sally and Jansher) showed clear evidence that their learning approaches had been stable for a long time prior to the transition to independent graduate study and were now being challenged by significant changes of environment. Thus, we feel it is reasonable to say that all the students had a clearly discernible cognitive style, except Paul. He appeared to be able to adopt a number of different approaches depending on the value he attributed to the material he was trying to learn. If he was interested in gaining information he might adopt the characteristics of an alien learner:

> ...you sit through a lecture, you'll write notes down, but quite often to be honest, very often in lectures, I didn't pay that much attention to the mathematics itself. I was more interested in getting it down off the board, and regularly over a semester course what I would do is, quite often, lose the plot, by about week three or four but not worry too much about that.

However, if Paul needed a better understanding of the underlying relationships between concepts, natural learning might be adopted:

> I mean for example this is how I've been learning recently, and I like to be able to view the object in my head and quite often work with it up there rather than going straight at pen and paper and reading through masses of theory and stuff. I can play about with it for a while in my head as an object.

Indeed, we even saw characteristics of our new cognitive style, 'coherence':

> I'll go through a fairly big overview so I've got an idea of where everything's going first, and that, when I then come back to include these things in detail, that helps me to be able to see where they are all going.

This flexible cognitive style, then, seems to adopt facets of the other cognitive styles as strategies (in the sense of Riding and Rayner (1998)). Unlike the other students who talked of a single general tendency towards learning new material, Paul considered the value to be gained from new material before deciding how to approach the learning of it. This fits well with Zazkis, Dubinsky, and Dautermann (1996) who suggest that some of the most successful learners may use different learning strategies, or combinations of those strategies, at different times.

## Theoretical Perspectives

Except for Paul, all the students in the study clearly demonstrated that they had a set way of approaching the learning of new material that was challenged by becoming a PhD student. In some cases (the majority of natural learners), the challenge was dealt with easily, perhaps with some sense of relief. In other cases (the majority of the alien learners), the challenge required a fundamental change in cognitive style. The resulting cognitive styles seem to match well with the descriptions Smith and Hungwe (1998) gave of the practices of young mathematicians who focussed either on structure and foundations or on insight and understanding. The question naturally arises, then, about what aspects of the transition to independent graduate study lie at the heart of this challenge.

The transition (in the context of this particular study: a middle ranking UK university) is characterized by many changes, which are intimately intertwined. Clearly there are obvious changes to the teaching/learning situation as students move from lecture classes of 100, with little planned regular contact with their lecturers, to a one-to-one relationship with a supervisor. The undergraduate syllabuses are clearly defined, the pace with which new material is presented is set, and the majority of the problems posed to students have predetermined solutions. In contrast, the aim of the PhD in mathematics is original research, which has no set syllabus, no set pace and no pre-determined solutions.

At the same time as these pedagogical changes are occurring, there is probably a similar upheaval in the students' social lives as they either move from one university to another, or their friends from their undergraduate days move on. Social groups that may have formed around working together on undergraduate mathematical problems are likely to have broken up, and, as postgraduate students tend to work in their own, highly specialized areas, the opportunities for collaboration are diminished. With the students in this sample, the social relationship with mathematics faculty members changed across the transition: the teacher-learner role pair that had existed as undergraduates seems to have been replaced with a senior-junior colleagues relationship.

It is difficult to untangle from this mass of differences the individual forces that lead to the kinds of changes in cognitive style that we have mapped above. To help us account for the changes, we might adapt Skemp's 'ecological' model of learning viability, as a response to the environmental change as a whole (Skemp, 1979). He proposed a model of learning based on the notion of goals and director systems (combinations of goals, sensors and operators): "learning is a goal-directed change of state of a director system towards states which, for the assumed environment, make possible optimal functioning." He described two types of director system - one that operates on the environment ($\Delta_1$) and one that operates on other director systems ($\Delta_2$). Skemp suggested that a director system is deemed to be *viable* in its environment if it both *functions* and is *capable*, That is, the director system is in an environment from which it can move to its goal (it can function) and the director system's operators are able to move it to its goal (it is capable).

The difference between a cognitive strategy (how a learner solves a particular problem) and a cognitive style (longer term approaches to learning new material) can be seen as analogous to $\Delta_1$ and $\Delta_2$ systems. We can consider cognitive style to be a $\Delta_2$ director system: it is a system which suggests *how* we go about learning rather than a system which learns. This has clear links with Curry's (1991) 'onion model' of layers of learning ('instructional preferences' wrapped around 'information processing style,' wrapped around 'cognitive personality style') the innermost of which is least affected by the environment and changes most slowly over time. When the learner moves to a new environment, the viability of these layers of director systems changes and, with a sufficiently radical change of environment, the only way to retain this viability is to change not just the surface systems, but the underlying systems. That is, the learner has to change cognitive style.

As we have indicated, the transition to independent graduate study can be an extreme change of environment for the learner. However, for the *natural* learner, some of the changes, such as the change of pace, may enhance the viability of their cognitive style while others, such as the loss of externally set syllabuses, will probably not do any damage to their cognitive style's viability. Thus, the natural cognitive style is stable across this transition. Indeed, the supposed increased viability in the new environment might account for Oliver's sense of relief at the ways he was allowed to learn once he became a PhD student.

Alien cognitive styles, however, are unstable. They need an environment in which the material to be learned is provided for them by an external authority, so the loss of an externally set syllabus is likely to make their style less viable. Similarly, the lack of an externally set pace is unlikely to make their style more viable. So, for these students, as they continue to use a cognitive style in an environment in which it is no longer viable, they will find themselves less able to reach learning goals.[2]

The adaptations alien learners make to retain viability in this new environment take them in one of two directions. One option is to begin to adopt the more viable natural cognitive style: searching for ways of connecting the new mathematics they are trying to learn to their existing understandings of mathematics. The other option is to retain a core of an alien cognitive style: requiring that an indication of the goal they need to achieve be given to them by an external authority – that is, being given a clear sense of 'where it is going' – while retaining a sense of the procedural, but changing focus to building a sense of the logical structure of the new mathematics.

Of course, in the most extreme cases of loss of viability in this new environment, it may be that a learner's cognitive style cannot adapt at all, and the learner may drop out of the PhD program. While we had no students in that situation, Herzig's study of participation in US mathematics doctoral programs hints at this difficulty to adapt as one of the reasons for the apparently high attrition rates (Herzig, 2002).

It is worth noting that all the students in our survey showed a clear awareness of a changed environment and changed expectations and one might suggest that this awareness is a necessary (but not sufficient) prerequisite for a movement to a more viable cognitive style.

The 'ecological' explanation of the changes in cognitive styles that we have mapped suggests some ways in which the transition to independent graduate study might be dealt with to ease some of the problems. Clearly, the most obvious ways in which departments can affect the transition is to address the two different environments and lessen the differences between them. Opportunities to engage in longer term research projects, with less emphasis on the pace of learning new material and on pre-determined goals at the undergraduate levels may both prepare students with alien cognitive styles for the change they will face and, perhaps, lessen the feelings of frustration felt by undergraduates with natural cognitive styles. One might argue that this approach merely pushes the transitional problems back into undergraduate years. However, rather than 'filtering out' students with alien cognitive styles, the approach might do much to help 'filter in' more students with

---

[2]Indeed, one might argue that the indicators of achieving learning goals have become more obscure for alien learners: the high scores and the teacher's check marks have disappeared. How are they now able to determine if they have reached their goals?

natural (and therefore more stable) cognitive styles. To counter this, one might also suggest to PhD supervisors of students with alien learning styles that they might initially provide more direction, or support for developing a sense of mathematical structure for new material.

However, perhaps the most telling aspect of the part that can be played by those involved in the initiation of new researchers into their role as practicing mathematicians could come from considering Burton's description of *their own* practices as working mathematicians (Burton, 1999):

> ...there is also a clear recognition that, although you *think* you know when you know, often you are wrong or you find unexpected gaps which have to be plugged before you are ready to offer your solution to the community at large. Nonetheless, the rich sense of pleasure gained from achieving an Aha!, even one which is later followed by an Oops! permeates their descriptions as does the sense that every new journey is, indeed, a new journey and you face again all the hazards that have been faced before. You might accumulate experience which helps you on your way but, at root, you are an explorer reliant upon your own strategies, expectations and fallacies, and those of your fellow travellers. There is a chasm between this perspective on coming to know, and the transmission pedagogy of the classroom dependent as it is on acquiring the knowledge of the expert. There is another chasm between the language of 'boring' heard from some pupils... and the language of fun, of 'euphoria', of excitement, of personal struggle and achievement of these mathematicians.

The opportunity for students to experience the challenge, the ups and downs, the successes and frustrations before the achievement of eventual success must surely be invaluable to their future as research mathematicians.

## References

Brousseau, G. (1997). *Theory of didactical situations in mathematics*. Dordrecht, The Netherlands: Kluwer Academic Publishers.

Burgess, R. (1988). Conversations with a purpose: The ethnographic interview in educational research. In R. Burgess (Ed.), *Studies in qualitative methodology*. London: JAI Press.

Burton, L. (1999). The practices of mathematicians: What do they tell us about coming to know mathematics? *Educational Studies in Mathematics, 37*, 121–143.

Curry, L. (1991). Patterns of learning style across selected medical specialists. *Educational Psychology, 11*, 247–278.

Daskalogianni, K., & Simpson, A. (2000). Towards a definition of attitude: The relationship between the affective and cognitive in pre-university students. In T. Nakahara & M. Koyama (Eds.), *Proceedings of the 24th conference of the International Group for the Psychology of Mathematics Education, Vol. 2* (pp. 217–224). Hiroshima, Japan: Hiroshima University.

Duffin, J., & Simpson, A. (1993). Natural, conflicting and alien. *Journal of Mathematical Behavior, 12*, 313–328.

Evans, J., & Over, D. (1996). *Rationality and reasoning.* Hove, UK: Psychology Press.

Fodor, J. (1983). *Modularity of mind.* Cambridge, MA: MIT Press.

Glaser, B., & Strauss, A. (1967). *The discovery of grounded theory.* Chicago: Aldine.

Herzig, A. (2002). Where have all the students gone? Participation of doctoral students in authentic mathematical activity as a necessary condition for persistence towards the PhD. *Educational Studies in Mathematics, 50,* 177–212.

Hiebert, J., & Lefevre, P. (1986). Conceptual and procedural knowledge in mathematics. In J. Hiebert (Ed.), *Conceptual and procedural knowledge: The case of mathematics.* Hillsdale, NJ: Erlbaum.

Messick, S. (1984). The nature of cognitive styles: Problems and promise in educational practice. *Educational Psychologist, 19,* 59–74.

Perry, W. (1970). *Forms of intellectual and ethical development in the college years: A scheme.* New York: Holt, Rinehart, and Winston.

Piaget, J. (1975). *The equilibration of cognitive structures: The central problem of intellectual development.* Chicago: University of Chicago Press.

Pinto, M. (1998). *Students' understanding of real analysis.* Unpublished doctoral dissertation, Warwick University.

Pole, C., Sprokkereef, A., Burgess, R., & Lakin, E. (1997). Supervision of doctoral students in the natural sciences: Expectations and experiences. *Assessment and Evaluation in Higher Education, 22,* 49–63.

Riding, R., & Cheema, I. (1991). Cognitive styles - an overview and integration. *Educational Psychology, 11,* 193–215.

Riding, R., & Rayner, S. (1998). *Cognitive styles and learning strategies.* London: Fulton.

Skemp, R. (1976). Relational understanding and instrumental understanding. *Mathematics Teaching, 77,* 20–26.

Skemp, R. (1979). *Intelligence, learning, and action: A foundation for theory and practice in education.* Chichester, UK: John Wiley and Sons.

Sloman, S. (1996). The empirical case for two systems of reasoning. *Psychological Bulletin, 119,* 3–22.

Smith, J., & Hungwe, K. (1998). Conjecture and verification in research and teaching: Conversations with young mathematicians. *For the Learning of Mathematics, 18.*

Stanovich, K., & West, R. (2000). Individual differences in reasoning: Implications for the rationality debate. *Behavioral and Brain Sciences, 23,* 645–726.

Tall, D. O. (1991). *Advanced mathematical thinking.* Dordrecht, The Netherlands: Kluwer.

Wason, P. (1966). Reasoning. In B. Foss (Ed.), *New horizons in psychology.* Harmondsworth, UK: Penguin.

Zazkis, R., Dubinsky, E., & Dautermann, J. (1996). Coordinating visual and analytic strategies: A study of students' understanding of the group $D_4$. *Journal for Research in Mathematics Education, 27,* 435–457.

School of Mathematics, University of Hull, Hull, HU6 7RX, UK.

School of Education, Durham University, Durham, DH1 1TA, UK.
*E-mail address*: a.p.simpson@durham.ac.uk

# Research in Collegiate Mathematics Education

## Editorial Policy

The papers published in these volumes will serve both pure and applied purposes, contributing to the field of research in collegiate mathematics education and informing the direct improvement of post-secondary mathematics instruction. The dual purposes imply dual but overlapping audiences and articles will vary in their relationship to these purposes. The best papers, however, will interest both audiences and serve both purposes.

**Content.** We invite papers reporting on research that addresses any and all aspects of collegiate mathematics education. Research may focus on learning within particular mathematical domains. It may be concerned with more general cognitive processes such as problem solving, skill acquisition, conceptual development, mathematical creativity, cognitive styles, etc. Research reports may deal with issues associated with variations in teaching methods, classroom or laboratory contexts, or discourse patterns. More broadly, research may be concerned with institutional arrangements intended to support learning and teaching, e.g. curriculum design, assessment practices, or strategies for faculty development.

**Method.** We expect and encourage a broad spectrum of research methods ranging from traditional statistically-oriented studies of populations, or even surveys, to close studies of individuals, both short and long term. Empirical studies may well be supplemented by historical, ethnographic, or theoretical analyses focusing directly on the educational matter at hand. Theoretical analyses may illuminate or otherwise organize empirically based work by the author or that of others, or perhaps give specific direction to future work. In all cases, we expect that published work will acknowledge and build upon that of others—not necessarily to agree with or accept others' work, but to take that work into account as part of the process of building the integrated body of reliable knowledge, perspective, and method that constitutes the field of research in collegiate mathematics education.

**Review Procedures.** All papers, including invited submissions, will be evaluated by a minimum of three referees, one of whom will be a volume editor. Papers will be judged on the basis of their originality, intellectual quality, readability by a diverse audience, and the extent to which they serve the pure and applied purposes identified earlier.

**Submissions.** Papers of any reasonable length will be considered, but the likelihood of acceptance will be smaller for very large manuscripts. Manuscripts should have citations and bibliographies according to the format of the American Psychological Association as described in the fifth edition of the *Publication Manual of the American Psychological Association.*

Note that the *RCME* volumes are produced for electronic submission to the AMS. Accepted manuscripts should be prepared using AMS-LaTeX and the CBMS author packages available from the AMS web site, www.ams.org. Illustrations should also be prepared in a form suitable for electronic submission (namely, encapsulated postscript files). Additional information for final formatting will be provided upon acceptance of a work for publication.

For further information, see the Issues in Mathematics Education section of the Conference Board of the Mathematical Sciences web site at www.cbmsweb.org.

**Correspondence.** Before submitting a manuscript, send an abstract to one of the current editors of *RCME*:

| Fernando Hitt | Derek Holton | Pat Thompson |
|---|---|---|
| ferhitt@yahoo.com | dholton@maths.otago.ac.nz | pat.thompson@asu.edu |

Subsequent correspondence may be with the production editor or with the volume editor who has been assigned primary responsibility for decisions regarding the manuscript.

# Titles in This Series

13 **Fernando Hitt, Guershon Harel, and Annie Selden, Editors,** Research in collegiate mathematics education. VI, 2006

12 **Annie Selden, Ed Dubinsky, Guershon Harel, and Fernando Hitt, Editors,** Research in collegiate mathematics education. V, 2003

11 **Conference Board of the Mathematical Sciences,** The mathematical education of teachers, 2001

10 **Solomon Friedberg et al.,** Teaching mathematics in colleges and universities: Case studies for today's classroom. Available in student and faculty editions, 2001

9 **Robert Reys and Jeremy Kilpatrick, Editors,** One field, many paths: U. S. doctoral programs in mathematics education, 2001

8 **Ed Dubinsky, Alan H. Schoenfeld, and Jim Kaput, Editors,** Research in collegiate mathematics education. IV, 2001

7 **Alan H. Schoenfeld, Jim Kaput, and Ed Dubinsky, Editors,** Research in collegiate mathematics education. III, 1998

6 **Jim Kaput, Alan H. Schoenfeld, and Ed Dubinsky, Editors,** Research in collegiate mathematics education. II, 1996

5 **Naomi D. Fisher, Harvey B. Keynes, and Philip D. Wagreich, Editors,** Changing the culture: Mathematics education in the research community, 1995

4 **Ed Dubinsky, Alan H. Schoenfeld, and Jim Kaput, Editors,** Research in collegiate mathematics education. I, 1994

3 **Naomi D. Fisher, Harvey B. Keynes, and Philip D. Wagreich, Editors,** Mathematicians and education reform 1990–1991, 1993

2 **Naomi D. Fisher, Harvey B. Keynes, and Philip D. Wagreich, Editors,** Mathematicians and education reform 1989–1990, 1991

1 **Naomi D. Fisher, Harvey B. Keynes, and Philip D. Wagreich, Editors,** Mathematicians and education reform: Proceedings of the July 6–8, 1988 workshop, 1990